W9-CKM-310

Musculoskeletal Soft-Tissue Aging: Impact on Mobility

American Academy of
Orthopaedic Surgeons
Symposium

Musculoskeletal Soft-Tissue Aging: Impact on Mobility

Edited by
Joseph A. Buckwalter, MD
Professor, Department of Orthopaedic Surgery
University of Iowa Hospitals
Department of Orthopaedics
Iowa City, Iowa

Victor M. Goldberg, MD
Charles H. Herndon Professor and Chairman
Department of Orthopaedics
Case Western Reserve University
Cleveland, Ohio

Savio L-Y. Woo, PhD
Ferguson Professor and Vice Chairman for Research
Department of Orthopaedic Surgery
Professor of Mechanical Engineering
University of Pittsburgh
Pittsburgh, Pennsylvania

with 80 illustrations

Workshop
Colorado Springs, Colorado
November 1992

Supported by the
American Academy of Orthopaedic Surgeons

and the
National Institute of Arthritis and Musculoskeletal and Skin Diseases

and the
National Institute on Aging

American Academy of Orthopaedic Surgeons
6300 North River Road
Rosemont, IL 60018

American Academy of Orthopaedic Surgeons

Musculoskeletal Soft-Tissue Aging: Impact on Mobility

Staff

Executive Director:
Thomas C. Nelson

Director, Division of Education:
Mark W. Wieting

Director, Department of Publications:
Marilyn L. Fox, PhD

Senior Editor:
Bruce Davis

Associate Senior Editors:
Joan Abern
Jane Baque

Production Manager:
Loraine Edwalds

Assistant Production Manager:
Kathy M. Brouillette

Editorial Assistants:
Susan Baim
Sharon Duffy
Katy O'Brien
Sophie Tosta

Publications Secretaries:
Geraldine Dubberke
Em Lee Lambos

The material presented in *Musculoskeletal Soft-Tissue Aging: Impact on Mobility* has been made available by the American Academy of Orthopaedic Surgeons for educational purposes only. This material is not intended to present the only, or necessarily best, methods or procedures for the medical situations discussed, but rather is intended to represent an approach, view, statement, or opinion of the author(s) or producer(s), which may be helpful to others who face similar situations.

Some drugs and medical devices demonstrated in Academy courses or described in Academy print or electronic publications have FDA clearance for use for specific purposes or for use only in restricted settings. The FDA has stated that it is the responsibility of the physician to determine the FDA status of each drug or device he or she wishes to use in clinical practice, and to use the products with appropriate patient consent and in compliance with applicable law.

Furthermore, any statements about commercial products are solely the opinion(s) of the author(s) and do not represent an Academy endorsement or evaluation of these products. These statements may not be used in advertising or for any commercial purpose.

Library of Congress Cataloging-in-Publication Data

Musculoskeletal soft-tissue aging : impact on mobility / edited by Joseph A. Buckwalter, Victor M. Goldberg, and Savio L-Y. Woo.
p. cm.
"Workshop, Colorado Springs, Colorado, November 1992, supported by the American Academy of Orthopaedic Surgeons and the National Institute of Arthritis and Musculoskeletal and Skin Diseases and the National Institute on Aging."
Includes bibliographical references and index.
ISBN 0-89203-086-0
1. Muscles—Aging—Congresses. 2. Connective tissues—Aging—Congresses. I. Buckwalter, Joseph A. II. Goldberg, Victor M. III. Woo, Savio L-Y. IV. American Academy of Orthopaedic Surgeons
QP321.M899 1993
612.7'4--dc20 93-41652
DNLM/DLC CIP
for Library of Congress

Contributors and Participants

Wayne H. Akeson, MD†
Professor and Chairman, Department of Orthopaedics
University of California San Diego
San Diego, California

Edward Amento, MD*
Genetech Inc. Rheumatology Department
South San Francisco, California

David Amiel, PhD†
Professor of Orthopaedics
University of California San Diego
San Diego, California

Kai-Nan An, PhD*†
Chair, Division of Orthopaedic Research
Professor of Bioengineering
Mayo Medical School
Rochester, Minnesota

Gunnar B.J. Andersson, MD, PhD*†
Professor and Associate Chairman
Department of Orthopaedic Surgery
Rush-Presbyterian-St. Luke's Medical Center
Chicago, Illinois

Frank W. Booth, PhD*†
Professor of Physiology and Cell Biology
University of Texas Medical School at Houston
Houston, Texas

Frank U. Brauer, MD†
Fellow, Department of Pathology
The Hospital for Special Surgery
New York, New York

Susan V. Brooks, PhD†
Postdoctoral Research Fellow
University of Michigan
Ann Arbor, Michigan

Marybeth Brown, PT, PhD*†
Assistant Professor
Program in Physical Therapy
Washington University School of Medicine
St. Louis, Missouri

Joseph A. Buckwalter, MD*†
Professor, Department of Orthopaedic Surgery
University of Iowa Hospitals
Department of Orthopaedics
Iowa City, Iowa

Kitty Buckwalter, PhD*
Professor, College of Nursing
University of Iowa
Iowa City, Iowa

Peter G. Bullough, MB, ChB*†
Director of Laboratory Medicine
The Hospital for Special Surgery
New York, New York

David L. Butler, PhD*†
Professor, Department of Aerospace Engineering and Engineering Mechanics
Noyes-Giannestras Biomechanics Laboratories
Cincinnati, Ohio

Judith Campisi, PhD*†
Senior Scientist, Department of Cell and Molecular Biology
Lawrence Berkeley Laboratory
Berkeley, California

Joseph Cannon, MD*
U.S. Department of Agriculture
Boston, Massachusetts

Arnold I. Caplan, PhD†
Professor/Director, Skeletal Research Center
Biology Department
Case Western Reserve University
Cleveland, Ohio

David R. Clemmons, MD*†
Professor of Medicine
University of North Carolina at Chapel Hill
Chapel Hill, North Carolina

Francesco DeMayo, PhD†
Assistant Professor, Department of Cell Biology
Baylor College of Medicine
Houston, Texas

William J. Evans, PhD†
Director, Human Performance Laboratory
The Pennsylvania State University
University Park, Pennsylvania

David Eyre, PhD*†
Ernest M. Burgess Professor
Depatent of Orthopaedics
University of Washington
Seattle, Washington

John A. Faulkner, PhD*†
Professor of Physiology
University of Michigan
Ann Arbor, Michigan

Cyril Frank, MD*†
Professor, Department of Surgery
McCaig Centre for Joint Injury and Arthritis Research
Faculty of Medicine
University of Calgary
Calgary, Alberta, Canada

Richard H. Gelberman, MD*†
Professor, Orthopaedic Surgery
Harvard Medical School
Boston, Massachusetts

Vijay Goel, MD*
Department of Biomedical Engineering
University of Iowa
Iowa City, Iowa

Victor M. Goldberg, MD*†
Charles H. Herndon Professor and Chairman
Department of Orthopaedics
Case Western Reserve University
Cleveland, Ohio

Stephen L. Gordon, PhD*
Chief, Musculoskeletal Diseases Branch
National Institute of Arthritis and Musculoskeletal and Skin Diseases
National Institutes of Health
Bethesda, Maryland

Alan J. Grodzinsky, ScD*†
Professor of Electrical, Mechanical, and Bioengineering
Department of Electrical Engineering and Computer Science
Massachusetts Institute of Technology
Cambridge, Massachusetts

Krishan L. Gupta, MD†
Assistant Professor, Department of Medicine
Division of Geriatrics
Baylor College of Medicine
Houston, Texas

Evan C. Hadley, MD*†
Associate Director (Geriatrics)
National Institute on Aging, NIH
Bethesda, Maryland

David A. Hart, PhD†
Professor, Department of Microbiology and Infectious Diseases
McCaig Centre for Joint Injury and Arthritis Research
Faculty of Medicine
University of Calgary
Calgary, Alberta, Canada

Vincent Hascall*
National Institutes of Health
Bethesda, Maryland

Stephen E. Haynesworth, PhD*†
Assistant Professor, Department of Biology
Case Western Reserve University
Cleveland, Ohio

Dick Heinegård, MD, PhD*†
Professor, Department of Medical and Physiological Chemistry
Lund University
Lund, Sweden

Raymond Hintz, MD†
Department of Pediatrics
Stanford University Medical Center
Stanford, California

Alan M. Jette, PT, PhD*†
Senior Research Scientist
New England Research Institute
Watertown, Massachusetts

Z.V. Kendrick, PhD†
Director, Biokinetics Research Laboratory
Temple University
Philadelphia, Pennsylvania

Klaus E. Kuettner, PhD*
Professor and Chairman
Department of Biochemistry
Rush Medical College
Rush Presbyterian-St. Luke's Medical Center
Chicago, Illinois

Cato T. Laurencin, MD, PhD*†
Instructor of Biochemical Engineering
Division of Health Sciences and Technology
Massachusetts Institute of Technology
Cambridge, Massachusetts

Heung Man Lee†
Department of Cell Biology
Baylor College of Medicine
Houston, Texas

Stefan Lohmander, MD, PhD*†
Associate Professor, Department of Orthopaedics
University Hospital
Lund, Sweden

Pilar Lorenzo†
Department of Medical and Physiological Chemistry
University of Lund
Lund, Sweden

Martin Lotz, MD†
Associate Professor, Department of Medicine
University of California San Diego
San Diego, California

David T. Lowenthal, MD, PhD*†
VA Medical Center
University of Florida College of Medicine
Gainesville, Florida

John Matyas, PhD†
McCaig Centre for Joint Injury and Arthritis Research
University of Calgary
Health Sciences Centre
Calgary, Alberta, Canada

Vincent M. Monnier, MD*†
Professor of Pathology
Institute of Pathology
Case Western Reserve University
Cleveland, Ohio

Roland W. Moskowitz, MD†
Professor of Medicine
Case Western Reserve University
Cleveland, Ohio

Van C. Mow, PhD*
Professor of Mechanical Engineering and Orthopaedic Bioengineering
Departments of Mechanical Engineering and Orthopaedic Surgery
Columbia University
New York, New York

Theodore R. Oegema Jr, PhD*†
Professor, Departments of Orthopaedic Surgery and Biochemistry
University of Minnesota
Minneapolis, Minnesota

Richard H. Pearce, PhD*†
Professor Emeritus, Department of Pathology
Faculty of Medicine
University of British Columbia
Vancouver, British Columbia, Canada

H. K. Pokharna, PhD†
Postdoctoral Fellow
Case Western Reserve University
Cleveland, Ohio

Malcolm H. Pope, Dr Med Sc, PhD*†
Professor, Department of Orthopaedics
University of Vermont
Burlington, Vermont

David Powell, MD†
Associate Professor of Pediatrics
Baylor College of Medicine
Texas Children's Hospital
Houston, Texas

Scott Powers, PhD, EdD†
Center for Exercise Science
University of Florida
Gainesville, Florida

Finn P. Reinholt, MD, PhD†
Associate Professor, Department of Pathology
Karolinska Institutet
Huddinge Hospital
Huddinge, Sweden

Peter J. Roughley, PhD*†
Professor, Department of Surgery
McGill University
Montreal, Quebec, Canada

Daniel Rudman, MD*†
Professor of Medicine
Medical College of Wisconsin
Milwaukee, Wisconsin

Paul Schreck, MD†
Department of Orthopaedics
University of California San Diego
San Diego, California

Albert B. Schultz, PhD*†
Vennema Professor of Mechanical Engineering and Applied Mechanics
University of Michigan
Ann Arbor, Michigan

Robert J. Schwartz, PhD†
Professor, Department of Cell Biology
Baylor College of Medicine
Houston, Texas

David R. Sell, PhD*†
Assistant Professor
Case Western Reserve University
Cleveland, Ohio

Kaup R. Shetty, MD†
Associate Professor of Medicine
Medical College of Wisconsin
Milwaukee, Wisconsin

Yngve Sommarin, PhD†
Department of Medical and Physiological Chemistry
Lund University
Lund, Sweden

Carol Teitz, MD*
Associate Professor, Department of Orthopaedic Surgery
University of Washington
Seattle, Washington

L.E. Underwood, MD†
Professor of Pediatrics
University of North Carolina at Chapel Hill
Chapel Hill, North Carolina

Jill Urban, PhD*†
Senior Research Fellow
Physiology Laboratory
Oxford University
Oxford, England

Kathryn G. Vogel, PhD*†
Professor, Department of Biology
The University of New Mexico
Albuquerque, New Mexico

Steven H. Weeden, BA†
Medical Student, 3rd year
University of Texas Medical School at Houston
Houston, Texas

Alan H. Wilde, MD*
Cleveland Center for Joint Reconstruction
Cleveland, Ohio

Savio L-Y. Woo, PhD*†
Ferguson Professor and Vice Chairman for Research
Department of Orthopaedic Surgery
Professor of Mechanical Engineering
University of Pittsburgh
Pittsburgh, Pennsylvania

Virgil L. Woods Jr, MD*†
Associate Professor of Medicine and Orthopedics
University of California San Diego Medical Center
San Diego, California

Kuo Chang Yin, MS†
Houston, Texas

---

* Workshop Participant
† Contributor to Volume

Contents

Preface

The need to improve the quality of life for people who are middle aged and older, and who suffer from age-related musculoskeletal impairments, including weakness and stiffness that lead to loss of mobility, will significantly increase in the next several decades. In 1980, 12% of the country's population was over 65, with the fastest growing segment made up of persons in their 70s and 80s. By 2050, the percentage of elderly Americans is expected to grow to 22% of the population, a total of 68 million people. Between 1990 and 2010 the number of people over age 45 will increase by more than 40 million, from 82 to 124 million. For these people, declining musculoskeletal function is a common cause of decreasing mobility. Many, if not most, of these people will want to, and should, remain physically active. Therefore, we must increase our understanding of the age-related changes in the musculoskeletal system and seek ways to maintain and, if possible, improve or restore musculoskeletal function and mobility for older individuals.

Despite the clinical importance of age-related changes in the musculoskeletal soft tissues, including muscle, cartilage, intervertebral disk, tendons, ligaments, and joint capsules, the recent advances in understanding of these changes in the soft tissues have not been widely disseminated to the scientists and clinicians who work on problems of the musculoskeletal system, and very few investigators have focused their efforts on defining the relationships between the tissue changes and age-related deterioration of musculoskeletal function. Furthermore, there has not been a previous effort to synthesize new basic scientific information with clinical observations and to define the most productive directions for future investigation in this increasingly important area.

For these reasons, the workshop "Age-Related Changes in the Musculoskeletal Soft Tissues" held November 14-17, 1992, was dedicated to evaluating current understanding of age-related changes in the musculoskeletal soft tissues, to identifying important future directions for investigation, and to encouraging future basic, clinical, and collaborative research in this important area. Workshop sponsors included the National Institute of Arthritis and Musculoskeletal and Skin Diseases, the National Institute on Aging, and the American Academy of Orthopaedic Surgeons.

This book summarizes the recent studies of age-related changes in musculoskeletal function and the musculoskeletal soft tissue presented and discussed at the workshop. It consists of five sections: Age-Related Changes in Function and the Mechanisms of Age Changes, Articular Cartilage, Skeletal Muscle, Tendon and Ligament, and Intervertebral Disk. In addition to individual chapters, each section contains an overview and a discussion of future research directions.

The workshop participants, listed on pages v-viii, included physicians, bioengineers, biochemists, cell biologists, molecular biologists, and pathologists.

Joseph A. Buckwalter, MD
Victor M. Goldberg, MD
Savio L-Y. Woo, PhD

Introduction

The Impact on Mobility of Age-Related Changes in the Musculoskeletal Soft Tissues

Joseph A. Buckwalter, MD

Impairments of musculoskeletal function, primarily stiffness and weakness, are among the most prevalent and symptomatic complaints of middle and old age.*

With increasing age, musculoskeletal soft-tissue injuries and diseases, including sprains and strains; arthritis; stiffness of muscles, tendons and joints; muscle weakness, soreness, and fatigue; and back pain and back stiffness, can progressively limit mobility and thereby compromise the quality of life. In addition, recovery from acute and repetitive soft-tissue trauma and from surgical repair or reconstruction of bone, synovial joint, tendon, and ligament takes longer with increasing age. Although these problems may be most severe in the elderly, many people notice their effects by middle age or before.

By decreasing strength, restricting activity, and causing pain, musculoskeletal impairments prevent older people from making full use of their abilities and opportunities for leisure or work and from participating in the regular physical activity necessary to maintain both optimal function of the musculoskeletal system and their general health. Many of these problems result from age-related changes in the musculoskeletal system. These age-related changes in the tissues adversely alter tissue responses to loads and to injury and, thereby, lead to further disability.

Bone fractures are a well-recognized cause of loss of mobility in the elderly. However, bone fractures do not cause impairment for as many older people as do weakness and restriction of joint motion.* In addition, age-related loss of spine motion clearly is associated with changes in the musculoskeletal soft tissues: skeletal muscle, ligaments, tendons, articular cartilage, and intervertebral disks. Furthermore, clinical experience suggests that musculoskeletal

*The concepts developed at the workshop are summarized in Buckwalter JA, Woo SL-Y, Goldberg VM, et al: Soft tissue aging and musculoskeletal function: A current concepts review. *J Bone Joint Surg* 1993;75A:1533-1548.

soft-tissue injuries occur more frequently and heal more slowly with increasing age.

Recently, a number of investigators have shown potentially important age-related changes in the musculoskeletal soft tissues. Those investigating the composition of the musculoskeletal soft-tissue matrices, especially the cartilage, dense fibrous tissue, and intervertebral disk matrices, have found significant changes with age. Other investigators have clearly shown age-related declines in the mechanical properties of ligaments and increasing stiffness of the human spine that presumably results from alterations in the spinal ligaments and the intervertebral disks. Data from experimental studies confirm the clinical evidence of delayed healing of the soft tissues with increasing age and suggest that the age-related alteration of healing may be at least partially reversible.

The concept of the relationship between age-related changes in musculoskeletal soft tissues and impairment developed at the workshop "Age-Related Changes in the Musculoskeletal Soft Tissues"* is that: with advancing age musculoskeletal soft-tissue function declines, susceptibility to disease and injury increases, and the ability to recover from disease or injury declines; these changes increase the probability of impairment (Fig. 1, *top*). However, the function of individual cells, tissues or organ systems may remain stable or even improve temporarily with age, and a number of interventions have the potential to maintain or improve musculoskeletal function (Fig. 1, *bottom*).

Recent work has identified age-related changes that probably occur in all tissues to some degree and specific changes in individual musculoskeletal soft tissues. The age-related changes that affect most tissues include diminished proliferative and synthetic capacities of differentiated cells, decreased numbers of mesenchymal stem cells, post-translational protein modification, accumulation of degraded matrix molecules, decreased levels of trophic hormones, and impaired healing. Study of skeletal muscle; articular cartilage, tendons, ligaments, and joint capsules; and intervertebral disks demonstrates significant age-related alterations in the cells, matrices, and function of these tissues. Yet the relationships between aging processes in the tissues and age-related musculoskeletal impairment remain unclear. Attempts to define these relationships have shown them to be exceedingly complex.

Despite these limitations, the information presented in the following chapters and in the discussions at the workshop support the general conclusions listed below concerning age-related changes in musculoskeletal function and the soft tissues. (1) Age-related musculoskeletal impairments (weakness, stiffness, pain with motion), along with increased susceptibility to injury and disease and slower or less effective healing, cause increasing disability with increasing age. (2) In many instances, aging changes (changes that occur as a direct result of increasing age) in the soft tissues appear to be synonymous with degeneration (tissue disintegration or disorganization

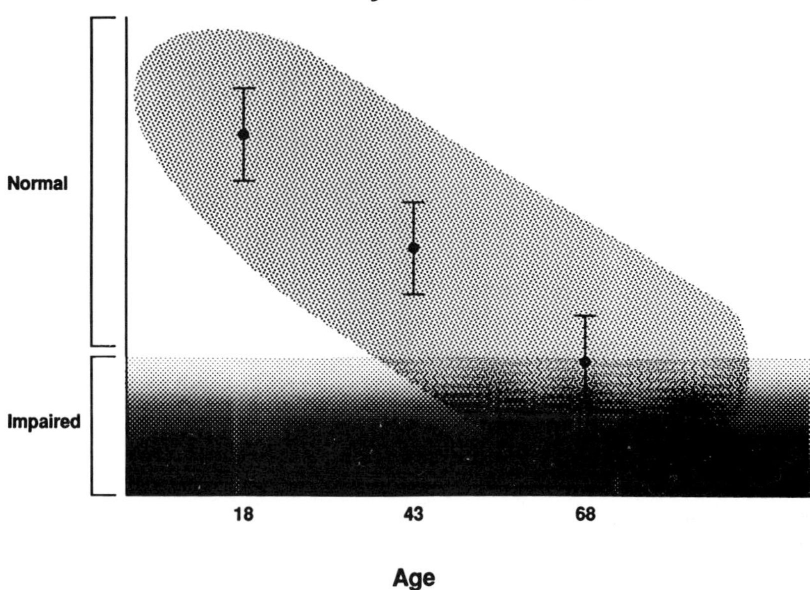

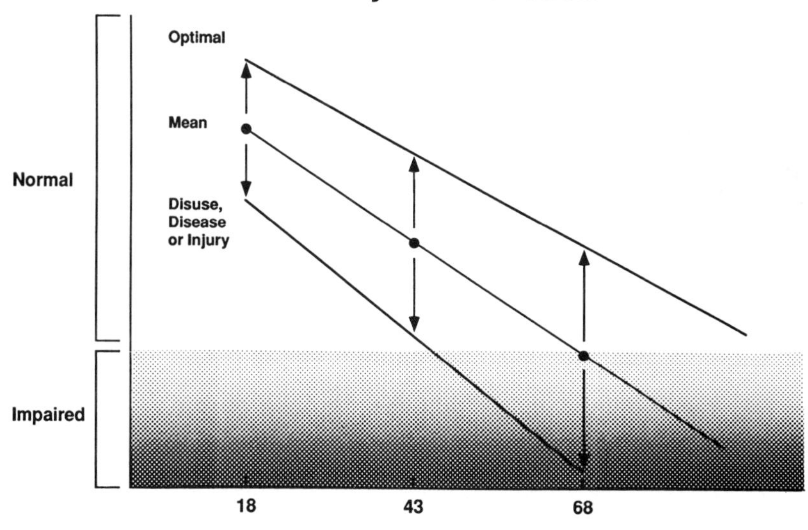

Fig. 1 *Schematic representations of the change in cell, tissue or musculoskeletal function with age.* ***Top****, The dots and error bars at 18, 43, and 68 years of age represent the mean values and standard errors for function at each age. The shaded elongated oval area represents the range of values.* ***Bottom****, The level of function at the ages represented can decrease because of disuse, disease or injury. A number of studies indicated that function can also be improved at any age.*

that results in a decline in function). Disease, decreased use, or injury can accelerate age-related decline of soft-tissue organization and function, but the effects of recognized age-related diseases and injuries have not yet been clearly distinguished from changes that occur as a direct result of increasing age or decreasing use. Therefore, the functional changes and tissue changes that are associated with increasing age are referred to as age-related changes in this book, rather than aging changes. (3) Although the primary concerns of the workshop were the changes in musculoskeletal tissue composition, structure, and function that occur following skeletal maturity, in some instances the tissue changes that occur prior to skeletal maturity have been more thoroughly studied or appear to be more extensive. Therefore, some of the chapters include cell and extracellular changes that occur during skeletal maturation. (4) In addition to age-related tissue changes, other factors adversely affect musculoskeletal function with age. These include factors directly related to the musculoskeletal system, including joint proprioception, reaction time, and postural balance. Other general factors that may be important include mental status, vision, systemic disease, living environment, economic status, and psychological factors. (5) In some tissues, increasing age is associated with a significant decline in cell proliferative capacity and the ability of differentiated cells to perform important functions, such as matrix synthesis, and possibly in the population of mesenchymal stem cells. Other important age-related changes include accumulation of degraded matrix molecules, posttranslational modification of matrix macromolecules, and alterations in matrix composition, matrix mechanical properties, tissue healing, trophic hormone levels, and overall musculoskeletal function. (6) Age-related changes in cells, tissues, and musculoskeletal function are not necessarily unidirectional or uniform among individuals, organ systems, tissues, or cells, nor are they necessarily irreversible. Progressive resistance and range of motion exercise can decrease the age-related loss of strength and may help maintain or restore flexibility. However, exercise programs can also cause injury. Older individuals should have a careful medical evaluation before starting an exercise program, and the program should be selected based on this evaluation. This is especially important for individuals with systemic illness and for individuals at greater risk of musculoskeletal injury, including people with previous joint injuries, obesity, osteoarthritis, joint deformity, weakness, or restricted joint motion. Trophic hormone replacement may also modify age-related changes in the soft tissues, including loss of strength. Systemic or local use of growth factors and cell transplantation after expanding the population of mesenchymal stem cells in culture could improve healing in older people. Potential future methods of slowing or reversing age-related deterioration of the soft tissues include inhibition of post-translational modification of matrix proteins, inducing production of selected cell and matrix components through hormonal or mechanical stimuli, and altering the activity of selected transcription factors. In particular, better understanding of the role of transcription factors in cell senescence may make it possible to devise methods of delaying or reversing age-related loss of cell proliferative and synthetic capacity.

The following chapters show that there is a clear need for increased understanding of the age-related changes in the musculoskeletal cells and tissues and their relationships to musculoskeletal impairment. Based on this knowledge, it will be important to develop clinically applicable interventions that modify age-related changes and to assess their benefits in terms of improved quality of life versus their costs. Progress in accomplishing this goal requires investment in investigator-initiated research and in the research models, methods, and clinical information necessary to provide background for this research. Examples of the type of work needed include development of better animal models of aging, in vitro model systems (especially of human tissues), autopsy and biopsy studies of human tissues, descriptive longitudinal population studies, and improved methods of measuring musculoskeletal function. By identifying methods of maintaining or restoring mobility, these investigations of the age-related changes in musculoskeletal function and soft tissues have the potential to maximize the quality of life and independence for the increasing number of people who are middle-aged and older.

Section One

Age-Related Changes in Function

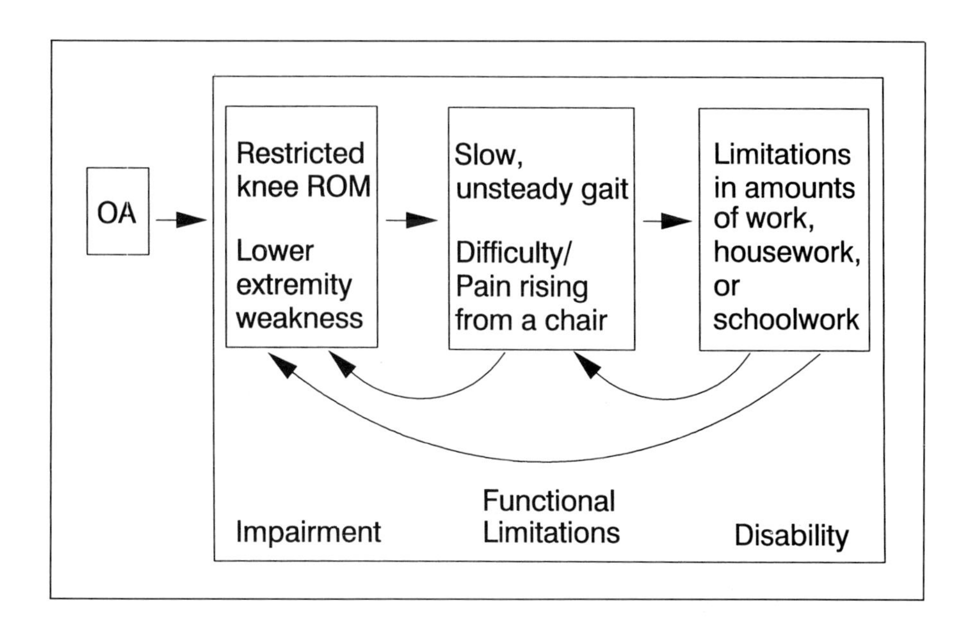

Section Leader

Evan C. Hadley, MD

Section Contributors

Judith Campisi, PhD
Arnold I. Caplan, PhD
Victor M. Goldberg, MD
Krishan L. Gupta, MD
Evan C. Hadley, MD
Stephen E. Haynesworth, PhD
Alan M. Jette, PhD
Vincent M. Monnier, MD
Roland W. Moskowitz, MD
H.K. Pokharna, PhD
Daniel Rudman, MD
Albert B. Schultz, PhD
David R. Sell, PhD
Kaup R. Shetty, MD

Overview

This section addresses the impact of musculoskeletal diseases in older persons, biomechanical factors in impaired physical performance abilities in old age, and methodologic problems in elucidating the role of aging processes in diseases of aging.

As discussed in Dr. Jette's chapter, physical disability is a major consequence of chronic diseases, and unless projected demographic trends are altered, there will be a substantial increase in the prevalence of disability in future elderly cohorts. Epidemiologic data reveal the high prevalence of musculoskeletal impairments among older persons. These impairments are a major causal factor leading from chronic disease to disability in old age.

The impact of musculoskeletal impairments on disability is specific to the site and severity of impairment. Little is known, however, about the medical, social, psychologic, behavioral, and economic factors that influence the extent to which musculoskeletal impairments lead to disability in an individual.

Age-related changes in musculoskeletal soft tissues are particularly likely to have some responsibility for the high frequency of mobility impairments in older persons. Conversely, mobility impairments almost certainly contribute to changes in these tissues in many older persons. Mobility impairments ultimately express themselves as changes in the biomechanics of physical task performance. Dr. Schultz' review notes that, although age-related differences in several biomechanical parameters relating to strength, balance, and specific performance tasks have been described, it is not clear which, if any, of those studied to date contribute to significant physical disability.

Although age-related musculoskeletal impairments are major causes of disability, there is much to learn about how other factors influence responses to these impairments and, thus, the degree of disability. Specifically, we especially need longitudinal data on factors that mediate the relationship between musculoskeletal impairments and specific disabilities, better field measures of these impairments, and development of secondary prevention strategies to detect and halt, slow, or reverse their progression.

Critical aspects of mobility task performance must be identified in order to assess and intervene against mobility impairments effectively. In particular, we need to learn how age-related changes in joint ranges of motion, joint torque strength, and rate of strength development relate to impaired mobility. There is also a need for better assessments of risk for mobility impairments, to permit detection at an early stage of development.

In the search for connections between aging processes and age-related morbidity, methodologic issues arise from the interplay of several aging-related phenomena: heterogeneity among the aging population of levels and rates of change in physiologic functions; high prevalence of chronic disease in old age; presence of many still-undefined pathologies of aging; the long time span of human aging changes compared to that of individual research careers; and differences in age-related pathologies between experimental animals and humans.

Because of these phenomena, the potential contribution of an age-related change to age-related morbidity would be clarified by studies in which the distribution, as well as the mean, of age-related differences is considered; the proportion of the population in whom age-related change enters a "morbid range" (levels at which it may contribute to disease) is estimated; and ages spanning as much of the life span as possible, including the "oldest old" are included. The contributions of an age-related change to morbidity are also clarified by appropriate comparisons between subjects who do and do not have the pathology of interest, other potentially related pathologies, and differences in such other related variables as levels of physical activity. Artifactual errors in animal studies can be avoided by attention to differences in age-related pathology between animal models and humans.

Several aging processes and age-related changes have great potential relevance to age-related changes in the musculoskeletal soft tissues. These include post-translational protein modification, increased prevalence of low levels of trophic hormone, limited cell proliferative capacity and associated effects on expression of differentiated function, and changes in stem cell populations. The chapters described below present explorations of the mechanisms underlying these phenomena, their impact on musculoskeletal tissues, and, in some cases, results of interventions to reverse age-related changes.

As noted in Dr. Monnier's chapter, collagen-rich tissues such as tendons and ligaments undergo progressive cross-linking changes, which are not adequately explained by enzymic reactions. In particular, glycosylation products, including pentosidine cross-links, increase with age and appear to contribute to joint stiffness in some cases. These glycosylation products may also stimulate macrophages and chondrocytes to release cytokines and proteases, which may play a role in osteoarthritis.

Dr. Rudman's chapter presents results of a clinical trial based on the observation that a large proportion of older persons have low levels of growth hormone. Older persons with these low levels appear to lose lean body mass faster than the general population of the same age. Administration of growth hormone to older growth hormone-deficient subjects appears to reverse several age-associated changes to which growth hormone deficiency may contribute, including loss of muscle mass. Excessive elevation of growth hormone levels, however, may produce musculoskeletal complications such as carpal tunnel syndrome.

The limited in vitro proliferative lifespan of human fibroblasts has been actively studied as an aging model. Because the onset of proliferative senescence also appears to be associated with changes in the expression of differentiated function, this phenomenon may have important implications for musculoskeletal soft tissues. Dr. Campisi's chapter presents important recent discoveries of transcription factors involved in loss of proliferative capacity, which raise the possibility of identifying one or a few genes responsible for ultimate control of this phenomenon.

A second issue relating to cell kinetics that has great potential relevance to musculoskeletal soft tissues is loss or change in behavior of stem cells. The chapter by Drs. Haynesworth, Goldberg, and Caplan presents results from a new model system, which indicate that rats do not appear to have major age-related differences in the ability of mesenchymal stem cells to generate cartilage-forming colonies, although old rat marrow appears to yield fewer mesenchymal stem cell colonies.

In all the above lines of research, many questions remain unanswered. We still need to know the extent of the contribution of damage from glycosylation products to degeneration of disks and cartilage, and whether inhibitors of the formation of these products could prevent conditions such as osteoarthritis. The effects of growth hormone replacement therapy on the full range of target organs affected by growth hormone and the effects of long-term therapy remain to be determined. Identification of transcription factors controlling proliferation and differentiation-specific gene expression in musculoskeletal tissues that undergo age-related degenerative change would permit studies of age-related changes in their expression and explorations of ways to modify such changes. Studies of changes in mesenchymal stem cell populations and their ability to yield functional differentiated progeny need to be expanded to human tissues. The role of age-related changes in production of other autocrine and paracrine factors besides TGF-β, and in tissue responses to these factors, is a very promising topic.

The above lines of research reflect only some of the mechanisms of aging processes which could contribute to age-related musculoskeletal soft tissue changes. Others, including unprogrammed cell death and oxyradical damage, have equally important potential effects. The contribution of all these aging processes to age-related musculoskeletal changes and disorders is a topic ready for energetic exploration.

Chapter 1

Musculoskeletal Impairments and Associated Physical Disability in the Elderly: Insights from Epidemiological Research

Alan M. Jette, PT, PhD

Introduction

Prevalence of Disability in Modern Society

The population of the United States is aging—an estimated 12.7% of the total U.S. population in 1990 (31.7 million persons) were 65 years of age or older, and this segment of the population is forecast to grow by the year 2020 to 51.4 million, 17.3% of the population.[1] The prevalence rates of most chronic conditions rise with age,[2] and the individual and social burdens of chronic diseases are rising accordingly.

Physical disability is the chief consequence of chronic diseases and, because most chronic diseases are lifelong, subsequent disability can trouble the individual throughout his or her remaining life span. Existing data show all too clearly that unless projected trends are changed, changing demographics will substantially increase the incidence and prevalence of disability in the coming years. In 1988, 10.6 million people 65 and older in the United States (37% of the non-institutionalized population) reported some limitation of their activities as a result of chronic disease or impairment; 14.4% were limited but not in their major activity; 12.1% were limited in the amount or kind of their major activity; and 10.5% were unable to carry on their major activity.[3] Roughly 40% of the elderly population have some activity limitation, and about 17% need assistance in basic life activities.[4] The age gradient for activities of daily living disability is substantial. By age 85, the risk of significant disablement approaches 50/50.

Assuming that age- and sex-specific disability prevalence rates stay the same, some analysts have estimated that the elderly population with chronic disabling conditions could grow by 31% to 7.2 million between 1985 and 2000.[5] This figure compares with a projected increase of 20% in the nondisabled population. In 2010, as the baby boom generation reaches their aged years, the number of people 65 and older with chronic disabling conditions could exceed 15 million.[2,5]

These projections have led to an increased emphasis on the epidemiology of disability—an investigation and understanding of the major mechanisms or pathways through which chronic disease, injury, and accident impact on the older person's ability to act in necessary, expected, and personally desired ways in society.[2] As this review of some of the epidemiologic literature will show, musculoskeletal disease and resultant impairments are among the most common and disabling of chronic conditions, and they deserve increased attention as approaches are developed to stem the rising tide of disability among our aging population.

This review is organized into three parts. In the first section, the prevalence of arthritis is reviewed, along with the research evidence linking it to various forms of physical disability. The second section reviews the research that has linked specific types and sites of musculoskeletal impairment to subsequent physical disability. In the final section, a model of disablement is presented with specific definition of terms within the model and suggestions for fruitful areas of future research.

Arthritis and Disability

Arthritis is the most prevalent chronic condition affecting the middle-aged and elderly.[6,7] Among the working-age, adult population in the United States, 3.7 million men and 5.6 million women report having arthritis.[8] Among men, 1 in 5 between the ages of 45 and 64 have arthritis, the second most common cause of activity limitation among this age group. One of every three women between the ages of 45 and 64 report having arthritis, the primary cause of work loss and most common cause of activity limitation.[8] Arthritis is the leading condition for women of middle and old age with prevalence rates 20% to 25% higher than their second-ranked condition, high blood pressure. Arthritis ranks first among men aged 65 to 74 years and second among those 45 to 64 and 75 and older. Among the elderly, arthritis ranks as the most prevalent with almost half (49%) reporting this condition. Deformity or orthopaedic impairment ranks seventh, reported by 16.1% of the 65 and older population.[3] As the data in Table 1 show, prevalence rates of arthritis are higher for women than for men regardless of race, and generally, prevalence increases with advancing age.[9]

The link between musculoskeletal disease and disablement is quite clear. Musculoskeletal diseases are among the chronic medical conditions that most consistently predispose an older individual to disability. This finding has been documented in the National Health Interview Survey's Supplement on Aging, the Alameda County study, the Framingham Study, and the 70-Year-Old Study in Gothenburg, Sweden.[7,10-14] Arthritis achieves its dubious status as a disabling condition through its high prevalence as well as its impact

Table 1 Arthritis prevalence rates in the United States in 1984 (%)

Arthritis	Men	Women
White		
55-64	28.8	44.2
65-74	39.6	52.8
75-84	39.5	57.2
85 +	35.4	55.1
Non-White		
55-64	36.3	45.8
65-74	45.0	61.2
75-84	40.3	69.0
85 +	53.9	54.3

(Adapted with permission from Verbrugge L, Lepkowski J, Konkol L: Levels of disability among U.S. adults with arthritis. *J Gerontol Social Sciences* 1991;46:S71-S83.)

on musculoskeletal impairments. Furthermore, when arthritis is augmented with other chronic conditions, disability levels increase considerably.[7]

Arthritis is reported most often as the principal cause of limitations (disability) by women middle-aged and older, and as the second-ranked cause of limitations among men of those same ages (after heart diseases).[7,15] In comparing arthritis with 12 other conditions of public health importance, Verbrugge and associates[9] showed that arthritis produces a moderate disability impact. People with arthritis experience more disability in their physical, personal care, and household care activities than do those without the disease.[9] This finding has been reported by others on regional samples (Table 2).[16]

Prevalence of Musculoskeletal Impairments

As has been shown by Guccione and associates[17] for arthritis, severity of disease has a profound impact on risk of disablement. Yet, most previous research focused on the relationship of specific diseases such as arthritis with disability and have all but ignored the type and severity of musculoskeletal impairments that intervene between a disease and subsequent disablement. Direct measurements of musculoskeletal impairments are required to understand better how disease (or pathology) leads to subsequent disability.[18] Such information is required to guide clinicians in the development of intervention strategies to prevent or retard physical disablement.

In the United States, Cunningham and Kelsey[19] have documented the overall prevalence of musculoskeletal impairments among non-institutionalized adults using the United States Health and Nutrition Examination Survey (HANES I) data (1971-1975). Musculoskeletal impairments were measured by physician-observed signs and symptoms of musculoskeletal disorders (such as joint tenderness, joint swelling, limitation of motion, and pain on motion) and self-report of musculoskeletal symptoms (such as pain in back, neck, hip, knee,

Table 2 Prevalence of disability for noninstitutionalized persons in the United States, aged 55 and older: With and without arthritis

	Arthritis (%)	Nonarthritis (%)
Difficulty in Physical Actions:		
walking	22.4	8.3
getting outside	10.1	4.2
walking 1/4 mile	34.8	14.9
standing 2 hours	41.9	17.4
stooping, crouching, kneeling	49.8	19.6
reaching over head	22.0	7.6
using fingers to grasp	16.5	4.0
lifting 25-lb weight	43.8	20.0
Difficulty in Personal Care:		
bathing	10.6	4.8
dressing	7.2	3.1
eating	1.6	1.1
transfers	10.8	3.5
toileting	4.1	2.0
Difficulty in Household Management:		
preparing meals	6.8	3.5
shopping	11.2	5.4
managing money	3.8	2.8
using telephone	3.7	2.7
heavy housework	29.2	11.2

(Adapted with permission from Verbrugge L, Lepkowski J, Konkol L: Levels of disability among U.S. adults with arthritis. *J Gerontol Social Sciences* 1991;46:S71-S83.)

or other joints on most days, lasting at least a month; joint swelling and pain or morning stiffness lasting at least a month (Table 3).

An estimated 60% of the 65- to 74-year-old group had physician-observed musculoskeletal disorders, 39.6% of this group reported musculoskeletal symptoms. Musculoskeletal symptoms and observed abnormalities were more prevalent among individuals of lower income and lower education level, and among widowed persons. While women were more likely than men to report musculoskeletal symptoms, there were no gender differences in physician-observed musculoskeletal abnormalities. Back symptoms (17.2%) and knee symptoms (13.3%) were the most frequently reported musculoskeletal impairments.

In Sweden, Bagge and associates[20] documented the prevalence of joint complaints in a cohort at ages 70, 75, and 79. In this study, subjects were asked if they had presently or ever had any problems with their joints, such as pain, stiffness, or swelling. Those who responded affirmatively were then asked to specify the site(s) of the complaint(s). As in the HANES I study, gender differences were seen in prevalence of joint complaints. Joint complaints were reported by 30% to 43% of the women and by 15% to 25% of the men. The prevalence of joint complaints increased significantly between the ages of 70 and 75 for both men and women. After age 75 no significant age increase was seen (Table 4).

Complaints of the lower extremities were more common than complaints of the upper extremities. The most common site of com-

Table 3 Prevalence of Musculoskeletal Impairments in HANES I Examinees (n = 6,913)

	Musculoskeletal Symptoms %	Musculoskeletal Signs %
Age		
55-64	39.1	47.5
65-74	39.6	60.3
Sex		
male	28.6	34.1
female	32.2	35.9
Race		
white	30.8	35.0
non-white	29.0	35.8
Education		
<12 yrs.	36.7	46.6
12 yrs.	27.6	28.8
>12 yrs.	24.6	24.6
Income		
<$5,000/yr.	39.1	48.4
$5,000-$9,999	30.3	36.7
$10,000-$14,999	27.5	29.4
$15,000 +	24.1	23.4
Marital Status		
married	30.1	32.9
widowed	41.0	57.5

(Adapted with permission from Cunningham L, Kelsey J: Epidemiology of musculoskeletal impairments and associated disability. *Am J Public Health* 1984;74:574-579.)

plaint in both men and women was the knee joint. Women were much more likely than men to report complaints in the hands. Few subjects had complaints of both the upper and lower extremities. Unfortunately, findings were not presented for sociodemographic subgroups of the cohort (Table 5).

Musculoskeletal Impairments and Disability

What is known about the associations of musculoskeletal impairments and disability in the elderly? Clearly, among those elders with musculoskeletal impairments, some are disabled by their condition while others are not. Relatively little is known of the circumstances under which musculoskeletal impairments are associated with disability.

In the HANES I sample, analyses suggest that medical, social, and economic factors all play a role in determining whether a person with musculoskeletal impairments also reports disability related to his or her impairment. In HANES I, persons classified as having musculoskeletal disability were those who reported having their physical activity restricted to a marked degree, those who had to change their job status, and those who had lost five or more days from work in the past year because of their joint condition. Among persons 65 years of age and older who reported a history of musculoskeletal symptoms, 28.8% reported moderate to severe activity restriction, 20.5% reported a change in job status, and 9.2% re-

Table 4 Prevalence of Joint Complaints in Swedish Women and Men at Age 70, 75, and 79

Age	Women (n=620)		Men (n=422)	
	%	(95% CI)	%	(95% CI)
70	30.2	(26.6–33.9)	14.9	(11.7–18.7)
75	43.1**	(36.1–47.0)	25.1**	(21.0–29.5)
79	38.7	(34.8–42.7)	22.0	(19.0–23.3)

**p<.01 denoting increase with age
(Reproduced with permission from Bagge E, Bjelle A, Eden S: A longitudinal study of the occurrence of joint complaints in elderly people. *Age Ageing* 1992;21:160-167.)

Table 5 Site of Joint Complaints in Swedish Women and Men at Age 70, 75, 79

Age	Joint	Women (n=620)	Men (n=422)
70	hand	10.4**	2.3
75		14.4**	4.1
79		14.4**	6.1
70	elbow	4.8**	0.8
75		1.9	0.4
79		2.1	1.2
70	shoulder	5.6	2.8
75		9.6	8.5
79		9.1	5.7
70	hip	4.5	4.1
75		9.6	5.7
79		10.7	8.1
70	knee	10.7	9.7
75		17.1**	13.4
79		24.9**	11.3
70	feet	3.5	2.4
75		6.7	2.8
79		3.7	2.0

** p<.01 for gender differences
(Reproduced with permission from Bagge E, Bjelle A, Eden S: A longitudinal study of the occurrence of joint complaints in elderly people. *Age Ageing* 1992;21:160-167.)

ported having lost five or more days from work in the past year because of their joint condition.

As Table 6 indicates, persons who reported involvement of all areas were most likely to report physical disability. Persons with lower extremity and back/neck impairments were the next most likely to report a disability outcome. Upper extremity impairment alone was the musculoskeletal impairment least likely to be associated with moderate to severe disability.

The HANES I findings suggest a number of social reasons why musculoskeletal impairments may or may not lead to subsequent disablement. Non-whites, lower education, lower income, and widowhood were all related to one or more of these disability indicators. A lack of financial and/or social resources with which to deal with a musculoskeletal impairment appears to be an important factor associated with disablement. The cross-sectional design of HANES I did not permit an analysis of whether musculoskeletal impairments were associated with incidence of disability. Factors

Table 6 Site of Musculoskeletal Impairment and Associated Disability

Area of Impairment	(Number)	Mod./Sev. Activity Restriction
UE/LE/Back/Neck	(348)	37.8
UE/LE	(206)	19.1
UE/Back/Neck	(123)	20.0
LE/Back/Neck	(367)	31.0
UE Only	(208)	7.8
LE Only	(441)	13.6
Back/Neck Only	(389)	17.8

UE = Upper extremity
LE = Lower extremity

other than musculoskeletal impairments have been shown to be important mediating mechanisms for subsequent disability.[21] Social disadvantage is a hypothesis that deserves further investigation.

Specificity of Effect

As the few cross-sectional studies suggest and the one longitudinal study confirms, there appears to be a specific causal pathway between specific sites of musculoskeletal impairments and subsequent disablement. Restrictions or limitations in activities that involve similar types of action or that use the same parts of the body cluster together, and the degree of disablement relates to the severity and type of impairment. In Cunningham and Kelsey's HANES I analyses,[19] for example, persons who reported musculoskeletal impairments in multiple areas were most likely to report disability; upper extremity impairment was associated with the least disability.

Badley and associates,[22] in a sample of 95 adult patients with arthritis, found that mobility disability was highly correlated with limitation of knee joint motion, and bending activities were associated with range of hip flexion. Hip range was not associated with mobility. Flexion of the hand joints was highly correlated with disability in dexterity, bending arm, and reaching-up activities (Table 7).

In a Swedish investigation of 89 community-dwelling adults, who were 79 years of age, moderately high positive correlations were found between the performance of specific functional tasks and specific sites of musculoskeletal impairment (Table 8).[10] Cervical spine impairments were correlated with raising hands above the head, an earlobe test, and using public transportation. The need for walking aids was correlated with impaired hip range of motion. Ability to grasp the opposite earlobe was related to restricted range of motion of the upper extremities. Hip and knee restrictions were correlated with restricted use of public transportation. These investigators also reported that restricted range of motion was of greater importance to the occurrence of a disability than subjects' complaints of joint problems.[10]

Table 7 Relationship Between Range of Motion of Joints and Physical Disability (n=95)

ROM of:	Mobility	Bending Down	Dexterity	Bending Arm	Reaching Up
Knee Flexion	0.56	—	—	—	—
Hip Flexion	—	0.53	—	—	—
Shoulder Abduction	—	—	0.42	0.70	0.78
Wrist Extension	—	—	0.47	0.58	0.55
Thumb Circumduction	—	—	0.63	0.58	0.58
Proximal Interphalangeal Joint Flexion	—	—	0.61	0.47	0.46

(Adapted with permission from Badley EM, Wagstaff S, Wood PH: Measures of functional ability (disability) in arthritis in relation to impairment of range of joint movement. *Ann Rheum Dis* 1984;43:563-569.)

Table 8 Association of ROM of Joints and Physical Disability (n=89)

Restricted ROM of:	Hands Above Head	Ear-Lobe Test	Use Public Transport	Use Walking Aids
Cervical Spine	0.38**	0.34**	0.35**	—
Wrist (mod/severe)	0.28	0.33**	0.08	—
Fingers (severe)	0.37**	0.41**	0.05	—
Shoulder (active)	0.40**	0.36**	0.30**	—
Lumbar Spine	—	—	0.07	0.15
Hip	—	—	0.35**	0.27**
Knee	—	—	0.49**	0.20

**p<.05
(Adapted with permission from Bergstrom G, Aniansson A, Bjelle A, et al: Functional consequences of joint impairment at age 79. *Scand J Rehabil Med* 1985;17:183-190.)

Jette and Branch,[23] for a longitudinal study of a statewide sample of non-institutionalized elderly individuals in Massachusetts, trained field staff to assess musculoskeletal impairments among cohort members at each examination by observing and scoring subjects' performance of ten gross body movements designed to put the major body joints through their complete range of motion.[24] These investigators demonstrated a high prevalence of musculoskeletal impairment among this cohort of individuals 70 years of age and older (Table 9). Among the 85 and older subgroup, over half displayed impairments of hip and knee flexion, and/or wrist motion. As the correlations in Table 10 illustrate, the pattern of musculoskeletal impairments in this cohort was quite distinct. Correlations among areas of impairment were moderately positive.

In longitudinal analyses of this same cohort,[25] these investigators showed that the emergence of specific areas of joint and muscle impairment during a five-year follow-up period was a significant causative factor in explaining subsequent development of physical disability. In these analyses, physical disability was measured as subjects' self report of dependence on another person or as inability to perform ten different life activities.[18] In multivariate analyses, the progression of hand impairments for both men and women in the cohort was related to increasing disability in such basic activities of daily living as eating, bathing, and dressing. The progression of lower extremity impairments was associated with worsening disability in more complex activities of daily living, such as grocery

Table 9 Prevalence of Musculoskeletal Impairment in Massachusetts Elders

Body Motion	Percent With Limitation by Age 71-74 %	75-84 %	85+ %	Total %
Opposition	3	7	7	5
Complete Fist	7	12	15	11
Wrist Extension**	20	28	52	28
Wrist Flexion**	28	32	59	33
Arms Above Head**	15	21	38	21
Hip and Knee Flexion**	16	24	51	24
(n)	(255)	(380)	(76)	(711)

(Reproduced with permission from Jette A, Branch CG: Musculoskeletal impairment among the non-institutionalized aged. *Int Rehabil Med* 1984;6:157-161.)
**p<.05 for age difference

Table 10 Correlation Across Site of Musculoskeletal Impairment

	(n)	1	2	3	4
1. Lower Extremity	(699)				
2. Upper Extremity	(701)	0.50			
3. Wrist	(707)	0.34	0.60		
4. Hand	(711)	0.24	0.37	0.31	

(Reproduced with permission from Jette A, Branch CG: Musculoskeletal impairment among the non-institutionalized aged. *Int Rehabil Med* 1984;6:157-161.)

shopping, housekeeping, and food preparation. Women were more likely than men to experience worsening disability in these more complex life activities.

Decrement in hand function was a significant musculoskeletal mechanism that led to observable disablement in basic activities of daily living. This finding is most likely a reflection of the fine motor control needed to perform many of the basic life activities such as dressing and bathing. In contrast, the progression of lower extremity impairments led to instrumental disabilities, reflecting the important role of the lower extremities in complex, strenuous activities such as housekeeping, food preparation, and driving an automobile.

The longitudinal design of Jette and associates[25] should allay fears about the proper sequence in the causal chain. It is the progression of musculoskeletal impairments that leads to disablement and not the reverse. Impairments clearly precede the progression of disability in specific life activities. Further longitudinal research is needed on the specific causal pathways for different basic and instrumental life activities.

A Model of Disablement

What has been learned from existing epidemiologic research? What are some of the most promising directions for future research regarding the relationship of musculoskeletal impairments and disablement?

First, existing epidemiologic data clearly demonstrate the high prevalence of musculoskeletal impairments among the elderly. This is an area of great concern to health providers and today's elderly and is likely to become increasingly important among future generations of the elderly.

As the epidemiologic data demonstrate, musculoskeletal impairments in the elderly have serious consequences. The available evidence shows that they are a major cause of physical disability among the elderly. The data further show that the relationship between musculoskeletal impairments and disability is very specific. The site of an impairment is a critical factor in understanding if and how it will relate to subsequent disablement. This finding, of course, is consistent with clinical experience. As this review has illustrated, musculoskeletal impairments represent one of the major causal mechanisms leading to disablement in older persons.

What research has not shown are the critical factors that mediate the effects of musculoskeletal impairments on disability. This is an area of great importance. What determines whether or not arthritis leads to impairment and impairment to disablement? Future research needs to be directed toward (1) longitudinal research that tracks the progression of musculoskeletal impairments and their impact on physical disablement to better identify the key modifying factors associated with change; and (2) the development of promising secondary and tertiary prevention intervention strategies needed to halt the onset of disability among those with musculoskeletal impairments.

The terms impairment and disablement are used with some imprecision by different professions and in previous disability research, leading to some confusion. The situation is compounded further by a general lack of agreement on how these concepts should be measured. Future disability research needs to define clearly the major disablement concepts and to adhere to these concepts within their research. The following are offered as suggestions for future musculoskeletal disablement research.

Impairments are abnormalities in specific organs or body systems, which usually represent a response to a specific pathologic process, injury, or accident. An impairment can be anatomic, physiologic, mental, or emotional. The emphasis and focus in this review, of course, are on impairments that affect the musculoskeletal system. Examples of musculoskeletal impairments include restricted range of motion and muscle weakness. Typically, musculoskeletal impairments are abnormalities in soft tissue that result from some pathologic process, such as osteoarthritis, or are residual abnormalities that remain after a pathology has been controlled or eliminated, for example, a healed amputation.[26] Common ways of measuring musculoskeletal impairments include the standard clinical examination (e.g., number of tender, inflamed, or swollen joints), patient history, symptom reports (e.g., pain, morning stiffness), or standardized clinical tests or protocols (e.g., isokinetic strength testing, range of motion).

Functional limitations, which represent the individual's inability or limitations in the performance of specific tasks and actions considered normal, are frequent consequences of musculoskeletal impairment. A gait deviation is a common functional limitation seen in older persons. The final step in the disablement process is the concept of physical disability, which refers to the impact chronic conditions have on the behavior of an older person in necessary, expected, and personally desired roles and activities in society. This definition of disablement is consistent with others seen in the literature.[2,26-28]

Physical disabilities are frequently discussed in terms of basic and instrumental activities of daily living. The term, basic activity of daily living, refers to basic physical tasks of life, such as eating, dressing, bathing, toileting, transferring, and walking.[27,29] Instrumental activities of daily living encompass the range of more complex life activities, and include meal preparation, shopping, outside mobility, housework, traveling, handling finances, and taking medications.[30]

Figure 1 provides a musculoskeletal example of the cycle of disablement. A pathologic process, in this example osteoarthritis, can have various consequences or effects. Commonly seen impairments include restricted range of joint motion and lower extremity joint weakness. These impairments, in turn, may result in specific functional limitations—in this example, a slow, unsteady gait and difficulty and pain in rising from a chair. The resultant physical disability includes limitation in the amount of work or housework performed by the individual.

An important element that is included in this disablement cycle is an understanding that the disabling process is not a unidirectional progression. The feedback loops shown throughout the model indicate the potential for primary elements in the model to have secondary consequences for the older person. For example, inactivity due to musculoskeletal impairments can lead to functional limitations, which in turn result in further inactivity leading to secondary cardiac and pulmonary deconditioning effects.

Disability Prevention

Traditionally, the public health field has focused its efforts on primary prevention—averting the onset of pathologic processes by reducing susceptibility, controlling exposure to disease-causing agents, and eliminating, or at least minimizing, behavioral and environmental factors that increase the risk of disease or injury. Work to understand the underlying mechanism of musculoskeletal disease in old age is critical and will eventually lead to important advances in primary prevention of potentially disabling disease. Not all potentially disabling conditions (such as arthritis) however, can be pre-

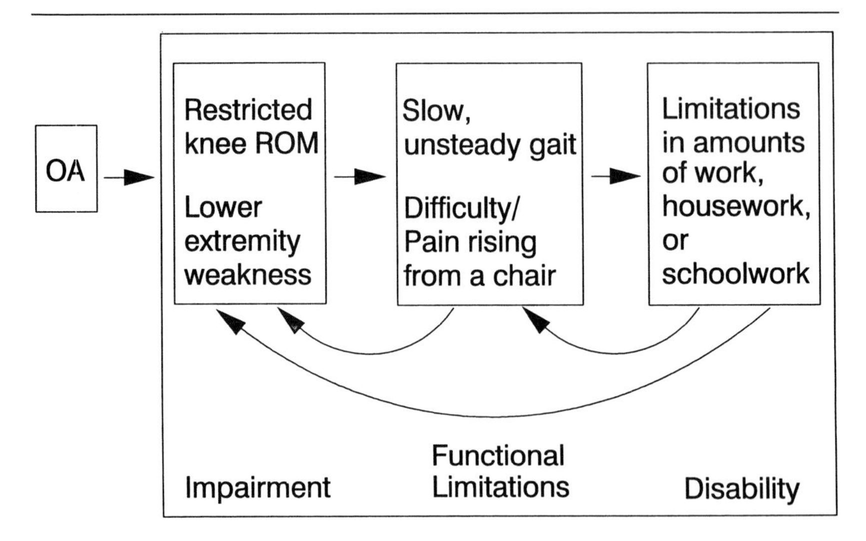

Fig. 1 *Musculoskeletal disablement cycle.*

vented given current limitations in our understanding of the etiology of these diseases. Nevertheless, much can be accomplished by focusing attention on means of preventing the consequences of chronic diseases. To prevent disablement, much more emphasis must be directed toward the needs of people who already have potentially disabling impairments—secondary prevention, and those who are already disabled—tertiary prevention (Fig. 2).

Secondary prevention includes efforts designed for early detection of potentially disabling conditions (such as arthritis), followed by the implementation of interventions designed to halt, reverse, or slow the progression of resultant impairments.[2] Some analysts strongly advocate secondary prevention,[31] arguing that we must attempt to control or eliminate risk factors that affect the onset or severity of impairments and resultant disablement rather than try to prevent the onset of the underlying pathologic process itself. Therapeutic exercise has proved to be of value in reducing musculoskeletal impairments and disability among those with various types of arthritis.[32,33]

Tertiary prevention of disability attempts to restore functional ability and prevent further functional limitations and disability among those who are already disabled. This includes therapeutic efforts traditionally referred to as geriatric rehabilitation. A growing body of literature on geriatric rehabilitation has begun to document its positive impact on functional ability and mortality.[34] Further work is needed to extend the scope of tertiary prevention beyond the limitations of traditional health interventions to encompass community-wide efforts to change behavior and/or modify the environment to reduce physical disability, prevent secondary com-

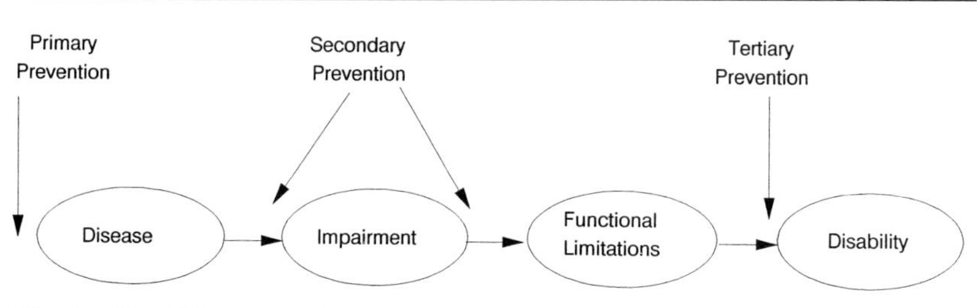

Fig. 2 *Disability prevention strategies.*

plications, and/or restore actual function among older persons who are already disabled.[2]

The models that have guided previous epidemiologic research, although useful, are too simplistic for the research needs outlined above. As Figure 1 illustrates, musculoskeletal impairments have been posited as the major pathway through which specific diseases like arthritis lead to physical disability. In future research, this basic model needs to be expanded to include relevant demographic, biologic, social, psychological, or behavioral attributes or conditions that either elevate or reduce the chances that a specific pathology will lead to impairments, functional limitations, and/or physical disabilities. The elucidation of important modifying factors will become a source of new secondary and tertiary prevention interventions needed to halt the cycle of disablement among older persons.

Figure 3 illustrates how various modifying attributes can be classified into a more fully developed model. Risk factors, one type of modifying factor, are those phenomena, present either before or after the onset of a chronic condition, that increase the chances of disablement. Risk factors such as low education or low income may exist before the onset of a chronic condition; others, such as iatrogenic complications from treatment, alcohol abuse, or anger may appear after the onset of a disease. Risk factors for disability can also be externally imposed—examples are architectural barriers, social prejudice, and inflexible work hours.

In contrast, buffers are modifying attributes that reduce risk of disability.[35] Professional interventions, including physical and occupational therapy, medicine and surgery, counseling, personal assistance and mechanical aids, coping, lifestyle changes and environmental accommodations, are examples of buffers that the affected individual or others may apply in an effort to avoid, retard, minimize, or reverse disability. Risk factors and buffers can be arrayed around the main concepts in this disablement model to illustrate how they can influence transitions from one state to another.

Disability researchers and health professionals currently have little existing epidemiologic data to guide their efforts to identify and test promising interventions on which to build chronic disease management plans. This simple disablement model can be used to gener-

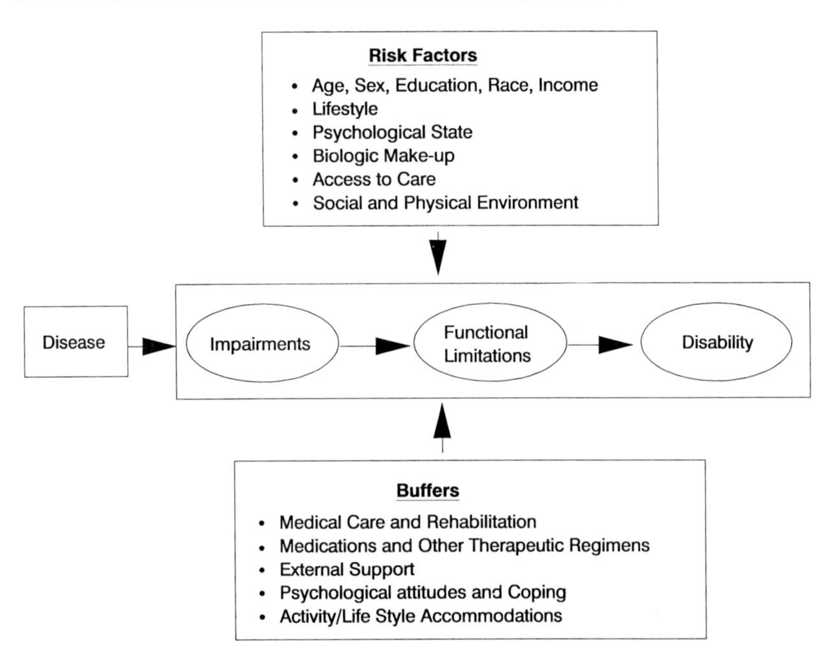

Fig. 3 *Relationship of risk factors and buffers to disablement.*

ate hypotheses for both patient- and population-based disability research. The model briefly outlined here can be used to generate hypotheses about the association of specific musculoskeletal diseases, associated impairments, and risk of disablement. It can help move the field beyond its current, mostly descriptive emphasis toward more analytic studies of disablement that can guide public health policies and programs and suggest interventions that can be applied either to treat or to prevent physical disability.

References

1. U.S. Bureau of the Census. Spencer G (ed): Projections of the Population of the United States, by Age, Sex and Race: 1983 - 2080. Washington, DC: 1984 Series: Current Population Reports; vol P-25, No. 952.
2. Pope AM, Tarlov AR (eds): *Disability in America: Toward a National Agenda for Prevention.* Washington, DC, National Academy Press, 1991.
3. National Center for Health Statistics. Hardy AM, Adams PF (eds): Current Estimates from the National Health Interview Survey, 1988. Vital and Health Statistics, Series 10, No. 173. Washington, DC, 1989; vol DHHS Pub. No. (PHS) 89-1501.
4. LaPlante M: Disability risks of chronic illnesses and impairments. Disability Statistics Report 2. San Francisco, University of California, Institute for Health and Aging, 1989.
5. Manton KG: Epidemiological, demographic, and social correlates of disability among the elderly. *Milbank Q* 1989;67(Suppl 2, Part I):13-58.

6. Dawson DA, Adams PF (eds): Current estimates from the National Health Interview Survey, United States, 1986. Hyattsville, MD: National Center for Health Statistics, 1987 Vital and Health Statistics, Series 10, No. 164; vol DHHS Publ. No. 87-1592.
7. Verbrugge LM, Lepkowski JM, Imanaka Y: Comorbidity and its impact on disability. *Milbank Q* 1989;67:450-484.
8. Yelin E: Arthritis: The cumulative impact of a common chronic condition. *Arthritis Rheum* 1992;35:489-497.
9. Verbrugge LM, Lepkowski JM, Konkol LL: Levels of disability among U.S. adults with arthritis. *J Gerontol* 1991;46:S71-S83.
10. Bergstrom G, Bjelle A, Sundh V, et al: Joint disorders at ages 70, 75 and 79 years: A cross-sectional comparison. *Br J Rheumatol* 1986; 25:333-341.
11. Jette AM, Pinsky JL, Branch LG, et al: The Framingham Disability Study: Physical disability among community-dwelling survivors of stroke. *J Clin Epidemiol* 1988;41:719-726.
12. Harris T, Kovar MG, Suzman R, et al: Longitudinal study of physical ability in the oldest-old. *Am J Public Health* 1989;79:698-702.
13. Guralnik JM, Kaplan GA: Predictors of healthy aging: Prospective evidence from the Alameda County study. *Am J Public Health* 1989;79:703-708.
14. Verbrugge LM: Disability. *Rheum Dis Clin North Am* 1990;16(3): 741-761.
15. LaPlante MP: Data on Disability from the National Health Interview Survey 1983-1985: An Info Use Report. Washington, DC, National Institute on Disability and Rehabilitation Research, 1988.
16. Yelin E, Lubeck D, Holman H, et al: The impact of rheumatoid arthritis and osteoarthritis: The activities of patients with rheumatoid arthritis and osteoarthritis compared to controls. *J Rheumatol* 1987;14:710-717.
17. Guccione AA, Felson DT, Anderson JJ: Defining arthritis and measuring functional status in elders: Methodological issues in the study of disease and physical disability. *Am J Public Health* 1990;80:945-949.
18. Jette AM, Branch LG: Impairment and disability in the aged. *J Chron Dis* 1985;38:59-65.
19. Cunningham LS, Kelsey JL: Epidemiology of musculoskeletal impairments and associated disability. *Am J Public Health* 1984; 74:574-579.
20. Bagge E, Bjelle A, Eden S, et al: A longitudinal study of the occurrence of joint complaints in elderly people. *Age Ageing* 1992;21:160-167.
21. Summers MN, Haley WE, Reveille JD, et al: Radiographic assessment and psychologic variables as predictors of pain and functional impairment in osteoarthritis of the knee or hip. *Arthritis Rheum* 1988;31:204-209.
22. Badley EM, Wagstaff S, Wood PHN: Measures of functional ability (disability) in arthritis in relation to impairment of range of joint movement. *Ann Rheum Dis* 1984;43:563-569.
23. Jette AM, Branch LG: Musculoskeletal impairment among the non-institutionalized aged. *Int Rehabil Med* 1984;6:157-161.
24. Eberl DR, Fasching V, Rahlfs V, et al: Repeatability and objectivity of various measurements in rheumatoid arthritis: A comparative study. *Arthritis Rheum* 1976;19:1278-1286.
25. Jette AM, Branch LG, Berlin J: Musculoskeletal impairments and physical disablement among the aged. *J Gerontol* 1990;45:M203-M208.

26. Nagi SZ: Appendix A: Disability concepts revisited: Implications for prevention, in Pope AM, Tarlov AR, (eds): *Disability in America: National Agenda for Prevention*. Washington DC, National Academy Press, 1991, pp 309-327.
27. Mahoney FI, Barthel DW: Functional evaluation: The Barthel index. *Md Med J* 1965;14:61-65.
28. Haber L: Issues in the definition of disability and the use of disability survey data, in *Disability statistics, an assessment: Report of a workshop*. Washington, DC, National Academy Press, 1990.
29. Katz S, Ford AB, Moskowitz RW, et al: Studies of illness in the aged: The index of ADL: A standardized measure of biological and psychosocial function. *JAMA* 1963;185:914-919.
30. Lawton MP: Assessing the competence of older people, in Kent DP, Kastenbaum R, Sherwood S (eds): *Research Planning and Action for the Elderly: The Power and Potential of Social Science*. New York, NY: Behavioral Publications, 1972, pp 122-143.
31. Fries JF, Crapo LM: *Vitality and Aging: Implications of the Rectangular Curve*. San Francisco, CA: WH Freeman, 1981.
32. Fries JF: Aging, illness, and health policy: Implications of the compression of morbidity. *Perspect Biol Med* 1988;31:407-428.
33. Kovar PA, Allegrante JP, MacKenzie CR, et al: Supervised fitness walking in patients with osteoarthritis of the knee: A randomized, controlled trial. *Ann Intern Med* 1992;116:529-534.
34. Rubenstein L, Stuck A, Siu A, et al: Impacts of geriatric evaluation and management programs on defined outcomes: Overview of the evidence. *J Am Geriatr Soc* 1991;39:8S-16S; discussion 17S-18S.
35. Verbrugge L, Jette A: The disablement process. *Soc Sci Med*, in press, 1993.

Chapter 2

Biomechanics of Mobility Impairment in the Elderly

Albert B. Schultz, PhD

Introduction

The size of the present population of elderly adults, its projected growth, and the extent of musculoskeletal impairments and physical disability among that population were outlined in an earlier chapter. Age-related changes in musculoskeletal soft tissues likely have some responsibility for the high frequency of mobility impairments in the elderly, and mobility impairments almost certainly have a role in causing changes in the musculoskeletal soft tissues of many elderly people. Studies of mobility impairments in the elderly are thus relevant to the concerns of this Workshop. In turn, mobility impairments ultimately express themselves in the form of changes in the biomechanics of physical task performance. Research on the biomechanics of mobility impairment in the elderly is needed to quantify impairment magnitudes, to determine what elements are critical to those impairments, to improve therapies for remediation and to design more effective programs for prevention. Present evaluations of impairment are largely qualitative. They often serve poorly as indicators of subtle changes in functioning over time, yet there are major needs to understand those changes and their relationship to age-related changes in soft tissues. Therapies and programs for prevention of physical disability can be made more effective once the specific factors important to the maintenance or restoration of function are better understood. Biomechanics research is needed to help achieve that understanding. The following review of some of the relevant issues has been adapted from a recent article.[1]

Age Effects on Musculoskeletal System Components

Anthropometry

The anthropometry of the elderly differs somewhat from that of young adults.[2-5] Stoudt[6] points out that, in the mean, persons 65 to

74 years old are approximately 3% shorter than those 18 to 24 years old. Elderly males are slightly lighter and elderly females approximately 11% heavier than young adult males and females. These differences result both from trends in the population and from biologic changes inherent in the aging process. In illustration of the latter factor, most of the decrement in stature takes place in the trunk and results from a flattening of the intervertebral disks and vertebral bodies and changes in kyphosis.

Joint Ranges of Motion

Ranges of motion (ROM) of body joints generally diminish with age. For example, Allander and associates[7] reported a decline of approximately 20% in hip rotation among 411 subjects aged 45 to 70 years, and declines of 10% in wrist and shoulder ROM. A comparison of the ROM of lower extremity joints reported by Boone and Azen[8] and Roaas and Andersson[9] for young and middle-aged adults with those reported by Walker and associates[10] for elderly adults shows declines that ranged from negligible to 57%. Svanborg[11] found that, at age 79, one-fifth of a large group of subjects had restricted knee joint motion and two-thirds had restricted hip joint motion. Battié and associates[12] tested 3,020 blue-collar workers and found a decline of approximately 25% in ability to bend to the side and a 45% decline in shoulder motion over the period from 20 to 60 years of age. Einkauf and associates,[13] who tested 109 healthy females, found declines of 25% to 50% in various ranges of motion of the lumbar spine between ages of approximately 20 years to 80 years. However, Walker and associates[10] reported no significant differences in 28 different joint ROM between two groups with mean ages of approximately 65 and 80 years.

The effects of decreased ROM on abilities to perform activities of daily living are less well studied, not only in older people but even in young adults. Badley and associates[14] found, when studying 95 subjects aged 28 to 84 years who had arthritis, that ability to move around in one's environment correlated significantly with ROM in knee flexion, ability to bend down correlated with hip flexion ROM, and abilities in activities requiring use of hands and arms correlated significantly with ROM of the upper extremities. Bergström and associates[15] reported that restricted knee motion in 79-year-old subjects correlated with disability in entering public transport vehicles. However, they found that joint impairment was not generally associated with commitment to institutional care. Bergström and associates[16] found that the majority of a group of 134 people, aged 79 years, had enough spinal mobility to perform common activities of daily living. Biomechanics research further detailing relationships between decreased ROM and mobility impairment seems merited.

Muscle Function

The changes in muscle function that accompany aging are described elsewhere. A prevalent casual belief is that many of the mobility impairments that arise in the elderly are caused by a decline in muscular strength. This belief warrants careful consideration. For example, the joint torques needed to maintain postural balance[17] and even to rise from a chair[18] are often well below available joint torque strengths. Further, comprehensive biomechanical studies are needed of the muscle strength requirements of all common activities of daily living in order to relate strength requirements, strength availabilities, and performance in those activities.

Reaction Times

Reaction times, the delays that occur between the onset of a stimulus and the onset of response to it, are of fundamental importance to biomechanical analyses of any tasks performed rapidly. Reaction times are often categorized into simple reaction times (SRT) and choice reaction times (CRT). The two differ in that the latter involves selection among several possible response options and, thus, presumably involves more central processing of motor commands. Usually at issue in studies of the slowing of performance with age have been questions of to what extent that slowing results from "central" mechanisms rather than "neuromuscular mechanisms."[19,20] CRT are thought to associate more with the former and SRT more with the latter.

The data Welford[19,20] reviewed suggest that the SRT of old adults are approximately 20% longer than those for young adults. Salthouse[21] reviewed 11 studies of the effects of age on SRT. Uniformly, age and reaction time show a positive linear correlation, but the correlation coefficients range only from 0.19 to 0.47. Gottsdanker[22] found SRT to increase with age at a rate of only 2 ms per decade.

Lack of precise definitions, lack of standards for experimental methods, and the frequent use of test conditions that are biomechanically complex can make data on reaction times difficult to interpret for the purposes of biomechanical analyses of rapid task performances. For example, Welford's review article[19] tabulates mean simple myoelectric latencies in young adults that range from 109 to 255 ms. More data are needed on age effects in SRT when attempts are made to move only single body joints. Studies of the effects of different SRT on abilities to perform complex tasks rapidly are also needed.

Proprioception

Relatively few studies have examined changes in proprioception with aging.[23] Skinner and associates,[24] who studied 29 subjects over a 62 year age span, found that joint position sense in the knee deteriorated with age. In approximate terms, joint angles could be reproduced to within 2° by 20 year olds, but only to within 6° by 80 year olds. Twenty year olds could detect passive joint motions of 4°, but 80 year olds could detect only motions larger than 7°. Kokmen and associates[25] compared joint motion sensation in 52 subjects over 60 years old with that in ten young adult subjects. In contrast to the findings of Skinner and associates,[24] they found no major decline with age in motion perception in finger and toe joints.

Mobility Impairment Biomechanics

Evaluation of Mobility Impairments in the Elderly

Impairments are anatomic or physiologic abnormalities of the body; disabilities are declines in an individual's functional performance, including physical and social performance. Jette and Branch[26] found musculoskeletal and visual impairments to be strongly related to physical but not to social disability in the elderly.

Biomechanics research is needed to improve evaluations of mobility impairment in the elderly. Guralnik and associates[27] pointed out that physical function level has generally been assessed through self- or proxy-report and that measures of physical performance are needed to supplement these. Tinetti[28] noted that locomotor, sensory, and cognitive functioning are each intimately related to mobility. Assessments of function in the elderly have often relied solely on a disease-oriented approach. Clinically practical, performance-oriented assessments of mobility are also needed. Tinetti and Ginter[29] found that currently used neuromuscular examinations for identifying mobility problems are not adequately able to identify those problems, and that systematic evaluations of position changes and balance maneuvers are needed to identify specific problems.

These authors and others have outlined the need for better ways to quantify degree of impairment largely for purposes of clinical evaluations of patient status. A more fundamental rationale for quantifying impairments in biomechanical terms is to make it possible to determine which elements of a mobility impairment are the critical ones. Quantifying impairments in biomechanical terms is a major requirement for the success of this search.

Postural Balance

Many studies of the ways aging affects the ability to maintain postural balance have been reported. Horak and associates,[30] in a recent review of this topic, pointed to the ample evidence of deterioration in many sensorimotor systems underlying postural control, even in elderly populations without obvious signs of disease. They concluded, however, that aging alone cannot account for the heterogeneity of postural control problems in the elderly. Winter and associates[31] noted that, because responses to postural perturbations are task and perturbation specific, no single assessment technique can serve as a true indicator of the overall integrity of the balance control system.

Reports of age effects on sway while standing generally show that, after maturity, sway increases with age. The studies of Sheldon,[32] Murray and associates,[33] and Hasselkus and Shambes[34] were among the earliest to report this. More recently, Era and Heikkinen[35] and Maki and associates[36] confirmed that standing postural sway is more pronounced in the elderly, although Era and Heikkinen found disturbed postural sway to be the same in the young, the middle-aged, and old adults.

Reports of myoelectric responses to postural disturbances include those of Woollacott and associates,[37] Inglin and Woollacott,[38] and Manchester and associates.[39] Woollacott and associates[37] found a statistically significant difference between young and old adults in mean ankle muscle latencies, but this difference was only 7 ms in the approximately 100 ms latencies. The biomechanical significance of such small differences is unknown, and latencies values vary depending on the methods used to determine them. Moreover, Manchester and associates[39] found that during involuntary responses to postural disturbances muscle latencies did not differ across age groups. Inglin and Woollacott[38] found that when arm movements were carried out voluntarily while standing, lower leg muscle response onset latencies were often significantly longer in old adults than in young, and arm muscle response latencies showed even more pronounced age differences. Manchester and associates[39] found that old adults used more antagonistic muscle contractions than did young adults during involuntary responses to postural disturbances, and that the sequence in which the leg muscles were contracted differed from that of young adults in five out of 12 old adults.

Few studies to date have attempted an in-depth analysis of whole-body biomechanics or even body segment motions or joint torques developed in response to postural disturbances. Allum and associates[40] and Manchester and associates[39] made limited measurements indicative of ankle torques, and Horak and Nashner[41] and Keshner and associates[42] made limited measurements of body movements. Two more comprehensive studies of the motions and torques needed to maintain balance are those of Alexander and associates[43] and Gu and associates.[17] The required motions and torques are generally modest compared to the literature-reported capacities of healthy old adults.

Bohannon and associates[44] reported on the ability to balance for 30 seconds on one leg with eyes closed. They tested 184 healthy subjects between ages 20 and 79. Every subject younger than 30 could balance for at least 22 seconds. No subject over 70 could balance for more than 13 seconds. It would be useful to learn why this occurs. Does it result from reduced ankle joint lateral bending muscle strength or endurance, increased muscle latency times, decreased cutaneous or joint proprioception, or decreased willingness to allow the center of the floor reaction to deviate from the centroid of the foot support area? These are questions well suited to biomechanics inquiry. The answers to them are likely to be of fundamental importance in addressing problems of mobility impairment in the elderly.

Gait

In their pioneering studies of gait in the elderly, Murray and associates[45] found that, compared to younger males, men in their 60s demonstrated significantly shorter step and stride lengths and decreased ankle extension and pelvic rotation. Murray and associates[46] found men 65 to 87 years old, when compared to younger men, to have significantly different stride length and cadence, and head, shoulder, and ankle movements. More recent studies have confirmed many of those findings,[47-49], but not without exception. Blanke and Hageman[50] found no significant differences between age groups in step and stride length, velocity, and movements of the ankles, pelvis, and total body mass center. Gabell and Nayak[51] concluded from their study that increased variability in gait should not be regarded as a normal concomitant of old age. These various study outcomes are perhaps a good illustration of the effects of subject inclusion criteria on the outcome of studies of mobility problems in the elderly.

Cunningham and associates[47] reported normal walking speed to be associated with maximum aerobic power independent of age. Bassey and associates[52] found the normal walking speed of elderly subjects to be significantly associated with their muscle strength, while Bendall and associates[53] found that among elderly subjects more than 40% of the variance in normal walking speed can be accounted for by differences in height, calf muscle strength, and the presence of leg pain or other health problems.

The relationships between difficulties in walking and tendencies to fall are not yet clear. Wolfson and associates[54] found stride length and walking speed to be significantly impaired in elderly fallers compared to controls, but Heitmann and associates[55] found no corresponding significant differences in step width.

Falls

One of the most serious problems of mobility impairment is the tendency of old adults to fall. During a one-year prospective study of 336 community-dwelling persons 75 years of age or older, 32% fell at least once. Of those who fell, 24% sustained serious injuries, and more than 5% experienced fractures.[56,57] Among the elderly, falls account for 87% of all fractures.[58] Retrospective interviews found that 35% of 1,042 community-dwelling persons 65 or more years old reported one or more falls in the preceding year.[59] Death rates from falls per 100,000 persons in 1984 were 1.5 for those younger than 65 and 147.0 for those 85 or more years old.[60]

While the mechanisms related to increased falls in the elderly remain unclear, tendencies to fall and increases in postural sway when standing are associated. Elderly individuals sway more than young adults when standing, as already noted, and elderly persons with a history of falls sway more when standing than elderly persons without such a history.[61,62]

There is an urgent need for further studies of the biomechanics of maintaining standing posture and responding to fall-provoking stimuli. Why do old adults sway more and experience fall-related injuries more often than young adults? Does the apparent motor response deterioration with age result from reduced muscular strengths and joint ROM, increased reaction times, inappropriate body segment motion sequencing, or combinations of these or other mechanisms? Relatively little is known about these issues, and quantitative rather than qualitative studies would seem needed to arrive at an understanding of them.

Fractures

Fractures lead directly to mobility impairments, and fractures frequently occur among the elderly. Of 70-year-old females, 5% to 20% have had fractures of their vertebrae.[63] The risk of hip fracture increases dramatically after age 50, and almost 50% of hip fractures occur in persons who are 80 years old or older. In 1987, approximately 220,000 hip fractures occurred in the United States. That number will grow to over 400,000 by the year 2020. Approximately 20% of women who experience a hip fracture do not survive the first year after the fracture. Another 20% do not regain the ability to walk without assistance.[64] Inability to walk without assistance often leads to institutionalization.

Cummings and associates[63] reviewed the literature on the relationship between osteoporosis and fractures. It is well known that bone strength is correlated closely with bone mass and mineral content, and that bone mass declines with age over the adult years. Nevertheless, the literature suggests that while osteoporosis is the primary determinant of vertebral fractures, other factors are important in hip and Colles' fractures. Cummings and associates[63] proposed that a combination of decreased bone mass and an increased frequency of trauma caused by falls may combine to produce the pattern of increased risk of fractures with age. In a later paper, Cummings and Nevitt[65] hypothesized that this combination alone does not explain the exponential rise of the incidence of hip fracture with age. They felt that other changes in neuromuscular function with age, including changes in gait speed, reaction times, and muscle strengths, may also be involved in the incidence of hip fractures.

Lotz and Hayes[66] found that the force needed to fracture an isolated femur is only about one-twentieth of the kinetic energy that can be developed in a typical fall. They suggest that the energy absorbed during falling and impact, rather than bone strength, is the dominant factor in the biomechanics of hip fracture. Cummings and associates[63] remarked that prevention of falls in the elderly may be the most promising way to prevent fractures.

New evidence suggests that hip fractures occur predominantly in falls to the side. Hayes and associates[67] reported findings from a falls surveillance study, which showed that elderly adults who fell and sustained a fracture more often fell to the side (57% versus 23%) and more often landed on the hip or side of the leg (70% versus 9%).

Closure

In its report, *Year 2000 Objectives for the Nation*, the Public Health Service[68] noted that a wide variation between chronological and physiological age is evident in the older population. A major research need is to "understand aging processes and how aging is distinct from disease, in order to better understand which changes are inevitable and which are open to modification." That report stated that a high priority should be placed on prevention of disability. This review has attempted to sketch out what biomechanics research probably must be done to help achieve that understanding and to learn how to prevent disability.

Acknowledgements

The support of PHS Grants AG 06621 and AG 08818 and the Vennema Endowment for the preparation of this paper is gratefully acknowledged.

This article has been modified with permission from Schultz AB: Mobility impairment in the elderly: Challenges for biomechanics research. *J Biomech* 1992;25:519-528.

References

1. Schultz AB: Mobility impairment in the elderly: Challenges for biomechanics research. *J Biomech* 1992;25;519-528.
2. Ward JS, Kirk NS: Anthropometry of elderly women. *Ergonomics* 1967;10:17-24.
3. Burr ML, Phillips KM: Anthropometric norms in the elderly. *Br J Nutr* 1984;51:165-169.
4. Shimokata H, Andres R, Coon PJ, et al: Studies in the distribution of body fat: II. Longitudinal effects of change in weight. *Int J Obes* 1989;13:455-464.
5. Kelly PL, Kroemer KH: Anthropometry of the elderly: Status and recommendations. *Hum Factors* 1990;32:571-595.
6. Stoudt HW: The anthropometry of the elderly. *Hum Factors* 1981;23:29-37.
7. Allander E, Björnsson OJ, Olafsson O, et al: Normal range of joint movements in shoulder, hip, wrist and thumb with special reference to side: A comparison between two populations. *Int J Epidemiol* 1974;3:253-261.

8. Boone DC, Azen SP: Normal range of motion of joints in male subjects. *J Bone Joint Surg* 1979;61A:756-759.
9. Roaas A, Andersson GB: Normal range of motion of the hip, knee and ankle joints in male subjects, 30-40 years of age. *Acta Orthop Scand* 1982;53:205-208.
10. Walker JM, Sue D, Miles-Elkousy N, et al: Active mobility of the extremities in older subjects. *Phys Ther* 1984;64:919-923.
11. Svanborg A: Practical and functional consequences of aging. *Gerontology* 1988;34(suppl 1):11-15.
12. Battié MC, Bigos SJ, Sheehy A, et al: Spinal flexibility and individual factors that influence it. *Phys Ther* 1987;67:653-658.
13. Einkauf DK, Gohdes ML, Jensen GM, et al: Changes in spinal mobility with increasing age in women. *Phys Ther* 1987;67:370-375.
14. Badley EM, Wagstaff S, Wood PH: Measures of functional ability (disability) in arthritis in relation to impairment of range of joint movement. *Ann Rheum Dis* 1984;43:563-569.
15. Bergström G, Aniansson A, Bjelle A, et al: Functional consequences of joint impairment at age 79. *Scand J Rehabil Med* 1985;17:183-190.
16. Bergström G, Bjelle A, Sorensen LB, et al: Prevalence of symptoms and signs of joint impairment at age 79. *Scand J Rehabil Med* 1985;17:173-182.
17. Gu MJ, Schultz AB, Shepard NT, et al: Postural control in young and elderly adults when stance is perturbed: Dynamics. *J Biomech*, in press.
18. Schultz AB, Alexander NB, Ashton-Miller JA: Biomechanical analyses of rising from a chair. *J Biomech* 1992;25:1383-1391.
19. Welford AT: Relations between bodily changes and performance: Some possible reasons for slowing with age. *Exp Aging Res* 1984;10:73-88.
20. Welford AT: Reaction time, speed of performance, and age. *Ann NY Acad Sci* 1988;515:1-17.
21. Salthouse TA: Speed of behavior and its implications for cognition, in Birren JE, Schaie KW (eds): *Handbook of the Psychology of Aging*. New York, Van Nostrand Reinhold, 1985, pp 400-426.
22. Gottsdanker R: Age and simple reaction time. *J Gerontol* 1982;37: 342-348.
23. Stelmach GE, Worringham CJ: Sensorimotor deficits related to postural stability: Implications for falling in the elderly. *Clin Geriatr Med* 1985;1:679-694.
24. Skinner HB, Barrack RL, Cook SD: Age-related decline in proprioception. *Clin Orthop* 1984;184:208-211.
25. Kokmen E, Bossemeyer RW Jr, Williams WJ: Quantitative evaluation of joint motion sensation in an aging population. *J Gerontol* 1978; 33:62-67.
26. Jette AM, Branch LG: The Framingham disability study: II. Physical disability among the aging. *Am J Public Health* 1981;71:1211-1216.
27. Guralnik JM, Branch LG, Cummings SR, et al: Review: Physical performance measures in aging research. *J Gerontol* 1989; 44:M141-M146.
28. Tinetti ME: Performance-oriented assessment of mobility problems in elderly patients. *J Am Geriatr Soc* 1986;34:119-126.
29. Tinetti ME, Ginter SF: Identifying mobility dysfunctions in elderly patients: Standard neuromuscular examination or direct assessment? *JAMA* 1988;259:1190-1193.
30. Horak FB, Shupert CL, Mirka A: Components of postural dyscontrol in the elderly: A review. *Neurobiol Aging* 1989;10:727-738.

31. Winter DA, Patla AE, Frank JS: Assessment of balance control in humans. *Med Prog Technol* 1990;16:31-51.
32. Sheldon JH: The effect of age on the control of sway. *Geront Clin (Basel)* 1963;5:129-138.
33. Murray MP, Seireg AA, Sepic SB: Normal postural stability and steadiness: Quantitative assessment. *J Bone Joint Surg* 1975; 57A:510-516.
34. Hasselkus BR, Shambes GM: Aging and postural sway in women. *J Gerontol* 1975;30:661-667.
35. Era P, Heikkinen E: Postural sway during standing and unexpected disturbance of balance in random samples of men of different ages. *J Gerontol* 1985;40:287-295.
36. Maki BE, Holliday PJ, Fernie GR: Aging and postural control: A comparison of spontaneous- and induced-sway balance tests. *J Am Geriatr Soc* 1990;38:1-9.
37. Woollacott MH, Shumway-Cook A, Nashner LM: Aging and posture control: Changes in sensory organization and muscular coordination. *Int J Aging Hum Dev* 1986;23:97-114.
38. Inglin B, Woollacott M: Age-related changes in anticipatory postural adjustments associated with arm movements. *J Gerontol* 1988;43: M105-M113.
39. Manchester D, Woollacott M, Zederbauer-Hylton N, et al: Visual, vestibular and somatosensory contributions to balance control in the older adult. *J Gerontol* 1989;44:M118-M127.
40. Allum JHJ, Keshner EA, Honegger F, et al: Organization of leg-trunk-head equilibrium movements in normals and patients with peripheral vestibular deficits. *Prog Brain Res* 1988;76:277-290.
41. Horak FB, Nashner LM: Central programming of postural movements: Adaptation to altered support-surface configurations. *J Neurophysiol* 1985;55:1369-1381.
42. Keshner EA, Allum JH, Pfaltz CR: Postural coactivation and adaptation in the sway stabilizing responses of normals and patients with bilateral vestibular deficit. *Exp Brain Res* 1987;69:77-92.
43. Alexander NB, Shepard NB, Gu MJ, et al: Postural control in young and elderly adults when stance is perturbed: Kinematics. *J Gerontol: Med Sci* 1992;47:M79-M87.
44. Bohannon RW, Larkin PA, Cook AC, et al: Decrease in timed balance test score with aging. *Phys Ther* 1984;64:1067-1070.
45. Murray MP, Drought AB, Kory RC: Walking patterns of normal men. *J Bone Joint Surg* 1964:46A:335-360.
46. Murray MP, Kory RC, Clarkson BH: Walking patterns in healthy old men. *J Gerontol* 1969;24:169-178.
47. Cunningham DA, Rechnitzer PA, Pearce ME, et al: Determinants of self-selected walking pace across ages 19 to 66. *J Gerontol* 1982;37: 560-564.
48. Hageman PA, Blanke DJ: Comparison of gait of young women and elderly women. *Phys Ther* 1986;66:1382-1387.
49. Himann JE, Cunningham DA, Rechnitzer PA, et al: Age-related changes in speed of walking. *Med Sci Sports Exerc* 1988;20:161-166.
50. Blanke DJ, Hageman PA: Comparison of gait of young men and elderly men. *Phys Ther* 1989;69:144-148.
51. Gabell A, Nayak US: The effect of age on variability in gait. *J Gerontol* 1984;39:662-666.
52. Bassey EJ, Bendall MJ, Pearson M: Muscle strength in the triceps surae and objectively measured customary walking activity in men and women over 65 years of age. *Clin Sci* 1988;74:85-89.

53. Bendall MJ, Bassey EJ, Pearson MB: Factors affecting walking speed of elderly people. *Age Ageing* 1989;18:327-332.
54. Wolfson L, Whipple R, Amerman P, et al: Gait assessment in the elderly: A gait abnormality rating scale and its relation to falls. *J Gerontol: Med Sci* 1990;45:M12-M19.
55. Heitmann DK, Gossman MR, Shaddeau SA, et al: Balance performance and step width in noninstitutionalized, elderly, female fallers and nonfallers. *Phys Ther* 1989;69:923-931.
56. Tinetti ME, Speechley M, Ginter SF: Risk factors for falls among elderly persons living in the community. *N Engl J Med* 1988;319: 1701-1707.
57. Tinetti ME, Speechley M: Prevention of falls among the elderly. *N Engl J Med* 1989;320:1055-1059.
58. Fife D, Barancik JI: Northeastern Ohio Trauma Study: III. Incidence of fractures. *Ann Emerg Med* 1985;14:244-248.
59. Blake AJ, Morgan K, Bendall MJ, et al: Falls by elderly people at home: Prevalence and associated factors. *Age Ageing* 1988;17:365-372.
60. Lambert DA, Sattin RW: Deaths from falls, 1978-1984. *MMWR CDC Surveill Summ* 1988;37:21-26.
61. Overstall PW, Exton-Smith AN, Imms FJ, et al: Falls in the elderly related to postural imbalance. *Brit Med J* 1977;1:261-264.
62. Fernie GR, Gryfe CI, Holliday PJ, et al: The relationship of postural sway in standing to the incidence of falls in geriatric subjects. *Age Ageing* 1982;11:11-16.
63. Cummings SR, Kelsey JL, Nevitt MC, et al: Epidemiology of osteoporosis and osteoporotic fractures. *Epidemiol Rev* 1985;7:178-208.
64. Schneider EL, Guralnik JM: The aging of America: Impact on health care costs. *JAMA* 1990;263:2335-2340.
65. Cummings SR, Nevitt MC: A hypothesis: The causes of hip fractures. *J Gerontol* 1989;44:M107-M111.
66. Lotz JC, Hayes WC: The use of quantitative computed tomography to estimate risk of fracture of the hip from falls. *J Bone Joint Surg* 1990;72A:689-700.
67. Hayes WC, Piazza SJ, Zysset PK: Biomechanics of fracture risk prediction of the hip and spine by quantitative computed tomography. *Radiol Clin North Am* 1991;29:1-18.
68. Public Health Service: *Promoting Health/Preventing Disease: Year 2000 Objectives for the Nation*. US Department of Health and Human Services, 1989.

Chapter 3

Morbid Consequences of Aging Processes: Methodologic Issues

Evan C. Hadley, MD

This paper focuses on methodologic issues in studies of the role of aging processes in age-related diseases. A potential role for many such processes, such as DNA damage, oxyradical damage, post-translational protein modification, neuroendocrine changes, and cell death, has been proposed; many more will almost certainly be suggested. Figure 1 illustrates the paradigm: A function (z) changes progressively with increasing age. (It could either increase or decrease; for simplicity, the change is always shown as a decrease in this chapter.) The function could be any parameter, ranging from the molecular to the behavioral level, that could affect morbidity. At some point the function enters a ''morbid range'' and begins to contribute to overt disease, which may increase in severity (indicated by deepening shading) as the function continues to change. The boundary of the morbid range may not be sharp, but the model does assume a transition from a tolerable ''healthy'' level to an undesirable ''morbid'' one.

A process like this, once identified, affords great opportunities for mechanistic and therapeutic research. One can learn what controls the rate of this change, and whether slowing it will delay the onset of morbidity. If each aging change occurred in all persons and progressively changed from the same starting level at the same rate, many methodologic issues discussed in this paper would be irrelevant. However, relationships like that shown in Figure 1 are often obscured by the great heterogeneity of individuals and their aging processes. The issues in this chapter focus on the challenges of finding these relationships amidst this heterogeneity. Once they are found, we can proceed with the task of learning how to control them.

This chapter contains simplifying assumptions about two methodologic issues relating to the model in Figure 1: The first is the problem of making inferences from cross-sectional data about aging changes within individuals. Because of the difficulties of longitudinal studies, much aging data are derived from cross-sectional com-

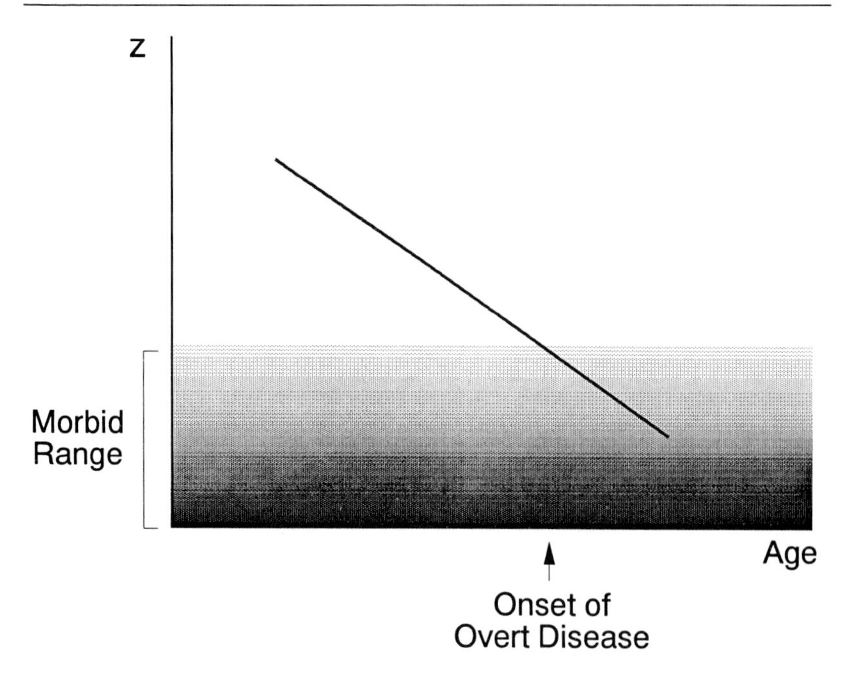

Fig. 1 *Basic model of an age-related change in a function (z) which progresses to a level (morbid range) where it contributes to overt disease. Deepening shading indicates increasing severity of disease.*

parisons, from which aging changes within individuals are inferred. The limitations of such inferences have been discussed elsewhere,[1,2] but in this chapter I assume that all differences observed between age groups actually reflect aging changes.

Second, there are numerous methodologic issues involved in finding the "morbid" range. In diseases with multiple causes, there is often no clear boundary for a morbid range for any one contributory factor—one can only speak of the factor affecting the level of risk for the disease. Further, in many cases there are no direct data establishing the morbid range. In this situation the investigator is faced with the choice of obtaining this information as part of the study, if possible; making the best estimate using other information, such as levels of the function in disorders similar to the one of interest; or simply guessing. For purposes of simplification, in this chapter I equate "morbid range" with "high-risk levels," and assume that this range has been identified or can be identified.

Aging Phenomena With Methodologic Implications

The interplay of six phenomena related to aging have particular implications for studies of the role of aging processes in morbidity.

(1) Age per se often explains little, even when aging has important effects. Figure 2 illustrates a common pattern seen in aging studies.

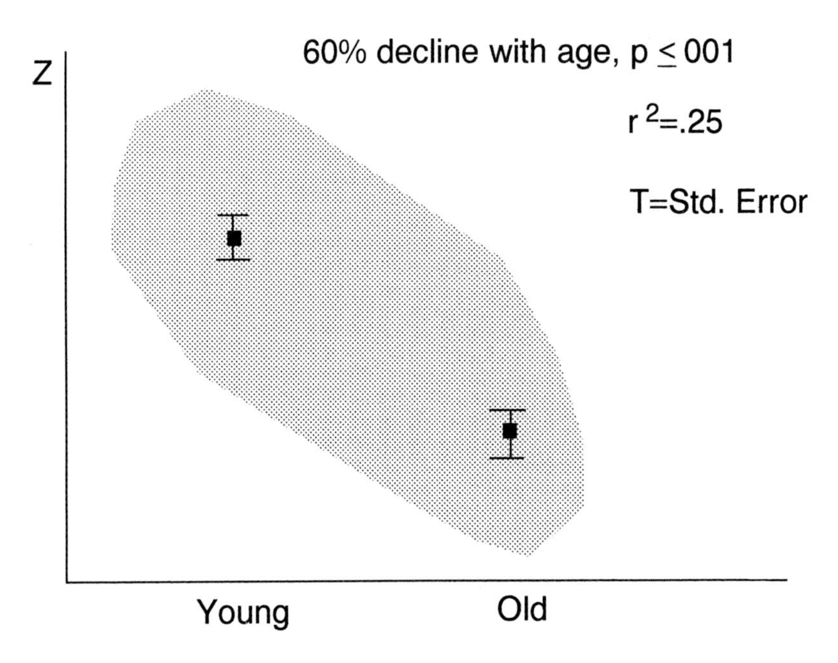

Fig. 2 *Distribution of levels of function (z) as a function of age. Stippled area represents range of levels in population.*

Although mean levels of a function may change dramatically and highly significantly with age, age itself often explains only a small share of the variance, and the function is higher in many old individuals than in many young individuals.

(2) Aging processes are not always universal. Although the mean level of a function may change with increasing age in a population, the direction of change is not the same in all individuals. For example, although it is common to state that creatinine clearance rate declines with age, a long-term longitudinal study of normal individuals found that it decreased in some, was unchanged in some, and actually increased in a substantial percentage.[3] Thus it is an oversimplification to say that such a function declines with age, when in fact both increases and decreases are within the normal range.

(3) There is a high prevalence of chronic diseases in old age. As shown in Table 1, numerous chronic diseases are common in the older population.[4] Further, they often coexist in the same person. Over half the people in the United States over the age of 70 have two or more chronic diseases, and less than one-fifth have none.[5] This fact complicates attempts to study either aging or diseases of aging in isolation, free of confounding effects of extraneous morbidity.

(4) Human aging changes occur over a longer time span than an individual research career. Although studies of the entire human life span are possible, they generally require corporate efforts with

Table 1 Prevalence of Chronic Conditions in Noninstitutionalized Persons Aged 65 and Over. US, 1982-1984 (%)

Arthritis	48.6
Hypertension	39.5
Ischemic heart disease	13.6
Diabetes	9.1
Cerebrovascular disease	5.8
Chronic bronchitis	5.8
Emphysema	4.1

(Reproduced with permission from National Center for Health Statistics, R Havlik, et al. Health Statistics on Older Persons. Vital and Health Statistics, Series 3, No. 25. Table 17, 1986.)

long-term commitments. The crucial scientific role of an individual in testing an original idea must still rely initially on studies that can be accomplished more quickly, employing cross-sectional comparisons, short-term longitudinal studies, or animal models whose life spans are shorter than those of humans.

(5) Not all pathologies of aging are defined yet. Many of the causes of morbidity in old age are still being elucidated. For example, another chapter in this volume discusses the possible role of growth hormone deficiency in several degenerative conditions of later life. Thus, one cannot conclude that an age-related change necessarily constitutes a primary aging process, even if it occurs in the absence of known disease. It may simply be secondary to a pathology we have yet to explore.

(6) There are important differences between experimental animals and humans in age-related pathologies. If an experimental animal developed the same spectrum of diseases as humans over a short life span, many methodologic problems would not exist. Unfortunately, for many diseases of human aging, animal models that spontaneously develop an identical or even similar disorder are lacking. Conversely, many established aging animal models develop age-related pathologies that differ from the major ones of humans, and which are highly prevalent in old animals.[6]

Given the complicating effects of the above factors, how can we judge whether a particular aging process is a likely contributor to a particular age-associated morbidity? One helpful strategy is to examine the distribution of aging change in relation to the morbid range. Looking closely at patterns of change with age can help separate likely from unlikely causes of pathology. It is useful to consider the distribution of age-related changes, in particular the degree to which they approach or enter the morbid range. Figure 3 illustrates two different patterns. The difference in means between young and old, the level of statistical significance, and the distance of the means from the morbid range, are identical in both. However, in the left-hand panel no persons fall within the morbid range, while a subset of older persons on the right-hand panel do so. Intuitively, one would suspect that the pattern of change on the right is more likely to reflect a process causing age-related morbidity.

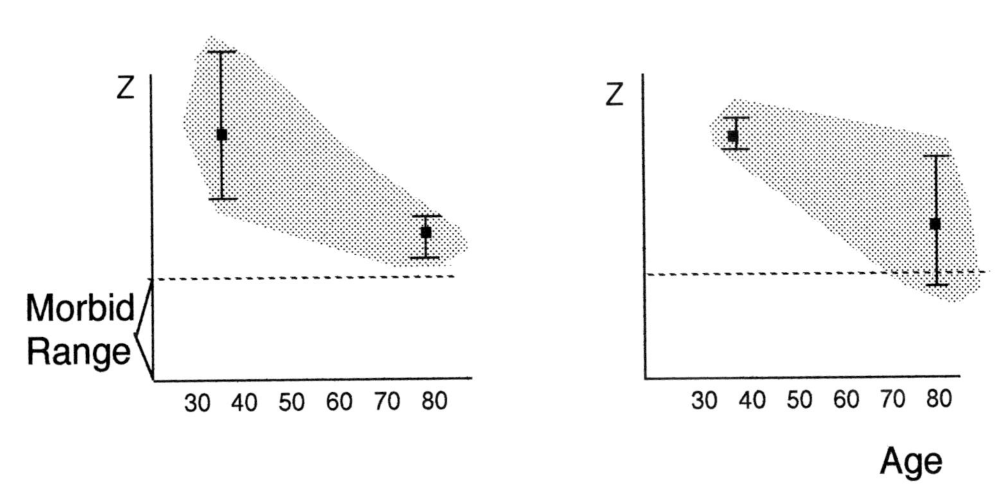

Fig. 3 *Differences in distribution affecting entry of function (z) into morbid range. Stippled area represents range of levels in population.*

An additional cautionary note relates to the age-composition of the older group in young versus old comparisons: Many changes progress throughout later life. If the study doesn't include enough subjects of very advanced age, it may miss the morbid effect of such changes. For example, had the study shown on the right of Figure 3 been conducted in 65 year olds it would have missed the morbid changes in the oldest subjects. The importance of including subjects of very advanced age is increasing in importance, because this group, which is the fastest growing segment of the United States population, is probably the least well characterized.

Interactions with Diseases and Other Sources of Variability

Comparisons of Subjects With and Without the Pathology of Interest

If Figure 1 represented the entire population, that is, if everyone followed the same pattern of change and developed the same pathology, many methodologic problems would not exist. Frequently, however, some individuals develop a particular pathology in old age and some don't. In looking for the possible etiologic role of an age-related change in a function, one can do case-control comparisons of the function in age-matched subjects with and without the disease (Fig. 4). The standard interpretation of a difference between cases and controls in the level of a possible etiologic factor is that the

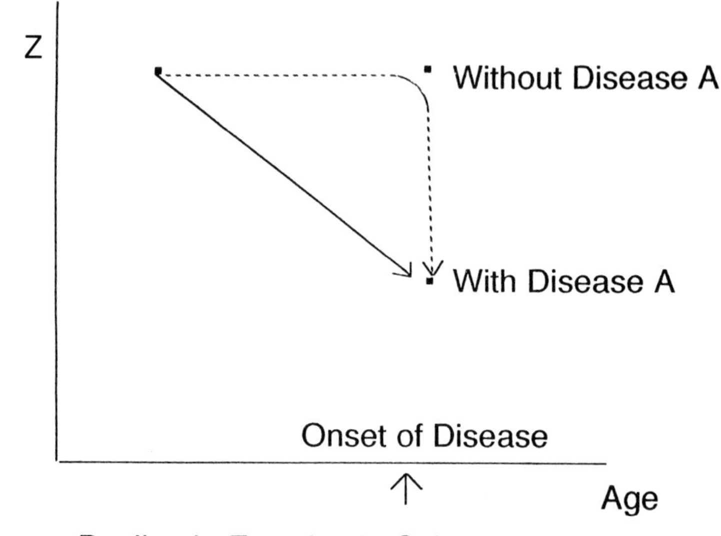

Fig. 4 *Alternative pathways to differences in level of function (z) in subjects with and without a particular disease.*

difference could be either the cause or the result of the onset of the pathology. The case-control study tells nothing about the path by which the differences between cases and controls appeared.

However, in some aging studies, the conclusion is drawn that absence of an age-related change in levels of a function in normal subjects indicates that aging changes in this function do not contribute to disease. This erroneous inference may stem from an implicit assumption that aging processes must be universal, and hence that only the pattern seen in the controls constitutes "normal" aging. However, as noted earlier, many important aging processes may occur only in subgroups of the population. Thus, if differences seen in case-control studies are not followed up with studies to determine whether the differences are causes or effects, one may fail to detect many aging processes with important pathologic consequences. Longitudinal studies are the most rigorous way to answer the question of causality, but cross-sectional studies spanning all age-ranges can provide clues. If in each successively older age group there is a subgroup with levels of the factor more and more closely approaching the morbid range, there is a good possibility that these age-related changes may contribute eventually to overt disease.

A converse problem exists when all older individuals undergo an age-related change in a function, whether or not they have the morbidity of interest. The debate about the relationship between bone density and hip fracture illustrates this problem. On one hand, the decline of femoral neck bone density with age, coupled with the rise in hip fracture rate with age, were cited as strong prima facie evidence for the role of age-related bone loss in hip fracture. On the other hand, similarities in bone density between hip fracture cases and age-matched controls were cited as evidence against an important relationship.

The debate illustrates two possible interpretations shown in Figure 5. Lack of a difference between cases and controls in age-related changes in a function may stem from the fact that the change is necessary but not sufficient to cause overt morbidity (illustrated by morbidity threshold A) and that one or more other conditions must also be present for morbidity to occur. For example, the occurrence of both low femoral neck density and a fall seems to be the prerequisite for most hip fractures. On the other hand, the age-related change may truly not contribute to the condition—the function remains outside the morbid range (illustrated by morbidity threshold B). Thus, again, the possibility of a morbid effect of an aging change cannot be dismissed until one determines which of these possibilities is occurring.

One way of discriminating between the two is by studying all age ranges to determine if at any age there is a difference in levels of the factor between cases and age-matched controls. In Figure 5, such differences would be seen in the age range in which persons crossed the morbidity threshold A. For example, in the case of hip fractures it was found that few fractures occurred in persons with bone densities of more than 1.0g/cm^2, and that this was so even at ages at which the population had substantial numbers with bone densities above and below that figure.[7] Multivariate techniques that correct for differences in other causal factors can also be applied to this approach.

However, if the level of the factor is very strongly correlated with age, the above approaches may be futile, and it may be impossible to distinguish effects of the factor from those of other age-related changes by observational studies alone. In such cases, an intervention study, if feasible, may be the best way to resolve the issue. Delaying the onset of morbidity by slowing the age-related change in the function is very compelling evidence of the change's morbid effect.

Inclusion of Subjects With Other Pathologies

When searching for an association between an age-related change and a particular pathology, one can be misled if the change is associated with another disease of high prevalence rather than being an "idiopathic" process. For example, there is evidence that diabetes mellitus may be a risk factor for osteoarthritis. Prevalence of both diseases increases with age. If subjects with diabetes are present in a study of a function potentially contributing to osteoarthritis, much of what is considered to be an age-related change in the function may actually be caused by diabetes and not occur in its absence.

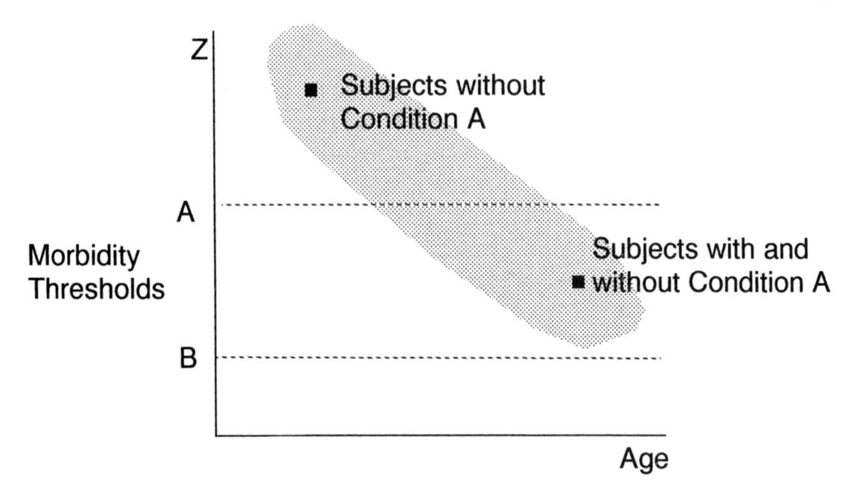

Fig. 5 *Age-related change in function (z) with respect to two possible morbidity thresholds A and B. Morbid ranges for the two thresholds extend from the dashed lines to the horizontal axis. Stippled area represents range of levels in the population.*

This type of situation is illustrated in Figure 6. In a total population study, entry of a function into a morbid range (D in upper panel) may be caused by the presence of other diseases. Thus, comorbidity is a crucial issue in aging studies, and there is a considerable literature indicating that many supposed aging changes diminish or disappear when subjects with various diseases are excluded. Thus, a common strategy is to exclude subjects with chronic diseases from the study, as illustrated in the lower panel of Figure 6.

However, there are problems with this approach as well. First, the extent of age-related diseases that could affect any physiologic system, including the musculoskeletal, is very wide. For example, it is not outlandish to think that ischemia related to cardiovascular disease could play a role in loss of muscle mass and other degenerative changes. A high proportion of persons over age 75 could have ischemia severe enough to have such an effect. One is not on safe ground simply excluding subjects with symptomatic disease: even asymptomatic disease can have significant effects.[8] The total number of older persons with chronic diseases that could affect the musculoskeletal system is probably a large majority of the population. Completely healthy older persons are the exception, not the rule. (This fact raises the philosophic issue of whether such persons are an elite minority who may differ in other ways from the rest of the population besides being free of disease.)

Further, even if one excludes all subjects with any sign of known chronic disease, there are probably many age-related pathologies yet undefined, and thus impossible to screen. In addition, there are physiologic and behavioral differences between young and old

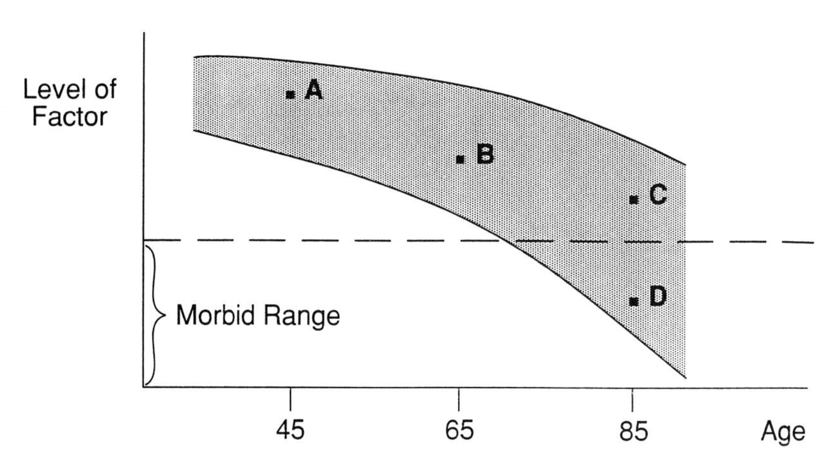

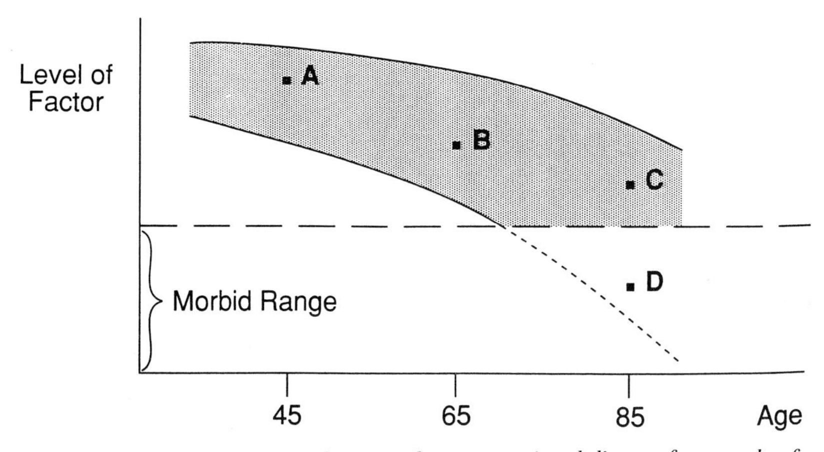

Fig. 6 *Impact of excluding subjects with age-associated disease from study of levels of a factor as a function of age. A, B, and C represent healthy subjects of different ages. D represents subjects with disease. Stippled area represents range of levels in the population.*

populations that may not reflect primary aging processes. The level of physical activity is one with clear implications for musculoskeletal research. For example, 57% of persons over age 65 engage in leisure physical activity, compared to 82% of those between ages 18 and 29.[9] Thus, differences between young and old populations in numerous musculoskeletal functions may result from differences in physical activity. Attempting to control for this by excluding inactive older subjects will further shrink the study population to an even more elite group. There are numerous such differences between young and old that raise analogous issues.

Further, excluding older subjects with diseases or other differences from younger subjects can obscure important contributions of fundamental aging processes to morbidity. Returning to the example of the possible association between diabetes and osteoarthritis: Figure 7 illustrates two alternative mechanisms that could cause such an association. In the pattern on the left, diabetes itself causes process Z, which in turn leads to osteoarthritis. On the left, however, process Z represents a fundamental aging process (for example, an abnormality in carbohydrate metabolism) which contributes to both diabetes and osteoarthritis. Excluding diabetics will impair the chances of detecting this change because it will exclude those in whom it is most pronounced and who are most likely to enter the morbid range.

The chances of detecting such relationships are better if one characterizes the subjects, and includes enough diabetics and non-diabetics so that one can determine whether the function enters the morbid range and, if so, whether diabetics constitute the bulk of the population in the morbid range (point D in Figure 6), while non-diabetics lie outside it (point C). If so, the task remains of determining whether the change is secondary to diabetes or a primary aging process. (Longitudinal studies can aid this.) The fundamental point is that the process would likely remain undetected in a study that excluded diabetics. Precisely because many fundamental aging processes may contribute to multiple diseases, their contribution to such diseases may be missed in studies of one disease that exclude all others.

From these considerations, one might infer that the only viable approach is a very large-scale study with sufficient power to assess relationships with all comorbid conditions and other variables that could contribute to differences in risk of morbidity. While the heterogeneity of the older population often imposes a need for larger sample sizes, a number-crunching study is not always necessary or even advisable. By judicious choices of the variables on which to accept heterogeneity and on which to limit it, one can still often do important and interpretable studies on a smaller scale.

Differences in Age-related Pathology Between Experimental Animals and Humans

Effects of Comorbid Endemic Age-related Pathologies in Experimental Animals

The species of experimental animals available for aging research develop age-related pathologies, but their major diseases of aging are not usually the major diseases of aging in humans. This is a particular problem with inbred strains because they typically develop one or a few pathologies that progress over the life span and

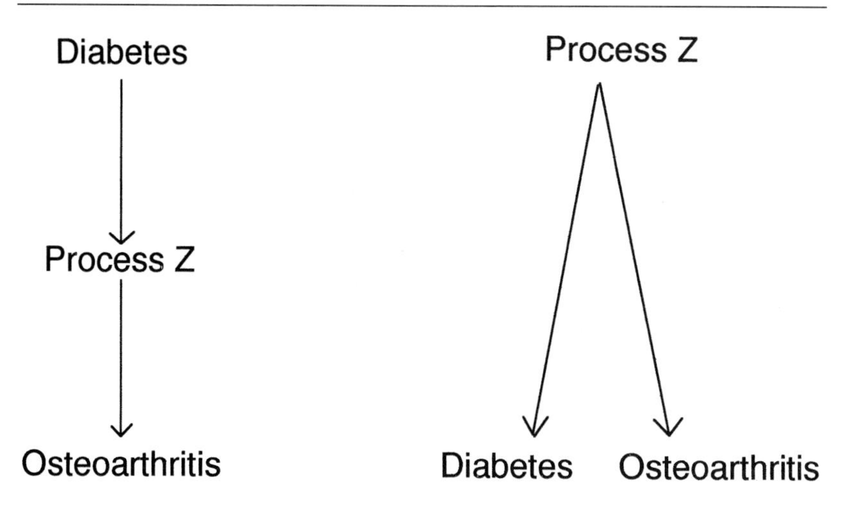

Fig. 7 *Two alternative mechanisms underlying a possible association between osteoarthritis and diabetes.* ***Left,*** *Diabetes causes intervening process contributing to osteoarthritis.* ***Right,*** *Common underlying process contributes to both diseases.*

become severe in a large majority of older animals in the strain. For example, the Fisher 344 rat, a commonly used strain, undergoes a progressive chronic glomerulonephropathy that progresses from a mild form early in life to severe disease in almost all animals aged 24 months—close to the mean life span, and a common age of "old" animals in aging studies.[10] Similar renal pathology is seen in other commonly used aging rat strains.[11] As noted by Weindruch and Masoro,[12] an endemic pathology such as this poses a major problem if it affects the function of interest in an aging study. For example, studies of age-related changes in muscle using a rat model face the problem that renal failure itself affects numerous muscle functions ranging from the tissue level to the molecular.[13-16] Thus, it is difficult, if not impossible, to determine, from a study confined to a species that develops endemic renal failure with increasing age, whether an age-related change in muscle is due to a primary aging process or is secondary to renal failure. The fact that other species develop different but also very common aging pathologies poses analogous methodologic problems for studies of any tissue at any level.[6]

There are ways to cope with this major problem. One obvious way is to avoid it by conducting the study in humans or human tissues whenever possible. Another is to choose a model whose pathology is unlikely to affect the process of interest. A third is to determine whether the aging process of interest occurs in several species or strains (preferably including humans), which develop very different age-related pathologies. If so, it is more likely that the change reflects a primary process rather than the artifact of one disease. In addition, the same longitudinal techniques that can help resolve the comorbidity issues discussed above for humans can also be applied to animal models.

Suitable Animal Models of the Pathology of Interest

A converse problem in animal studies concerns the age-related pathology for which one is trying to find an underlying aging mechanism. Ideally, one would wish for two strains of the same species, identical except that one develops the pathology with advancing age and the other does not. Comparison of aging processes between two such strains would be very valuable in identifying causal factors. However, in most cases (including musculoskeletal soft-tissue problems) we haven't developed such pairs of strains yet. Indeed, for many common age-related musculoskeletal problems in humans, there appears to be a lack of animal models that have been demonstrated to develop analogous pathology with advancing age. The identification or development of such models is of obvious value. Even in cases where parallels between humans and animal models have been identified, such as loss of muscle mass, it would be helpful to know more details about how closely the pattern in the animal model matches that in humans.

A common pitfall in animal studies is to do an animal study of age-related changes in some system, for example, cartilage, in order to learn more about the basis of an age-related pathology such as osteoarthritis in humans, but to do it in an animal that does not develop that pathology. If an age-related change is found in such a study, this fact is often used to support the idea that it might play a role in causing the disease in humans. This seems a non sequitur: if the change occurs in a strain that doesn't develop the disease, this implies that it doesn't necessarily have anything to do with the disease. (It might of course play a facilitating role in the presence of other factors not present in the animal model.) Thus, studies of age-related factors that contribute to a pathology of human aging are severely hampered when conducted in animal models that don't develop that pathology and would benefit from comparisons with models that do develop it.

Conclusion

The ideas in this chapter share a common theme: Aging doesn't occur in a vacuum. Numerous factors affect the progression and pathologic consequences of aging processes. Including and analyzing these sources of variability will reveal more about aging and age-related pathology than excluding or ignoring them. Viewed in this light, the complexities of research into the contributions of aging processes to disease, rather than being hindrances, are opportunities to elucidate major factors underlying the increase with age in the prevalence and severity of chronic diseases.

Acknowledgment

I am very grateful to Ms. Vivian Williams and Ms. Josephine Cruz for preparing the figures for this chapter.

References

1. Shock NW, Greulich RC, Andres R, et al: Normal human aging: The Baltimore Longitudinal Study of Aging. Washington DC, United States Government Printing Office, 1984.
2. Zimmer AW, Calkins E, Hadley E, et al: Conducting clinical research in geriatric populations. *Ann Intern Med* 1985;103:276-283.
3. Lindeman RD, Tobin J, Shock NW: Longitudinal studies on the rate of decline in renal function with age. *J Am Geriatr Soc* 1985;33:278-285.
4. National Center for Health Statistics, R Havlik, et al. Health Statistics on Older Persons. Vital and Health Statistics, Series 3, No. 25. Table 17. (1986).
5. Guralnik JM, LaCroix AZ, Everett DF, et al: Aging in the eighties: The prevalence of comorbidity and its association with disability. Advance data from vital and health statistics: No. 170. Hyattsville, Maryland: National Center for Health Statistics, 1989.
6. National Academy of Sciences. Committee on Animal Models for Research on Aging: Mammalian Models for Research on Aging. Washington DC, National Academy Press, 1981.
7. Riggs BL, Melton LJ: Involutional osteoporosis. *N Engl J Med* 1986;314:1676-1686.
8. Rodeheffer RJ, Gerstenblith G, Gecker LC, et al: Exercise cardiac output is maintained with advancing age in healthy human subjects: Cardiac dilatation and increased stroke volume compensate for a diminished heart rate. *Circulation* 1984;69:203-213.
9. Blair SN, Brill PA, Kohl HW: Physical activity patterns in older individuals, in Spirduso WW, Eckert HM (eds.): *Physical activity and aging*. American Academy of Physical Education Papers; No. 22. Champaign, IL, Human Kinetics Books, 1989, pp 120-139.
10. Maeda H, Gleiser CA, Masoro EJ, et al: Nutritional influences on aging of Fischer 344 rats: II. Pathology. *J Gerontol* 1985;40;671-688.
11. Bronson RT: Rate of occurrence of lesions in 20 inbred and hybrid genotypes of rats and mice sacrificed at 6 month intervals during the first years of life, in Harrison DE (ed): *Genetic Effects on Aging II*. Caldwell, NJ, Telford Press, 1991, pp 279-358.
12. Weindruch R, Masoro EJ: Concerns about rodent models for aging research. *J Gerontol* 1991;46:B87-B88.
13. Bradley JR, Anderson JR, Evans DB, et al: Impaired nutritive skeletal muscle blood flow in patients with chronic renal failure. *Clin Sci* 1990;78:239-245.
14. Bonilla S, Goecke IA, Bozzo S, et al: Effect of chronic renal failure on Na,K-ATPase alpha-1 and alpha-2 mRNA transcription in rat skeletal muscle. *J Clin Invest* 1991;88:2137-2141.
15. Bak JF, Schmitz O, Srensen SS, et al: Activity of insulin receptor kinase and glycogen synthase in skeletal muscle from patients with chronic renal failure. *Acta Endocrinol (Copenh)* 1989;121:744-750.
16. Bergstrom J, Alvestrand A, Furst P, et al: Sulphur amino acids in plasma and muscle in patients with chronic renal failure: Evidence for taurine depletion. *J Intern Med* 1989;226:189-194.

Chapter 4

Posttranslational Protein Modification By the Maillard Reaction: Relevance to Aging of Extracellular Matrix Molecules

Vincent M. Monnier, MD
David R. Sell, PhD
H.K. Pokharna, PhD
Roland W. Moskowitz, MD

Introduction

Collagens, lens crystallines, elastin, and proteoglycans are long-lived proteins, which are therefore susceptible to postsynthetic nonenzymatic modification. One striking observation is that the most significant measurable age-related changes occur in collagen-rich tissues, most of which lose their elasticity and become thickened or sclerosed with age.[1] At a physicochemical level, collagen becomes less soluble,[2] less digestible by collagenase,[3] increasingly thermostable,[4] and acquires covalently bound fluorophores.[5,6] Observations suggesting that these changes occur ubiquitously and at a faster rate in short-lived versus long-lived species led gerontologists like Bjorksten and Andrews[7] and Verzár[8] to propose a "crosslinking theory of aging," whereby proposed cross-linking agents included free radical reaction products and an array of endogenous and exogenous factors. Interestingly, Bensusan[9] proposed in 1965 that carbohydrates could lead to collagen cross-linking in a Maillard-type reaction, but the report, published in a book on the structure and function of connective and skeletal tissue, received little attention.

Considerable support for the notion that age-related postsynthetic modifications of collagen reflect metabolic processes of significance for basic aging research stems from work of Hamlin and associates,[10] who showed that the age-related loss in collagenase digestibility of tendons was accelerated in diabetes. Furthermore, dietary restriction, which is currently the only intervention known to prolong mean and maximum life-span, is also associated with decelerated aging rate of collagen.[4,11] Thus, the rationale for investigating the biochemical basis of collagen aging is strong, both from the perspective of basic aging process as well as from the perspective of diabetic complications, many of which are closely linked to changes in the extracellular matrix.[12]

Three mechanisms should be considered as a basis for the age-related cross-linking of collagen. The first involves borohydride reducible lysyl oxidase (LOX)-dependent cross-links, which undergo secondary changes to form trivalent and tetravalent cross-links such as histidine-hydroxylysinonorleucine (HIS-HLNL) and histidine-hydroxy-merodesmosmine (HHMD), respectively. Whereas the LOX-dependent cross-links undoubtedly play an important role in the stabilization of collagen during development and growth of the tissue, they do not appear to have a ubiquitous role in the age-related changes of collagen. HIS-HLNL, for example, increases in human and bovine skin to reach plateau levels at mid-life span.[13] Furthermore, its formation, like that of other LOX cross-links, is tissue specific,[14] in contrast to the loss of collagen digestibility and solubility, which is a ubiquitous phenomenon that progresses throughout life. However, LOX cross-linking may play a role in some of the changes that occur in the skin and other tissues of diabetic individuals. Increased LOX activity and LOX cross-linking was found in lungs,[15] granulation tissue,[16] and skin of diabetic animals or humans.[17] These changes are most likely associated with newly synthesized collagen, and a relationship of LOX with cross-linking of postmature collagen has not been demonstrated.

A second possibility for the introduction of collagen cross-links in aging tissues is based on free radical-mediated peroxidation of lipids and formation of malonyldialdehyde and alkenals such as 4-hydroxynonenal. Although their involvement in collagen aging is likely, evidence for their presence in collagen is based on nonspecific fluorescence assays.[18] Immunochemical demonstration that lipid peroxidation products are found in atherosclerosis was made,[19] but such products as collagen cross-links have been shown to be present in arteries and other tissues.

The third and, we believe, most plausible explanation for age- and diabetes-mediated modifications of postmature collagen involves Maillard reaction-mediated cross-linking through glycated residues, as well as sugars and ketoaldehydes that are generated during glycolysis and the advanced Maillard reaction.

The Maillard Reaction as a Basis for Age- and Diabetes-Related Postsynthetic Modification of Proteins

A simplified scheme of the Maillard reaction is depicted in Figure 1. The reaction is initiated by the nonenzymatic reaction of a reducing sugar with a primary amine in an amino-carbonyl type condensation reaction. The labile Schiff base undergoes an acid catalyzed rearrangement to form the Amadori product, the levels and stability of which depend on a variety of factors.[20] The rate of formation and degradation of the Amadori product generally follows the anomerization rate or percentage of aldehydic form of the sugar and, thus, is lowest for glucose and highest for trioses and glycolaldehyde.[21] This latter observation may be of fundamental importance for both intra- and extracellular modification of proteins and other molecules such as DNA.

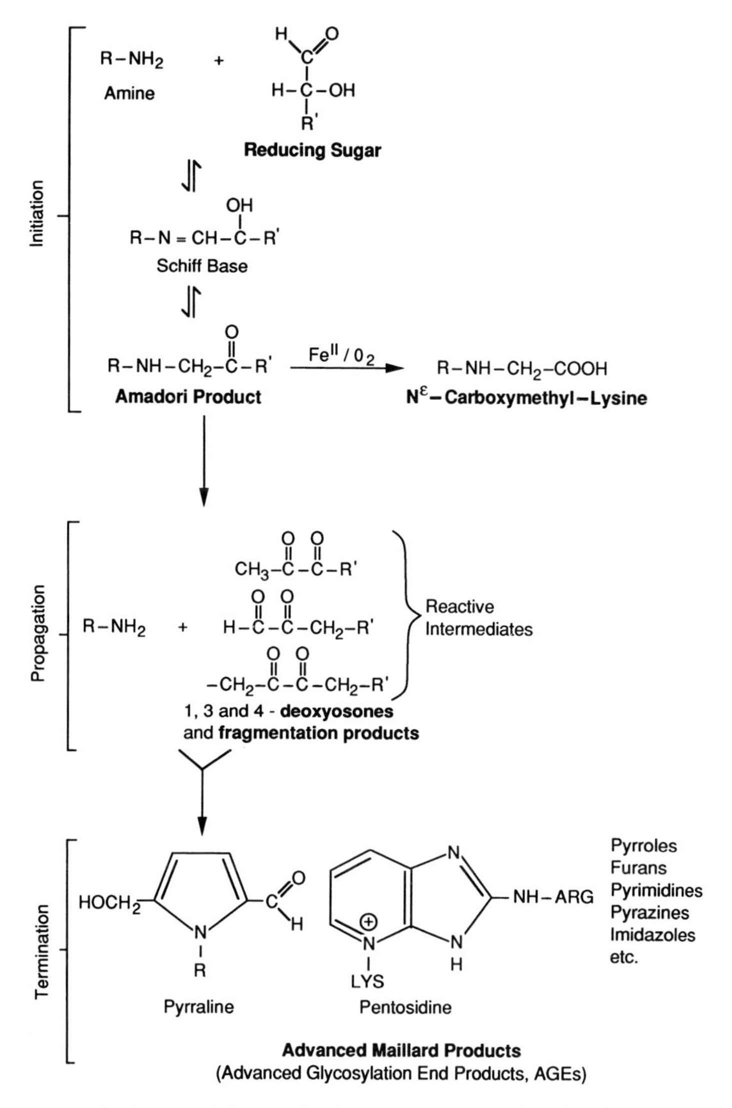

Fig. 1 *General scheme of the Maillard reaction. (Reproduced with permission from Monnier VM: Invited Minireview: Nonenzymatic glycosylation: The Maillard reaction and the aging processes.* J Gerontol *1990;45:B105-B111.)*

Depending on the pH, the Amadori product undergoes enolization reactions to form 1-deoxy-, 3-deoxy-, or 4-deoxyglucosones, which serve as precursors of advanced Maillard products. Hayase and associates[22] and Miyata and Monnier[23] showed that 3-deoxyglucosone is a precursor of pyrraline (Fig. 1), detected immunologically in proteins incubated with glucose, in diabetic plasma, and, using immunohistochemical methods, in human tissue sections.

A novel and important insight into the degradation of Amadori products came from the work of Ahmed and associates,[24,25] who showed that the Amadori product can be fragmented in presence of pro-oxidant conditions into carboxymethyl- (CML) and lactyl-lysine under formation of erythronic and glyceric acid, respectively. Carboxymethyl-lysine was detected in increased amounts in aging collagen and lens crystallines[26,27] and in the urine of diabetic individuals, suggesting that oxidative stress modulates the Maillard reaction in vivo.[28]

Further evidence for a tight relationship between oxidative modifications of carbohydrates and the Maillard reaction comes from studies by Wolff and associates,[29-32] who showed that formation of protein-linked fluorescence can occur during autoxidation of carbohydrates in presence of trace metals. Addition of chelating agents to the reaction mixture completely prevented formation of fluorescent molecules. More recently, Dyer and associates[33] demonstrated that complete removal of oxygen from incubation mixtures containing glycated proteins dramatically decreased polymerization through cross-linking. Their data suggest that the major part of the cross-link(s) forming in vitro from glucose appears to be nonfluorescent. Taken together, these studies clearly emphasize the necessity to consider nonfluorescent oxidation products of glycated proteins as a primary cause of collagen cross-linking in diabetes and aging.

Using a different approach to the question of Maillard reaction-mediated cross-linking in diabetes and aging, Brownlee and associates[34] discovered that aminoguanidine given to diabetic rats completely prevented the loss of solubility of collagen and the formation of collagen-linked fluorescence at 440 nm (excit. 370 nm). This landmark study was followed by an array of reports showing that aminoguanidine (AG) normalizes vascular permeability,[35] renal changes,[36-38] nerve conduction velocity and blood flow,[39] and retinal changes[40] in diabetic rats.

Aminoguanidine appears to block selectively the advanced Maillard reaction[41] by virtue of being able to trap deoxyosones (Fig. 1).[42] However, the puzzling observation that AG, which does not appear to be a significant inhibitor of lysyl oxidase,[43] has effects similar to those of rutin, an aldose reductase inhibitor,[44] suggests that it may also have direct effects on cellular metabolism that may be unrelated to scavenging of Maillard reaction intermediates. Aminoguanidine had no effect on the polyol pathway in nerve,[45] but a weak effect in lens.[46]

The complexity of the mechanism of collagen cross-linking in diabetes is further exemplified through intervention studies in streptozocin diabetic rats. Yue and associates[47] reported that both aspirin and sodium salicylate prevented the diabetes-induced rise in thermal breaking time but that only aspirin was able to block collagen glycation. Furthermore, they found that other inhibitors of cyclooxygenase, such as indomethacin and naproxen, also prevented the rise in thermal rupture of tail tendon collagen, thereby strongly suggesting that collagen cross-linking in diabetes is somehow related to abnormal cellular function.[48]

In our own studies, we have used collagen-linked fluorescence as a marker for the advanced Maillard reaction and studied its relationship to tail tendon breaking time as a marker for collagen cross-linking in hyperglycemia.[49,50] Following 12 months of dietary-induced galactosemia, tail tendon breaking time and fluorescence were increased whereas the LOX-dependent cross-link pyridinoline was unchanged. This suggested that advanced glycation by galactose was a possible factor in tendon cross-linking. However, when the identical study was repeated using aldose reductase inhibition, a much more significant decrease in tail tendon breaking time compared to fluorescence was noted, which suggests a partial dissociation in the biochemical mechanism responsible for cross-linking and fluorescence formation.[50]

A biochemical investigation of the nature of collagen changes in diabetic rat tail tendon collagen was performed by Brennan.[51,52] Her results support the presence of acetic acid stable cross-links but fluorescence was not increased in the acid stable fraction of diabetic collagen. In a second study, she showed that CNBr peptide maps were identical in diabetic rats and controls, suggesting that the number of cross-linked peptides did not increase in diabetes. In these studies, however, the utilization of ammonium- and formic acid-based solvents may have masked age-related fluorescence and cross-linking changes, respectively.

Structure of Collagen-Linked Fluorophores

By 1987, evidence supporting an involvement of the advanced Maillard reaction in vivo was based on the glucose-derived putative protein cross-link called 2-(2-furoyl)-4(5)-(2-furanyl)-1H-imidazole (FFI)[53] and on the presence of collagen-linked fluorescence that could be duplicated by incubating proteins, such as glucose, under physiologic conditions.[6] However, mechanistic studies revealed that FFI was a by-product of acid hydrolysis of glycated proteins.[54,55] Therefore, a detailed investigation of the biochemical nature of age-related collagen fluorescence was initiated, which led to isolation of two fluorescent peptides from enzymatic digest of aging human collagen. These fluorescent peptides were named "P" and "M" for their properties, which resembled pyridinium-type and Maillard-type fluorescence.[56]

Detailed characterization of the peptide-linked fluorophores P and M revealed the presence of a yellow and blue fluorophore, respectively.[56] Structure elucidation of fluorophore P revealed presence of an imidazo [4,5b] pyridinium compound bridging a lysyl residue with an arginyl residue (Fig. 1).[57] The fluorophore could be synthesized by the reaction of pentoses with L-lysine and L-arginine and was, therefore, named pentosidine.

Elucidation of the mechanism of formation of pentosidine revealed a number of unexpected findings.[58] First, long-term incubation of protein with glucose or fructose also led to pentosidine formation, albeit at a much slower rate than with D-ribose. Thus, obviously, spontaneous fragmentation of hexoses into pentoses must be occurring. Second, ascorbate was also a precursor of pentosidine, an observation that appears to be of paramount importance for lens aging (see below). Third, Amadori products were more active as pentosidine precursors than was the sugar from which they were made. Taken together, these observations suggested the formation of a common intermediate molecule in pentosidine synthesis.[58] Overall similar conclusions were reached by Dyer and associates.[33]

Relevance of Pentosidine to Aging and Disease

Pentosidine was assayed by reverse phase HPLC in a variety of human tissues obtained at autopsy. Three distinct patterns of pentosidine biosynthesis as a function of age were observed in dura mater, skin, and glomerular basement membranes (GBM). Pentosidine increased linearly in dura mater,[57] exponentially in skin,[59] and curvilinearly with plateau in GBMs.[60] Levels at the end of life span were 250, 75, and 40 pmol/mg protein, respectively.

Although it was not quite unexpected to find increased pentosidine levels in skin and glomerular basement membranes of diabetic subjects, it was very unexpected to find much higher levels of pentosidine in skin of subjects with end-stage renal disease on hemodialysis.[61] This latter observation suggests that high levels of a pentosidine precursor sugar are present in uremia, which, we hypothesized, should result in increased plasma pentosidine formation. Indeed, elevated pentosidine levels (10 to 30 ×) were found in red blood cells and plasma proteins from uremic individuals. Most interestingly, plasma pentosidine levels were unaffected by hemodialysis and were almost completely normalized following renal transplantation.[62] The significance of these findings will be discussed below.

The relationship between the severity of diabetic complications and pentosidine formation in skin biopsies from insulin-dependent subjects was also explored. Previous data from our laboratory indicated that collagen-linked fluorescence of unknown origin was higher in individuals with severe retinopathy, nephropathy, and arterial and joint stiffness than it was in individuals free of complications.[63] Essentially similar results were obtained for pentosidine, thereby providing evidence for an association between a metabolic abnormality reflected in increased pentosidine synthesis and the pathogenesis of severe complications of diabetes.[61]

Another interesting spin-off from the studies on pentosidine is the discovery of a high correlation between crystalline-linked pentosidine levels and the degree of browning and cross-linking of human lens crystallines during aging and cataractogenesis. In this case it appears that mechanisms responsible for reduction of dehydroascorbate (DHA) back to ascorbate weaken as a function of age, thus leading to spontaneous degradation of DHA into xylosone, which we found to be a pentosidine precursor sugar. Presence of pentosidine in the lens and skin was also found by Lyons and associates,[64,65] who also found that short-term control of glycemia did not correct pentosidine and carboxymethyl-lysine levels, although reversal of Amadori products was observed.

Relevance of the Advanced Maillard Reaction to Degenerative Disk Disease

Because of the extreme longevity of matrix molecules in the fibrocartilage of disk tissue, disks constitute a prime target for damage by the advanced Maillard reaction. The color of human disk tissue changes from white during early life to yellow by middle age, and, often, to a deep brown color by old age.[66] Preliminary evidence that such changes may be attributable to the Maillard reaction comes from work by Hormel and Eyre,[67] who found increased Maillard type fluorescence in disk tissue from elderly individuals. The fluorescence was linked to a lysyl residue from the $\alpha 1(II)$ CB12 peptide. Similarly, preliminary evidence for the presence of pentosidine in human cartilage was obtained.[57] Although both modifications are minor in quantitative terms, they reflect broader modifications of disk tissue proteins by the overall Maillard reaction.

It would be logical to expect that the ongoing damaging effects on disk and cartilage tissue proteins through cross-linking, insolubilization, and free radical mediated fragmentation would contribute to the senile degeneration of these tissues. We hypothesize that the activation of chondrocytes by AGE-proteins through a receptor pathway could lead these cells to release proteases and cytokines, such as IL-1 and TNFα. In osteoarthritis, such activation would lead to profound tissue remodeling and destruction of the joint. This proposition is based on the analogous properties of chondrocytes with macrophages, which have been shown to release these cytokines upon ingestion of AGE-proteins.[68]

References

1. Kirk E, Kvorning SA: Quantitative measurements of the elastic properties of the skin and subcutaneous tissue in young and old individuals. *J Gerontol* 1949;4:273-284.
2. Schnider SL, Kohn RR: Effects of age and diabetes mellitus on the solubility and nonenzymatic glucosylation of human skin collagen. *J Clin Invest* 1981;67:1630-1635.
3. Hamlin CR, Kohn RR: Evidence for progressive, age-related structural changes in postmature collagen. *Biochim Biophys Acta* 1971; 236:458-467.
4. Everitt AV: Food intake, growth and the ageing of collagen in rat tail tendon. *Gerontology* 1971;17:98-104.
5. Miksík I, Deyl Z: Change in the amount of ϵ-hexosyllysine, UV absorbance, and fluorescence of collagen with age in different animal species. *J Gerontol* 1991;46:B111-116.

6. Monnier VM, Kohn RR, Cerami A: Accelerated age-related browning of human collagen in diabetes mellitus. *Proc Natl Acad Sci USA* 1984;81:583-587.
7. Bjorksten J, Andrews F: Fundamentals of aging: A comparison of the mortality curve for humans with a viscosity curve of gelatin during cross-linking reaction. *J Am Geriatr Soc* 1960;8:632-637.
8. Verzár F: Aging of the collagen fiber. *Int Rev Connect Tissue Res* 1964;2:243-300.
9. Bensusan HB: A novel hypothesis for the mechanism of crosslinking of collagen, in Partridge SM, Tristram GR (eds): *Structure and Function of Connective and Skeletal Tissue*. London, Butterworths, 1965, pp 42-46.
10. Hamlin CR, Kohn RR, Luschin JH: Apparent accelerated aging of human collagen in diabetes mellitus. *Diabetes* 1975;24:902-904.
11. Harrison DE, Archer JR: Measurement of changes in mouse tail collagen with age: Temperature dependence and procedural details. *Exp Gerontol* 1978;13:75-82.
12. Kohn RR, Hamlin CR: Genetic effects on aging of collagen with special reference to diabetes mellitus. *Birth Defects* 1978;14:387-401.
13. Yamauchi M, Woodley DT, Mechanic GL: Aging and crosslinking of skin collagen. *Biochem Biophys Res Comm* 1988;152:898-903.
14. Reiser K, McCormick RJ, Rucker RB: Enzymatic and nonenzymatic cross-linking of collagen and elastin. *FASEB J* 1992;6:2439-2449.
15. Madia AM, Rozovski SJ, Kagan HM: Changes in lung lysyl oxidase activity in streptozotocin-diabetes and in starvation. *Biochim Biophys Acta* 1979;585:481-487.
16. Chang K, Uitto J, Rowold EA, et al: Increased collagen cross-linkages in experimental diabetes: Reversal by β-amino propionitrile and D-penicillamine. *Diabetes* 1980;29:778-781.
17. Buckingham B, Reiser KM: Relationship between the content of lysyl-oxidase-dependent cross-links in skin collagen, nonenzymatic glycosylation, and long-term complications in type I diabetes mellitus. *J Clin Invest* 1990;86:1046-1054.
18. Chio KS, Tappel AL: Synthesis and characterization of the fluorescent products derived from malonaldehyde and amino acids. *Biochemistry* 1969;8:2821-2826.
19. Haberland ME, Fong D, Cheng L: Malondialdehyde-altered protein occurs in atheroma of Watanabe heritable hyperlipidemic rabbits. *Science* 1988;241:215-218.
20. Baynes JW, Watkins NG, Fisher CI, et al: The Amadori production protein: Structure and reactions, in Baynes JW, Monnier VM (eds): *The Maillard Reaction in Aging, Diabetes and Nutrition*. New York, Alan R Liss, 1989, pp 43-67.
21. Bunn HF, Higgins PJ: Reaction of monosaccharides with proteins: Possible evolutionary significance. *Science* 1981;213:222-224.
22. Hayase F, Nagaraj RH, Miyata S, et al: Aging of proteins: Immunological detection of a glucose-derived pyrrole formed during Maillard reaction in vivo. *J Biol Chem* 1989;264:3758-3764.
23. Miyata S, Monnier VM: Immunohistochemical detection of advanced glycosylation end products in diabetic tissues using monoclonal antibody to pyrraline. *J Clin Invest* 1992;89:1102-1112.
24. Ahmed MU, Dunn JA, Walla MD, et al: Oxidative degradation of glucose adducts to protein: Formation of 3-(N^{ϵ}-lysino)-lactic acid from model compounds and glycated proteins. *J Biol Chem* 1988; 263:8816-8821.

25. Ahmed MU, Thorpe SR, Baynes JW: Identification of N^{ϵ}-carboxymethyllysine as a degradation product of fructoselysine in glycated protein. *J Biol Chem* 1986;261:4889-4894.
26. Dunn JA, McCance DR, Thorpe SR, et al: Age-dependent accumulation N^{ϵ}(carboxymethyl) lysine and N^{ϵ}-(carboxymethyl) hydroxylysine in human skin collagen. *Biochemistry* 1991;30:1205-1210.
27. Dunn JA, Patrick JS, Thorpe SR, et al: Oxidation of glycated proteins: Age-dependent accumulation of N^{ϵ}-(carboxymethyl) lysine in lens proteins. *Biochemistry* 1989;28:9464-9468.
28. Baynes JW: Role of oxidative stress in development of complications in diabetes. *Diabetes* 1991;40:405-412.
29. Wolff SP, Crabbe MJC, Thornalley PJ: The autoxidation of glyceraldehyde and other simple monosaccharides. *Experientia* 1984;40:244-246.
30. Hunt JV, Dean RT, Wolff SP: Hydroxyl radical production and autoxidative glycosylation: Glucose autoxidation as the cause of protein damage in the experimental glycation model of diabetes mellitus and aging. *Biochem J* 1988;256:205-212.
31. Hunt JV, Smith CC, Wolff SP: Autoxidative glycosylation and possible involvement of peroxides and free radicals in LDL modification by glucose. *Diabetes* 1990;39:1420-1424.
32. Thornalley P, Wolff S, Crabbe J, et al: The autoxidation of glyceraldehyde and other simple monosaccharides under physiological conditions catalysed by buffer ions. *Biochim Biophys Acta* 1984;797:276-287.
33. Dyer DG, Blackledge JA, Thorpe SR, et al: Formation of pentosidine during nonenzymatic browning of proteins by glucose: Identification of glucose and other carbohydrates as possible precursors of pentosidine in vivo. *J Biol Chem* 1991;266:11654-11660.
34. Brownlee M, Vlassara H, Kooney A, et al: Aminoguanidine prevents diabetes-induced arterial wall protein cross-linking. *Science* 1986;232:1629-1632.
35. Chang K, Allison W, Harlow J, et al: Aminoguanidine prevents glucose- and sorbitol-induced vascular dysfunction in skin chamber granulation tissue (Abstract). *Diabetes* 1991;40:210A.
36. Nicholls K, Mandel TE: Advanced glycosylation end-products in experimental murine diabetic nephropathy: Effect of islet isografting and of aminoguanidine. *Lab Invest* 1989;60:486-491.
37. Soulis-Liparota T, Cooper M, Papazoglou D, et al: Retardation by aminoguanidine of development of albuminuria, mesangial expansion, and tissue fluorescence in streptozocin-induced diabetic rat. *Diabetes* 1991;40:1328-1334.
38. Yagihashi S, Kamijo M, Baba M, et al: Effect of aminoguanidine on functional and structural abnormalities in peripheral nerve of STZ-induced diabetic rats. *Diabetes* 1992;41:47-52.
39. Kihara M, Schmelzer JD, Poduslo JF, et al: Aminoguanidine effects on nerve blood flow, vascular permeability, electrophysiology, and oxygen free radicals. *Proc Natl Acad Sci USA* 1991;88:6107-6111.
40. Hammes H-P, Martin S, Federlin K, et al: Aminoguanidine treatment inhibits the development of experimental diabetic retinopathy. *Proc Natl Acad Sci USA* 1991;88:11555-11558.
41. Edelstein D, Brownlee M: Mechanistic studies of advanced glycosylation endproduct inhibition by aminoguanidine. *Diabetes* 1992;41:26-29.
42. Hirsch J, Baynes JW, Blackledge JA, et al: The reaction of 3-deoxy-D-glycero-pentos-2-ulose (''3-deoxyxylosone'') with aminoguanidine. *Carbohydr Res* 1991;220:C5-7.

43. Oxlund H, Andreassen TT: Aminoguanidine treatment reduces the increase in collagen stability of rats with experimental diabetes mellitus. *Diabetologia* 1992;35:19-25.
44. Odetti PR, Borgoglio A, de Pascale A, et al: Prevention of diabetes-increased aging effect on rat collagen-linked fluorescence by aminoguanidine and rutin. *Diabetes* 1990;39:796-801.
45. Cameron NE, Cotter MA, Dines K, et al: Effect of aminoguanidine on peripheral nerve function and polyol pathway metabolities in streptozotocin-diabetic rats. *Diabetologia* 1992;35:946-950.
46. Kumari K, Umar S, Bansal V, et al: Inhibition of diabetes-associated complications by nucleophilic compounds. *Diabetes* 1991;40:1079-1084.
47. Yue DK, McLennan S, Handelsman DJ, et al: The effect of salicylates on nonenzymatic glycosylation and thermal stability of collagen in diabetic rats. *Diabetes* 1984;33:745-751.
48. Yue DK, McLennan S, Handelsman DJ, et al: The effects of cyclooxygenase and lipoxygenase inhibitors on the collagen abnormalities of diabetic rats. *Diabetes* 1985;34:74-78.
49. Monnier VM, Sell DR, Abdul-Karim FW, et al: Collagen browning and cross-linking are increased in chronic experimental hyperglycemia: Relevance to diabetes and aging. *Diabetes* 1988;37:867-872.
50. Richard S, Tamas C, Sell DR, et al: Tissue-specific effects of aldose reductase inhibition on fluorescence and cross-linking of extracellular matrix in chronic galactosemia. *Diabetes* 1991;40:1049-1056.
51. Brennan M: Changes in solubility, non-enzymatic glycation, and fluorescence of collagen in tail tendons from diabetic rats. *J Biol Chem* 1989;264:20947-20952.
52. Brennan M: Changes in the cross-linking of collagen from rat tail tendons due to diabetes. *J Biol Chem* 1989;264:20953-20960.
53. Pongor S, Ulrich PC, Bencsath FA, et al: Aging of proteins: Isolation and identification of a fluorescent chromophore from the reaction of polypeptides with glucose. *Proc Natl Acad Sci USA* 1984;81:2684-2688.
54. Njoroge FG, Fernandes AA, Monnier VM: Mechanism of formation of the putative advanced glycosylation end product and protein cross-link 2-(2-furoyl)-4(5)-(2-furanyl)-1H-imidazole. *J Biol Chem* 1988; 263:10646-10652.
55. Horiuchi S, Shiga M, Araki N, et al: Evidence against in vivo presence of 2-(2-furoyl)-4(5)-(2-furanyl)-1H-imidazole, a major fluorescent advanced end product generated by nonenzymatic glycosylation. *J Biol Chem* 1988;263:18821-18826.
56. Sell DR, Monnier VM: Isolation, purification and partial characterization of novel fluorophores from aging human insoluble collagen-rich tissue. *Connect Tissue Res* 1989;19:77-92.
57. Sell DR, Monnier VM: Structure elucidation of a senescence cross-link from human extracellular matrix: Implication of pentoses in the aging process. *J Biol Chem* 1989;264:21597-21602.
58. Grandhee SK, Monnier VM: Mechanism of formation of the Maillard protein cross-link: Glucose, fructose and ascorbate as pentosidine precursors. *J Biol Chem* 1991;266:11649-11653.
59. Sell DR, Monnier VM: End-stage renal disease and diabetes catalyze the formation of a pentose-derived crosslink from aging human collagen. *J Clin Invest* 1990;85:380-384.
60. Sell DR, Carlson EC, Monnier VM: Structure elucidation of a novel protein crosslink from human extracellular matrix: Role in complications of diabetes. *Diabetes* 1990;39:16A.
61. Odetti P, Fogarty JF, Sell DR, et al: Chromatographic quantitation of plasma and erythrocyte pentosidine in diabetic and uremic subjects. *Diabetes* 1992;41:153-159.

62. Hricik DE, Schulak JA, Fogarty J, et al: Pentose-derived glycation of plasma proteins after kidney and kidney-pancreas transplantation in patients with diabetic nephropathy. *Kidney Int* 1993;43:398-403.
63. Monnier VM, Vishwanath V, Frank KE, et al: Relation between complications of type I diabetes mellitus and collagen-linked fluorescence. *N Engl J Med* 1986;314:403-408.
64. Lyons TJ, Bailie KE, Dyer DG, et al: Decrease in skin collagen glycation with improved glycemic control in patients with insulin-dependent diabetes mellitus. *J Clin Invest* 1991;87:1910-1915.
65. Lyons TJ, Silvestri G, Dunn JA, et al: Role of glycation in modification of lens crystallins in diabetic and nondiabetic senile cataracts. *Diabetes* 1991;40:1010-1015.
66. Ghosh P (ed): *The Biology of Intervertebral Disc*. Boca Raton, FL, CRC Press, 1988, vol 2, pp 73-120.
67. Hormel SE, Eyre DR: Collagen in the ageing human intervertebral disc: An increase in covalently bound fluorophores and chromophores. *Biochem Biophys Acta* 1991;1078:243-250.
68. Vlassara H, Brownlee M, Manogue KR, et al: Cachectin/TNF and IL-1 induced by glucose-modified proteins: A role in normal tissue remodeling. *Science* 1988;240:1546-1548.

Chapter 5

New Clinical Applications for Recombinant Human Growth Hormone

Kaup R. Shetty, MD
Krishan L. Gupta, MD
Daniel Rudman, MD

Human growth hormone (hGH) is a single-chain polypeptide with 191 amino acid residues and a molecular weight of 22,000 daltons. The normal adult pituitary gland contains 5 to 10 mg of growth hormone (GH), which has a half-life of 20 to 25 minutes after secretion into the circulation. Two GH-regulating hypothalamic peptides, somatostatin and growth hormone-releasing hormone (GHRH), are released into the portal venous system under the control of neurotransmitters of the suprahypothalamic origin, and insulin-like growth factor I (IGF-I, also called somatomedin C) inhibits GH secretion by feedback at the pituitary level. The low basal level of GH secretion appears to reflect mostly the inhibitory action of somatostatin, while the pulses of GH secretion reflect mainly the stimulating effect of GHRH.

IGF-I is a 7,600-molecular-weight peptide and is the main mediator of the somatic effects of GH. The major site of IGF-I synthesis is the liver. Nevertheless, measurements of the tissue concentrations of the messenger RNA for IGF-I show that muscle, kidney, and other extrahepatic organs also produce this substance under the influence of GH. It is now believed that IGF-I, acting through endocrine, paracrine, and autocrine routes, mediates most of the effects of GH. For example, IGF-I stimulates whole body growth, renal blood flow, and muscle protein synthesis in the same manner as GH itself. Receptors to IGF-I have been demonstrated in chondrocytes, adipocytes, placental cells, hepatocytes, lymphocytes, kidney cells, and fibroblasts. Activation of IGF-I receptors stimulates cell growth in many types of cells and inhibits lipolysis in explanted adipocytes. Nevertheless, GH itself has a lipolytic effect and directly inhibits lipogenesis.

The operation of the hypothalamus/GH/IGF-I axis changes during the life cycle. The integrated serum GH level and the circulating IGF-I concentration rise to their highest values at puberty. Thereafter they gradually decline during the ensuing decades of the life span. Plasma IGF-I in healthy young men ranges from 0.5 to 1.5

units/ml. After about age 30, average peak nocturnal GH secretion decreases significantly.[1] At age 70 to 80, 40% of adults have a circulating IGF-I level less than 0.25 units/ml, the range found in GH-deficient children. IGF-I below 0.35 units/ml (about 50% of octogenarians) corresponds to the absence of significant nocturnal peaks of serum GH (<2.0 ng/ml). That the hyposomatomedinemia reflects cessation of GH secretion rather than hepatic unresponsiveness to the hormone is shown by the prompt rise in plasma IGF-I in response to exogenous hGH.

Human growth hormone has been available for therapeutic use for more than three decades, but the supply was restricted because it had to be extracted from the pituitary glands of cadavers. Moreover, there was concern about contamination of the natural product with the virus that causes Creutzfeldt-Jacob disease. Since 1985, with the advent of recombinant DNA technology, sufficient quantities of hGH have been available. In the United States at present there are two biosynthetic preparations of hGH on the market: Somatropin (Humatrope) and Somatrem (Protropin). Both preparations are identical to natural hGH in structure and function, except that Somatrem has an extra N-terminal methionine, which is related to the bacterial synthesis of the hormone. At present GH deficiency in children is the only therapeutic indication approved by the Food and Drug Administration, but novel clinical uses are being investigated. This review will focus on the newer uses of GH outside the field of pediatric growth retardation, with special emphasis on its clinical use in GH deficiency in adults and aged individuals. Outline 1 shows the various disorders concerning which investigative data are being accumulated for the potential use of hGH.

Effect of GH Replacement in GH-Deficient Adults

Body Composition

GH deficiency in adults with hypopituitarism is associated with increased overall mortality due to cardiovascular disease, despite adequate conventional replacement with thyroxine, cortisol, and sex steroids. It has been suggested that GH deficiency may be a contributory factor in the pathogenesis of cardiovascular disease because of its known effects on body composition and lipid metabolism. It is known that GH deficiency in children is associated with a decrease in lean body mass (LBM) and an increase in adipose mass (AM), both of which can be reversed with the administration of hGH. Moreover, hGH has been shown to promote a redistribution of fat from abdominal to peripheral regions, thereby changing the adipose topography from android to gynoid.

Studies of hGH replacement in conventional dosage have provided information on the physiological role of this hormone in adults. Salomon and associates,[2] in a double-blind placebo-controlled trial, replaced hGH at a dose of 0.07 units/kg body weight

Outline 1 Potential Therapeutic Uses of Growth Hormone

- Growth enhancement
  - Turner's syndrome
  - Post head and neck irradiation
  - Chronic renal failure
  - Skeletal dysplasias
  - Intrauterine growth retardation
  - Inflammatory bowel disease
  - Familial short stature
  - Constitutional delay in growth
- Anabolic effect
  - Growth hormone deficient adults
  - Normal aging
  - Malnourished older individuals
  - Postsurgical total parenteral nutrition
  - Burn stress
  - Osteoporosis and healing of fractures
  - Chronic obstructive pulmonary disease
  - Glucocorticoid treatment
- Miscellaneous
  - Obesity
  - Ovulation induction

given daily by the subcutaneous route in 12 adults with a mean age of 39 years, all of whom had GH deficiency caused by pituitary tumors. The body composition, as determined by total body potassium and skin fold thickness, showed decreased LBM and increased AM. In response to six months of hGH therapy, LBM increased by an average of 5.5 kg and fat mass decreased by 5.7 kg. Total body weight was unchanged, but the waist:hip ratio and the skin fold subcutaneous adipose thickness decreased. A similar effect on body composition was shown by Jorgensen and associates[3] in a placebo-controlled study of 22 GH-deficient adults with mean age of 23.8 years, who had previously received GH treatment as children. In response to four months of hGH therapy, there was an increase in muscle volume and a reduction in adipose mass as measured by computerized tomography of the thigh. These results have been further substantiated in an article published by Whitehead and associates.[4] Studies done for four to 12 months with hGH therapy have shown varying effects on bone mineral density. The results are consistent with the conclusion that GH deficiency in adults is associated with abnormal body composition, which can be partially reversed by hGH replacement therapy.

Functional Capacity

GH deficiency in adults is associated with muscle weakness, fatigue, decreased strength, and diminished exercise performance. Cuneo and associates[5,6] studied 24 GH-deficient adults with a mean age of 39 years in a double-blind placebo-controlled trial at a dose of 0.07 units/kg of body weight given daily as a subcutaneous injection for six months. Isometric strength was measured with a hand-held electrodynamometer, and exercise capacity with a cycle ergometer. There was significant improvement in muscle strength which correlated with an expansion of the LBM. Human growth hormone therapy improved and normalized the maximal exercise performance. Improved cardiac output was noted, which may also have contributed to the effect of hGH on exercise performance. These data have been confirmed in two separate studies.[4,7] It can be concluded that GH deficiency in adults is associated with subnormal muscle mass and exercise capacity, both of which show improvement with hGH replacement.

Psychosocial Well-being

It has been noted that GH-deficient adults who were treated with hGH during childhood have a higher than normal occurrence of muscle weakness and fatigue, and of psychosocial maladjustment associated with social isolation, unemployment, and unmarried status. There is a paucity of published work on the effect hGH replacement in adults has on their general well-being and quality of life. McGauley and associates,[8] in a double-blind, placebo-controlled, six-month replacement trial of hGH in 24 GH-deficient adults, 18 to 55 years old, found significantly impaired quality of life based on self-rating questionnaires. In response to hGH therapy, the subjects perceived less illness and a significant psychological improvement in their mood and energy level compared with the placebo group. The effect on well-being, mood, and cognition were variable in other studies. Longer treatment periods are necessary before the psychosocial aspects of hGH replacement can be properly assessed.

Effect of hGH on Body Composition in Elderly Men

It is well documented that body composition changes progressively in the second half of the human life cycle.[9] LBM shrinks with advancing age, and there is an average atrophy of about 30% from age 30 to age 75 years in the sizes of the liver, kidneys, brain, and pancreas. Body weight tends to remain stable because of a reciprocal expansion of the AM.

Gerontologists have concluded that the age-related changes in body composition are undesirable for at least three reasons: (1) Aerobic work capacity is directly proportional to LBM. (2) The geriatric atrophy of LBM organs is generally associated with diminished functional capacities in muscle strength, glomerular filtration rate, and renal blood flow as the aging muscles and kidneys shrink in size. (3) Increased adiposity predisposes to unfavorable changes in blood pressure, glucose clearance, and the plasma lipoprotein profile. Because GH causes expansion of the LBM and contraction of the AM, and because GH secretion tends to diminish in late adulthood, it has been postulated that geriatric hyposomatotropism contributes to the body composition changes described above.

Although the beneficial effects of hGH have been documented in GH-deficient children and young adults, the consequences of hyposomatotropism have not been defined in the elderly population. A short-term study for seven days showed that hGH caused positive nitrogen balance and an increase in the serum concentration of osteocalcin, a marker of bone synthesis.[10] Rudman and associates[11,12] reported that a one-year trial of hGH replacement in elderly men showed beneficial effect on body composition.

The latter study involved 45 independent men over 61 years old with plasma IGF-I level below 0.35 units/ml, indicating a low level of GH secretion. The 21-month protocol was as follows: Baseline period 0 to 6 months, experimental period 6 to 18 months, and post-experimental period 18 to 21 months (Fig. 1). During the experimental period, 26 men (group I) received approximately 0.03 mg/kg of biosynthetic human growth hormone subcutaneously three times a week, while 19 men (group II) received no treatment. Plasma IGF-I was measured monthly. The following outcome variables were measured at zero, six, 12, and 18 months: LBM, AM, skin thickness (dermis plus epidermis), sizes of the liver, spleen, and kidneys, the cross-sectional areas of ten muscle groups, and bone density at nine skeletal sites. LBM and AM were also measured at 21 months. In group I, hGH treatment raised the plasma IGF-I level and maintained it in the range 0.5 to 1.5 units/ml. Significant changes occurred in the following outcome variables, expressed as percent change at 18 months over baseline as shown in Table 1: LBM +6%, AM −15%, skin thickness +4%, liver volume +8%, spleen volume +23%, sum of ten muscle areas +11%. Three months after hGH treatment stopped, about one half of the hGH-induced increment in lean body mass had disappeared and about one third of the hGH-induced decrement in AM had reappeared. In group II, the IGF-I level remained below 0.35 units/ml. At the 18-month time point these untreated controls showed a significant decline in LBM to 96% of initial baseline and in skin thickness to 94% of initial baseline.

During hGH treatment, ten subjects developed carpal tunnel syndrome and four subjects developed gynecomastia. The adverse side effects disappeared spontaneously within three months after cessation of hormone treatment. Elevation of plasma IGF-I above 1.0 units/ml was found to be associated with substantial frequencies of carpal tunnel syndrome and gynecomastia. It was concluded that the desirable hormone effects in expanding LBM and reducing AM can be achieved, and the undesirable side effects avoided, by maintaining the mean IGF-I level in the range of 0.5 to 1.0 units/ml during the period of treatment.[13]

These observations were consistent with the hypothesis that diminished GH secretion in the later years of adulthood is a contributory cause to the changes in body composition that occur with advancing age. The study developed two types of evidence consistent with hypothesis. The first type of evidence was the relatively rapid

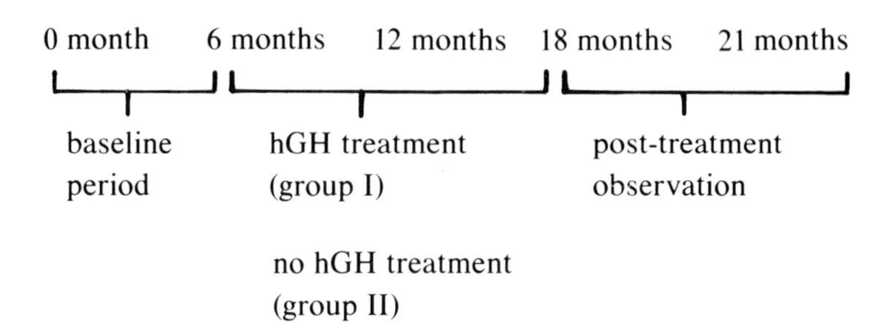

Fig. 1 *Experimental design of the clinical trial of hGH in hyposomatomedinemic elderly men. (Reproduced with permission from Rudman D, Feller AG, Cohn L, et al: Effects of human growth hormone on body composition in elderly men.* Horm Res *1991;36(suppl 1):73-81.)*

Table 1 Clinical trial of hGH in elderly hyposomatomedinemic men: Outcome variables as percent of initial baseline value

	Group	Month				
		0	6	12	18	21
Lean body mass	I*	100	99.0	104.8‡	105.7‡	102.7‡
	II†	100	99.8	99.1	96.0‡	91.7‡
Adipose mass	I	100	96.8	86.9‡	84.8‡	90.1‡
	II	100	98.2	102.2	97.8	105.5
Skin thickness (sum of four sites)	I	100	98.9	106.4‡	104.3‡	
	II	100	99.0	98.0	93.9‡	
Liver size	I	100	99.2	119‡	108‡	
	II	100	98.6	98.3	93.3	
Spleen size	I	100	95.1	116.6‡	123.0‡	
	II	100	95.2	101.9	93.2	
Sum of ten muscle areas	I	100	101.7	111.3‡	110.6‡	
	II	100	104.2	96.7	98.3	

*Group I, hGH treatment.
†Group II, no treatment.
‡$p < 0.05$ for change from initial baseline value by paired t test.
(Reproduced with permission from Rudman D, Feller AG, Cohn L, et al: Effects of human growth hormone on body composition in elderly men. *Horm Res* 1991;36(suppl 1):73-81.)

rate of decline in LBM in the untreated hyposomatotropic men compared to the general population. The second line of evidence was the reversal of body composition changes during the 12 months of hGH treatment, and the partial dissipation of these effects within three months after hGH was stopped.

Further investigations are necessary to assess the functional changes following GH replacement in the elderly. Future research objectives can now be stated as follows: (1) In hyposomatomedinemic elderly men who are treated with hGH, how do the individual organs within the LBM change in size? Which cell types within each organ participate in the response? What are the effects on cell size and cell number? (2) In the case of adipose tissue, how is the regional distribution influenced by hGH? (3) How do the functional capacities of the affected organs react to the hormone? (4) For those elderly men with a low versus a youthful plasma IGF- I, do body composition, organ sizes, and organ function differ in the manner predicted by GH deficiency? (5) Longitudinally, does the plasma IGF-I level of aging men correlate with the rate of decline in LBM and in the sizes of its constituent organs? (6) Can the gender difference in the rate of loss of LBM during mid and late adulthood be explained by a greater activity of the GH/IGF-I axis in older women than in older men?

Other Potential Uses

Preliminary results have suggested the potential for use of hGH in several additional disorders and clinical situations which are common in old age (Outline 1). These inquiries are briefly summarized below.

Postsurgical Stress

Several investigators have reported anabolic effects of hGH in postsurgical stress. In a comprehensive study, Jiang and associates[14] demonstrated that hGH given at a dose of 0.06 mg/kg/day, in a setting of postoperative hypocaloric total parenteral nutrition, reduced the loss of body weight and the degree of negative nitrogen balance, compared to placebo-treated patients. Moreover, stimulation of protein synthesis and preferential loss of fat in the hGH treated group were revealed by studies of labelled amino acid flux across the forearm and by compartmental analysis. Muscle strength as measured by hand grip remained stable in the treated group compared to a 10% reduction in the placebo group. Nevertheless, further studies of functional status and rate of recuperation in the postoperative period are necessary before GH therapy can be recommended for surgical patients.

Bone Density and Calcium Metabolism

GH enhances calcium and phosphate absorption from the gut either by increasing 1,25-dihydroxyvitamin D3 production, by increasing the sensitivity of the intestinal epithelium to this vitamin, or both. Remodeling in adult bone is a constant process that involves both formation and resorption; an array of growth factors and local hormones contribute to the modulation of this process. Osteoblasts contain IGF-I receptors, which proliferate in response to IGF-I and GH. IGF-I production in osteoblasts can be stimulated by GH, estradiol, calcitriol, or parathyroid hormone; IGF-I may be a critical determinant of bone turnover. A commentary on the role of GH in adult bone remodeling was recently written by Parfitt.[15] GH supplementation in GH-deficient adults is associated with significant increases in the various biochemical markers of bone formation (circulating levels of osteocalcin, procollagen I, and bone specific alkaline phosphatase) and of bone resorption (urinary hydroxyproline, calcium, and the more bone-specific collagen pyridinium cross-links). These studies indicate stimulation in whole body remodeling activity, resulting in increases of all components of bone turnover during GH therapy.

The published animal data investigating the effects of GH administration on bone mass and fracture repair are contradictory and variable, depending upon species, age, and hormone dosage. Children with GH deficiency have decreased bone mass and bone growth, which improve with GH replacement. Adults with hypopituitarism develop osteoporosis, but the role played by GH deficiency is difficult to ascertain.[16] Studies in acromegaly have shown an increase in total body mineral mass and skeletal size, especially in the areas of flat bones and in the shaft widths of long bones.[17] There is also enhancement in intestinal calcium absorption and in urinary calcium excretion. Dynamic histologic, radiokinetic, and biochemical bone markers indicate accentuation of bone turnover.[17] The greater amount of bone mass in the phase of increased bone turnover in acromegaly indicates that GH might mediate a positive bone balance at each remodelling cycle, making it a potential therapeutic agent for osteoporosis. Although bone mineral content is increased in the forearm of acromegalic patients, it is decreased in the spine.[18] This apparent discrepancy between the axial and appendicular skeletons suggests that GH excess cannot prevent the adverse effects of hypogonadism on the spine.

Although the lack of estrogen is a major pathogenetic factor in the postmenopausal osteoporosis of women, age-related GH deficiency may also have a role in the osteoporosis of the elderly population. A positive correlation has been reported between low serum IGF-I levels and symptomatic osteoporosis in middle-aged male patients.[19] Studies on the effects of hGH on the bone mineral content of GH-deficient non-elderly adults with isolated or panhypopituitarism are limited to three in number, for periods ranging from four to six months, and have shown either no change or only minimal improvement.[16,20,21] In a study by Rudman and associates,[11] 12 healthy elderly males with low IGF-I levels were treated for six months with hGH at 0.03 mg/kg of body weight given three times a week; a small but significant increase occurred in vertebral bone density, but there was no change in the radius or proximal femur. However, when hGH was continued for 12 months, no change in bone density was found at any of the nine sites, including the lumbar vertebrae.[12]

There are few studies on the treatment of primary osteoporosis with hGH. Aloia and associates[22-24] reported three clinical trials. In the first study, before the availability of the biosynthetic hormone, a group of eight osteoporotic patients were treated for six months at a lower dose of natural hGH, followed by another six months at a higher dose. Total body calcium remained unchanged, and the radial bone density declined. A two-year randomized parallel study of calcitonin alone, versus combined treatment with hGH plus calcitonin, was then carried out in two groups of women with postmenopausal osteoporosis. It was speculated that GH would improve bone formation, whereas calcitonin would inhibit the GH-mediated increase in bone resorption. Total body calcium content increased by the same amount in both groups, but the combined treatment group showed a progressive loss of bone mineral content in the distal radius. In a third study, a 24-month study of 14 postmenopausal women with osteoporosis, hGH was administered for two months followed by salmon calcitonin for three months. There was a significant increase of 2.3% per year in the total body calcium content, but bone mineral content in the distal radius did not change. Until now, no reports have been published on the effects of hGH on spinal and femoral bone density or the frequency of fractures in patients with primary osteoporosis.

In summary, in spite of significant activation of bone remodelling by hGH, consistent beneficial effects on skeletal mass and density have not been apparent in GH-deficient adults or in women with postmenopausal osteoporosis. Longer duration clinical trials, and perhaps the combination of hGH with other agents that inhibit bone resorption, such as calcitonin or bisphosphonates, are needed.

Miscellaneous Uses

In underweight subjects with chronic obstructive pulmonary disease, hGh administration for three weeks caused weight gain, nitrogen retention, and improvement in the strength of ventilatory muscles compared to pretreatment levels while on the same balanced diet.[25] In malnourished older individuals, hGH produced positive nitrogen balance,[26] and it decreased the protein catabolic effects of pharmacologic doses of corticosteroids in normal human volunteers.[27] Studies of women with amenorrhea have shown that cotreatment with hGH augments the gonadal response to gonadotropins and promotes the induction of ovulation.[28] The authors have noted lower levels of IGF-I in young quadriplegic men,[29] and in those polio survivors with residual muscle weakness and functional disability.[30,31] It is speculated that physical inactivity and an age-related impairment of the GH/IGF-I axis are factors causing the low IGF-I levels in these subjects. It is not known whether hGH replacement will be able to improve muscle mass and functional capacities in these chronic neuromuscular disorders. As with the anabolic androgenic steroids, there is a potential for the abuse of hGH in sports. The available data indicate that, although hGH therapy in athletes increases muscle size, it does not seem to affect strength or endurance.

Complications of hGH Therapy in Adults, Including the Elderly

The side effects of hGH when used in younger adults include glucose intolerance, transient edema, and hypertension. In the elderly, high incidences of carpal tunnel syndrome and gynecomastia have been noted, which disappear following discontinuation of treatment. The data indicate that these complications can be prevented, and the beneficial effects on body composition preserved, by excluding from treatment patients with early or subclinical signs of carpal tunnel syndrome, and by adjusting the hGH dosage so as to maintain the intratreatment IGF-I level in the range of 0.5 to 1.0 units/ml.[13] Elderly men seem to be more susceptible to carpal tunnel syndrome and gynecomastia side effects of hGH than are younger adults; the occurrence of these complications is strongly related to the dose of hGH.

Acknowledgments

This work was supported by the Department of Veterans Affairs Research Service. The authors wish to thank Judi Wildes for her expertise in manuscript preparation.

References

1. Rudman D, Kutner MH, Rogers CM, et al: Impaired growth hormone secretion in the adult population: Relation to age and adiposity. *J Clin Invest* 1981;67:1361-1369.
2. Salomon F, Cuneo RC, Hesp R, et al: The effects of treatment with recombinant human growth hormone on body composition and metabolism in adults with growth hormone deficiency. *N Engl J Med* 1989;321:1797-1803.
3. Jorgensen JO, Pedersen SA, Thuesen L, et al: Beneficial effects of growth hormone treatment in GH-deficient adults. *Lancet* 1989; 1:1221-1225.
4. Whitehead HM, Boreham C, McIlbrath EM, et al: Growth hormone treatment of adults with growth hormone deficiency: Results of a 13-month placebo controlled cross-over study. *Clin Endocrinol (Oxf)* 1992;36:45-52.
5. Cuneo RC, Salomon F, Wiles CM, et al: Growth hormone treatment in growth hormone-deficient adults: I. Effects on muscle mass and strength. *J Appl Physiol* 1991;70:688-694.
6. Cuneo RC, Salomon F, Wiles CM, et al: Growth hormone treatment in growth hormone-deficient adults: II. Effects on exercise performance. *J Appl Physiol* 1991;70:695-700.
7. Jorgensen JO, Pedersen SA, Thuesen L, et al: Long-term growth hormone treatment in growth hormone deficient adults. *Acta Endocrinol (Copenh)* 1991;125:449-453.
8. McGauley GA, Cuneo RC, Salomon F, et al: Psychological well-being before and after growth hormone treatment in adults with growth hormone deficiency. *Horm Res* 1990;33(suppl 4):52-54.

9. Rudman D: Growth hormone, body composition, and aging. *J Am Geriatr Soc* 1985;33:800-807.
10. Marcus R, Butterfield G, Holloway L, et al: Effects of short term administration of recombinant human growth hormone to elderly people. *J Clin Endocrinol Metab* 1990;70:519-527.
11. Rudman D, Feller AG, Nagraj HS, et al: Effects of human growth hormone in men over 60 years old. *N Engl J Med* 1990;323:1-6.
12. Rudman D, Feller AG, Cohn L, et al: Effects of human growth hormone on body composition in elderly men. *Horm Res* 1991;36(suppl 1):73-81.
13. Cohn L, Feller AG, Draper MW, et al: Carpal tunnel syndrome and gynecomastia during treatment of elderly hyposomatomedinemic men with human growth hormone, *Clin Endocrinol*, in press.
14. Jiang ZM, He GZ, Zhang SY, et al: Low-dose growth hormone and hypocaloric nutrition attenuate the protein-catabolic response after major operation. *Ann Surg* 1989;210:513-525.
15. Parfitt AM: Growth hormone and adult bone remodeling. *Clin Endocrinol (Oxf)* 1991;35:467-470.
16. Degerblad M, Almkvist O, Grunditz R, et al: Physical and psychological capabilities during substitution therapy with recombinant growth hormone in adults with growth hormone deficiency. *Acta Endocrinol (Copenh)* 1990;123:185-193.
17. Bouillon R: Growth hormone and bone. *Horm Res* 1990;36(suppl 1):49-55.
18. Diamond T, Nery L, Posen S: Spinal and peripheral bone mineral densities in acromegaly: The effects of excess growth hormone and hypogonadism. *Ann Intern Med* 1989;111:567-573.
19. Ljunghall S, Johansson AG, Burman P, et al: Low plasma levels of insulin-like growth factor I (IGF-I) in male patients with idiopathic osteoporosis. *J Intern Med* 1992;232:59-64.
20. van der Veen EA, Netelenbos JC: Growth hormone (replacement) therapy in adults: Bone and calcium metabolism. *Horm Res* 1990;33(suppl 4):65-68.
21. Binnerts A, Swart GR, Wilson JHP, et al: The effect of growth hormone administration in growth hormone deficient adults on bone, protein, carbohydrate and lipid homeostasis, as well as on body composition. *Clin Endocrinol* 1992;37:79-87.
22. Aloia JF, Zanzi I, Ellis K, et al: Effects of growth hormone in osteoporosis. *J Clin Endocrinol Metab* 1976;43:992-999.
23. Aloia JF, Vaswani A, Kapoor A, et al: Treatment of osteoporosis with calcitonin, with and without growth hormone. *Metabolism* 1985; 34:124-129.
24. Aloia JF, Vaswani A, Meunier PJ, et al: Coherence treatment of postmenopausal osteoporosis with growth hormone and calcitonin. *Calcif Tissue Int* 1987;40:253-259.
25. Pape GS, Friedman M, Underwood LE, et al: The effect of growth hormone on weight gain and pulmonary function in patients with chronic obstructive lung disease. *Chest* 1991;99:1495-1500.
26. Kaiser FE, Silver AJ, Morley JE: The effect of recombinant human growth hormone on malnourished older individuals. *J Am Geriatr Soc* 1991;39:235-240.
27. Horber FF, Haymond MW: Human growth hormone prevents the protein catabolic side effects of prednisone in humans. *J Clin Invest* 1990;86:265-272.

28. Jacobs HS, Bouchard P, Conway GS, et al: Role of growth hormone in infertility. *Horm Res* 1991;36(suppl 1):61-65.
29. Shetty KR, Sutton CH, Mattson DE, et al: Hyposomatomedinemia in quadriplegic men. *Am J Med Sci* 1993;305:95-100.
30. Shetty KR, Mattson DE, Rudman IW, et al: Hyposomatomedinemia in men with post-poliomyelitis syndrome. *J Am Geriatr Soc* 1991; 39:185-191.
31. Rao U, Shetty KR, Mattson DE, et al: Prevalence of low plasma IGF-I in poliomyelitis survivors. *J Am Geriatr Soc* 1993;41:697-702.

Chapter 6

Mechanisms of Aging That Might Contribute to Musculoskeletal Impairment: Diminished Cell Proliferative Capacity

Judith Campisi, PhD

Cellular Senescence

Most normal eukaryotic cells have an intrinsic limit to the number of cell divisions through which they can proceed.[1-3] This property has been termed the finite (proliferative) life span phenotype of cells, or cellular senescence. Cellular senescence has been most thoroughly studied using cells in culture, principally human fibroblasts. Limited data exist for other cell types and even fewer data exist for cells in vivo. Nonetheless, these data clearly indicate that a finite proliferative life span is by no means confined to fibroblastic cells, and that it is unlikely to be an artifact of cell culture.

Cells that can proliferate in vivo will, under appropriate conditions, initially proliferate well in culture. Such cultures are often termed early passage or "young cells." With increasing passage or population doublings (PD), there is a gradual and progressive decline in proliferative capacity of the culture. Eventually, the culture becomes composed entirely of viable but nondividing (senescent) cells. Senescent cells remain viable for long periods of time, during which they retain the capacity to synthesize RNA and protein; however, senescent cells fail to enter the S phase of the cell cycle (synthesize DNA) when provided with mitogens that readily stimulate early passage cells.

There is substantial, albeit indirect, evidence that cellular senescence may reflect either or both of two fundamentally important properties of cells.

First, senescence may reflect, at a cellular level, processes analogous to organismal aging. In support of this idea, the cells from old individuals or short-lived species senesce (in culture) after fewer cell divisions than cells from young individuals or long-lived species. In addition, human cells from donors with premature aging syndromes (eg, Werner's syndrome) reach the end of their proliferative life span after fewer divisions in culture than cells from age-matched controls.

A second, and not mutually exclusive, view holds that cellular senescence may constitute a tumor suppressive mechanism. In support of this idea, tumor cells often, although not always, have an immortal (infinite proliferative life span) phenotype, and a number of oncogenes appear to act primarily by overcoming processes responsible for cellular senescence. In addition, immortal cells are orders of magnitude more susceptible to tumorigenic transformation than cells having a finite proliferative life span, and senescent cells are virtually refractory to tumorigenic transformation.

Much of what is known about the mechanisms that control senescence relates to the proliferative failure that is characteristic of senescent cell.[2,3] However, senescent cells also show changes in differentiated characteristics.[4-6] The growth arrest associated with senescence has obvious implications for its relevance to tumor suppression. By contrast, both the growth arrest and the alterations in differentiation may be important in the relevance of the finite life span of cells to aging. Here, we focus on the relevance of cellular senescence to aging.

Control of Cell Proliferation

Cell proliferation in higher eukaryotes is controlled by such external factors as growth factors, extracellular matrix, or cell-cell contacts and by intrinsic programs of gene expression, such as the state of differentiation or stage in the cellular life span. For reasons of convenience and history, fibroblasts have been the most thoroughly characterized cell type with regard to the mechanisms that control cell proliferation, as well as the mechanisms that control cellular senescence.

Early passage human fibroblasts proliferate in response to a number of physiologic mitogens. In simple terms, mitogens act by inducing or repressing the expression of specific genes, some of which are necessary or permissive for the proliferative response. Senescent human fibroblasts do not proliferate in response to physiologic mitogens. However, a number of genes in senescent cells do remain mitogen inducible.[4,7] Thus, the growth arrest associated with senescence is not the result of a general breakdown of mitogen signalling pathways. Rather, the senescence-associated growth arrest appears to be caused by the selective repression of a few key cell cycle regulatory genes.

Fibroblasts in culture can be induced to enter a reversible, nondividing growth state, termed quiescence, by depriving them of growth factors for several days or by allowing them to grow to confluent cell densities. The cells can be induced to reenter the cell cycle and proliferate by providing them with growth factors or releasing them from cell-cell contacts. In vivo, many, but certainly not all, cells are quiescent, and they proliferate when the need for cell replacement arises, either because of normal cell loss or trauma.[8]

In either case, quiescent fibroblasts that have been stimulated to proliferate show an increase in the expression of many genes, some, but not all, of which are critical to the proliferative response. Of the genes that are essential for the proliferation of fibroblasts, an even smaller subset are truly regulatory. Cell cycle regulatory genes include a few proto-oncogenes and tumor suppressor genes, as well as a small number of other genes that have been highly conserved throughout evolution.

In fibroblasts, at least four proto-oncogenes appear to be regulatory and essential for the proliferative response to mitogens. These are the c-fos, c-jun, c-myc, and c-ras proto-oncogenes. In addition, at least two tumor suppressor genes—p53 and the retinoblastoma susceptibility gene product (Rb)—must be inactivated before proliferation can proceed. In the case of Rb, inactivation is effected by phosphorylation. Rb is phosphorylated by one or more members of a family of growth regulatory protein kinases, known as cdc2-like protein kinases, whose expression is induced by mitogens.

Proliferative Blocks in Senescent Cells

As noted earlier, several genes remain mitogen-inducible after human fibroblasts reach senescence. Among these are the c-myc, c-ras, and c-jun proto-oncogenes.[4,7,9] However, one regulatory proto-oncogene remains repressed in senescent cells. This is the c-fos proto-oncogene.[4] Because c-fos expression is essential for the proliferation of fibroblasts, repression of c-fos has the potential to account for the senescence-associated block to cell proliferation. We have strong circumstantial evidence that the repression of c-fos is clearly an important component of the proliferative block in senescent cells; however, it is also clearly not the sole cause of their failure to undergo cell division in response to mitogens.

A second cell cycle regulatory event that fails to occur in senescent cells is the phosphorylation (that is, inactivation) of the Rb tumor-suppressor protein.[10] Again, senescent cells express a number of protein kinase activities and mitogen-dependent phosphorylated proteins, but Rb remains underphosphorylated in senescent cells. Our data suggest that inactivation of Rb is another important component of the proliferative block in senescent cells.

The constitutive underphosphorylation of Rb that occurs in senescent cells may relate to a third cell cycle regulatory event that is altered by senescence. Cdc2, which is the protein kinase that is at least partially responsible for Rb phosphorylation, is repressed in senescent cells.

In summary, senescent human fibroblasts retain the ability to respond to mitogens in that they express a variety of mitogen-inducible genes, including several proto-oncogenes. However, the expression of at least three genes important in cell cycle progression is altered once the cells have reached the end of their proliferative life span.

Altered Transcription Factors in Senescent Cells

Is there a common theme to the genes discussed above that are altered by cellular senescence? This may well be the case.

The c-fos gene product is a nuclear-localized protein that is a critical activator component of the AP-1 transcription factor complex.[11] A number of genes are regulated by AP-1 activity, including those important in growth and differentiation. Thus, repression of c-fos suppresses the growth response of cells and alters the expression of some differentiation-specific genes.

Rb is a repressor component of the E2F transcription factor complex.[12] Like AP-1, E2F regulates the expression of a variety of genes, only some of which overlap with those regulated by AP-1. In fact, there is evidence to suggest that Rb may be a negative regulator of c-fos expression.[13]

Thus, constitutive activity of the Rb tumor suppressor in senescent cells (that is, maintenance in the unphosphorylated or suppressive form) is expected to have wide-ranging effects on the expression of several genes, including those needed for cell proliferation. One such gene may be c-fos, which, in turn, when repressed, will alter the expression of an additional subset of genes.

What of the mechanism responsible for the repression of cdc2, the protein kinase whose repression may be responsible for the constitutive activity of Rb? Preliminary data suggest that the expression of an as-yet-uncharacterized transcription factor may be responsible for the repression of cdc2 that occurs in senescent cells. Indirect evidence suggests that it belongs to a class of factors, termed helix-loop-helix (HLH) factors, that regulate yet another subset of genes, which is overlapping yet distinct from those regulated by AP-1 and E2F. Thus, there may be a hierarchy of changes in transcription factors that occur during cell senescence.

In summary, among the changes in gene expression that accompany cellular senescence are changes in the expression of selected transcription factors. The senescence-sensitive transcription factors are known, or postulated, to control progression through the cell cycle. Thus, the transcription factors that are altered by cellular senescence may hold the key to the proliferative failure of senescent cells, as well as to the changes in cell function that also occur with senescence. It is likely that the entire repertoire of changes in gene expression that characterize senescent cells may well be under the control of a small number of transcription factors. We have yet to understand how this process is initiated, or how it is amplified or maintained with each cell division.

Implications for Cells in Aging Tissues

As tissues and organisms age there is a gradual deterioration in overall function, an increased susceptibility to disease, and a diminished ability to recover from stress or trauma. Some of this deterioration may be caused by epigenetic factors. In addition, there is increasingly strong evidence that segments of the aged or senescent phenotype are under genetic control. Whatever the etiology of the deterioration, the effect is a decline in cell proliferative capacity and an accompanying change in cell function.

As discussed earlier, there is reason to believe that the proliferative senescence of individual cells may reflect some aspects of senescence or aging in intact organisms. Recent studies on the molecular biology of cellular senescence suggest that changes in the expression or activity of transcription factors may be the key to understanding the senescent phenotype. These studies show that there are changes in specific transcription factors as cells progress through their proliferative life span, and that the transcription factors that are altered in senescent cells are those that control both cell proliferation and cell function.

It is not unreasonable, then, to expect that a key to understanding aging at the level of the tissue or organism may lie in the transcription factors that control the growth and differentiation of specific cell types. If this idea proves valid, one might imagine, then, that specific age-related changes in tissue or organ function could be arrested or reversed by targeting specific transcription factors. If a hierarchy of changes in transcriptional control is evident, it may be desirable and perhaps more efficacious to target a downstream factor, rather than a "master" transcription factor. The notion of targeting select transcription factors is still more concept than reality, but it is certainly not beyond the realm of reality.

References

1. Stanulis-Praeger BM: Cellular senescence revisited: A review. *Mech Ageing Dev* 1987:38:1-48.
2. Goldstein S: Replicative senescence: The human fibroblast comes of age. *Science* 1990;249:1129-1133.
3. McCormick A, Campisi J: Cellular aging and senescence. *Curr Opin Cell Biol* 1991;3:230-234.
4. Seshadri T, Campisi J: Repression of c-*fos* transcription and an altered genetic program in senescent human fibroblasts. *Science* 1990; 247:205-209.
5. Bayreuther K, Rodemann HP, Hommel R, et al: Human skin fibroblasts in vitro differentiate along a terminal cell lineage. *Proc Natl Acad Sci USA* 1988;85:5112-5116.
6. Cheng CY, Ryan RF, Vo TP, et al: Cellular senescence involves stochastic processes causing loss of expression of differentiated function genes: Transfection with SV40 as a means for dissociating effects of senescence on growth and on differentiated function gene expression. *Exp Cell Res* 1989;180:49-62.
7. Rittling SR, Brooks KM, Cristofalo VJ, et al: Expression of cell cycle-dependent genes in young and senescent WI-38 fibroblasts. *Proc Natl Acad Sci USA* 1986;83:3316-3320.
8. Baserga R: *The Biology of Cell Reproduction*. Cambridge, MA, Harvard University Press, 1985.

9. Phillips PD, Pignolo RJ, Nishikura K, et al: Renewed DNA synthesis in senescent WI-38 cells by expression of an inducible chimeric c-fos construct. *J Cell Physiol* 1992;151:206-212.
10. Stein GH, Beeson M, Gordon L: Failure to phosphorylate the retinoblastoma gene product in senescent human fibroblasts. *Science* 1990;249:666-669.
11. Cohen DR, Curran T: The structure and function of the fos proto-oncogene. *Crit Rev Oncog* 1989:1:65-88.
12. Chellappan SP, Hiebert S, Mudryj M, et al: The E2F transcription factor is a cellular target for the RB protein. *Cell* 1991;65:1053-1061.
13. Robbins PD, Horowitz JM, Mulligan RC: Negative regulation of human c-fos expression by the retinoblastoma gene product. *Nature* 1990;346:668-671.

Chapter 7

Diminution of the Number of Mesenchymal Stem Cells as a Cause for Skeletal Aging

Stephen E. Haynesworth, PhD
Victor M. Goldberg, MD
Arnold I. Caplan, PhD

Background

Skeletal tissues differ in the rates in which they turn over and in their ability to repair themselves. For example, bone has the most rapid turnover of any skeletal tissue and the potential to completely regenerate itself after destructive injury. In contrast, cartilage turns over very slowly and its repair potential in the adult organism is very limited. The repair potentials and rates of turnover of tendon and ligament appear to be between the exhibited extremes of bone and cartilage.

One factor that correlates with and influences the rates of skeletal tissue turnover and repair is the availability of undifferentiated progenitor cells called mesenchymal stem cells, which can differentiate and fabricate new skeletal tissue matrix. For bone turnover and repair, mesenchymal stem cells are available in abundance from the surrounding periosteum and enclosed bone marrow. However, for cartilage repair few, if any, mesenchymal stem cells are normally available to differentiate into new chondrocytes. The availability of mesenchymal stem cells in tendon and ligament is not well established. However, in our laboratories, cells with mesenchymal stem cell characteristics have recently been isolated and culture-expanded from tendon and ligament tissue.

The differentiation process whereby mesenchymal stem cells form skeletal phenotypes has not been described in detail, but it appears to be similar for tissue turnover and repair. The process involves the following stages: mesenchymal stem cell recruitment; stimulation of cell division; and entrance into and progression through the differentiation pathway. Each of these stages is presumably mediated and controlled by bioactive protein factors, which are released by surrounding tissues and vasculature elements in a temporally and spatially controlled cascade.

A well-recognized consequence of aging is that the rates of repair of all skeletal tissues diminish with advancing age. The mechanisms that cause these decreased rates of repair are not understood and could potentially involve any or all of the three stages—cell recruitment, cell replication, and cell differentiation. For example, mesenchymal stem cells in older organisms may lack the cell surface receptors necessary to respond to the protein factors responsible for recruiting them to the site of injury or tissue replacement. Another possibility is that mesenchymal cells respond to recruitment factors, but do not respond to the mitogenic factors that stimulate them to divide. Still further, mesenchymal cells recruited to the site of injury may divide in response to local mitogenic factors, but they may lack the ability to differentiate in response to the local factors that stimulate their differentiation. Currently, no published data exist that address the question of whether mesenchymal stem cells from donors of different ages differ in their ability to respond to various bioactive protein factors involved in their differentiation into skeletal phenotypes.

Alternatively, there may be no differences in the ability of mesenchymal stem cells from young and old organisms to respond to local chemotactic, mitogenic, and morphogenic factors. Instead, the levels of secretion of one or more of these various factors may change with aging, thereby altering the processes of cell recruitment, cell division, or cell differentiation. Consistent with this mechanism, Syftestad and Urist[1] demonstrated a diminished capacity of bioactive factors from demineralized bone matrix (DBM) from older animals to induce bone formation when implanted into young synegenic host animals. In addition, Cesnjaj and associates[2] showed that DBM derived from ovariectomized rats does not stimulate bone formation as well as does DBM derived from normal age-matched controls.

Lastly, age-related decline in repair potential may be explained by a decrease in the number of mesenchymal stem cells available in the surrounding tissue, which can be recruited to the injury site and stimulated to divide and differentiate. A dramatic decline in the number of recruitable mesenchymal stem cells could cause a prolonged period of cell proliferation at the injury site, as fewer mesenchymal stem cells divide in response to local factors to fill the injured tissue. This mechanism is supported by several lines of data.[3-6] Irving and associates[3] showed that when DBM from young rats is implanted into six-week-old rats, new bone forms in 14 days. However, formation of new bone when DBM was implanted into two-year-old rats took 23 days; similar results were observed in a later study by Nishimoto and associates.[4] More directly, evidence is presented by Tsuji and associates,[7] who recently measured the number of bone nodules that form from adherent rat bone marrow stroma cells. They found a three-fold decline in the number of bone nodules and a 50% decline in the number of alkaline phosphatase positive cells in cultures derived from old donor rats as compared to young donor rats.

Although it is possible that all of the mechanisms described above play some role in the observed decline in the rate and potential of skeletal tissue repair, considerable recent data from our laboratories and from others described above support the concept that a dramatic decline in the number of recruitable mesenchymal stem cells with age is a primary factor. A review of recent studies from our laboratories, which contribute to the understanding of the role of mesenchymal stem cells in the age-related decline of skeletal tissue repair, is provided below.

Experimental Design

In the background section above, three general mechanisms involving mesenchymal stem cells were proposed to explain their role in the decline of skeletal tissue turnover and repair with age. The three mechanisms are: (1) diminished potential of mesenchymal stem cells to respond to the bioactive proteins that mediate their differentiation; (2) diminished or altered secretions in older animals of the bioactive proteins responsible for mediating differentiation of mesenchymal stem cells; and (3) a decline in the number of mesenchymal stem cells in skeletal tissues with advancing age. In order to identify which of these mechanisms are involved in the age-related decline of skeletal tissue formation, we developed the following experimental design, using mesenchymal stem cells derived from rat bone marrow.

In this experimental design, we assayed for the differentiation of culture-expanded marrow-derived rat mesenchymal stem cells into bone and cartilage. The optimal conditions for inducing these mesenchymal stem cells to differentiate into bone and cartilage in our in vivo assay system have been firmly established and developed.[8-10] The same experimental design may also be used to analyze the differentiation of marrow-derived mesenchymal stem cells into other skeletal phenotypes as assays for inducing stem cell differentiation into other skeletal tissues.

The experimental design is illustrated in Figure 1. Bone marrow cells from young adult rats (4 to 6 months of age) and old rats (27 to 32 months) were harvested from the tibiae and femurs of Fischer rats, converted into a homogeneous cell suspension and introduced into culture. Mesenchymal stem cells attach to and divide on the culture dishes; hematopoietic cells are removed by changing the culture medium. When the resultant mesenchymal stem cell colonies in primary cultures became near confluent, the cells from each plate were trypsinized and replated onto three new plates. As a rule, cultures generated from young adult rats generated significantly more fibroblastic colonies and had to be replated at an earlier timepoint than cultures generated from old donors. Current studies are being conducted to quantify these differences, however, our qualitative observations are that marrow from young adult rats contains significantly more mesenchymal cells than does marrow from old rats.

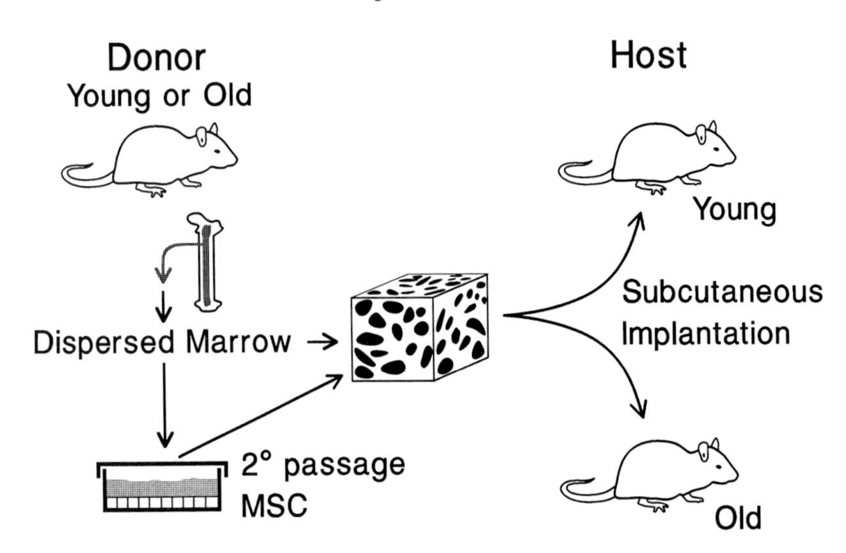

Fig. 1 *Experimental design for differentiation of culture-expanded, marrow-derived rat mesenchymal stem cells into bone and cartilage.*

Replated cells were allowed to divide until cultures again became near confluent, at which time the cells were again harvested with trypsin and counted. Harvested cells were loaded into porous cubes of ceramic made from tricalcium phosphate and hydroxyapatite. We have demonstrated in several studies that when loaded into the ceramic cubes and then implanted into synegenic host rats, mesenchymal stem cells from young adult rats differentiate primarily into osteoblasts and synthesize bone onto the walls of the pores in the ceramic cubes.[8,10] Less frequently, mesenchymal stem cells under these same conditions differentiate into chondrocytes and fabricate a cartilage matrix in the ceramic pores. The factor that determines whether the mesenchymal stem cells in a ceramic pore differentiate into osteoblasts or chondrocytes appears to be whether or not the pore becomes invaded by the host vasculature. In most pores, host capillaries enter rapidly, and, in such cases, bone formation is consistently observed with oriented secretion of bone matrix by the newly differentiated osteoblasts away from the vasculature onto the walls of the pores (Fig. 2, *top*). However, in some cases, pore geometry or a large bolus of loaded mesenchymal stem cells appear to prevent vasculature ingrowth, and cartilage formation is observed (Fig. 2, *bottom*).

In this study, mesenchymal stem cells from young adult rats and old rats were loaded at identical cell concentrations into separate ceramic cubes, which were then implanted subcutaneously into the

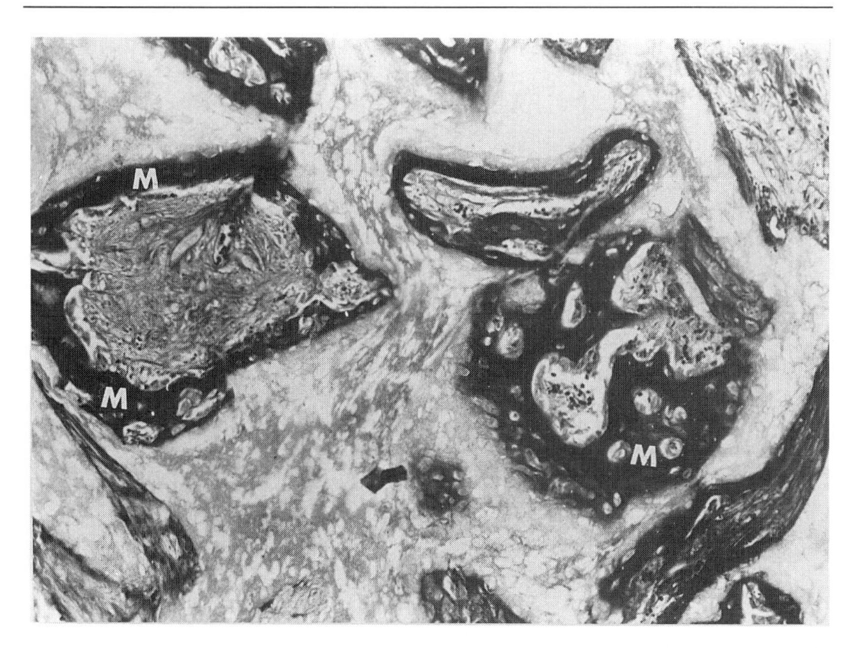

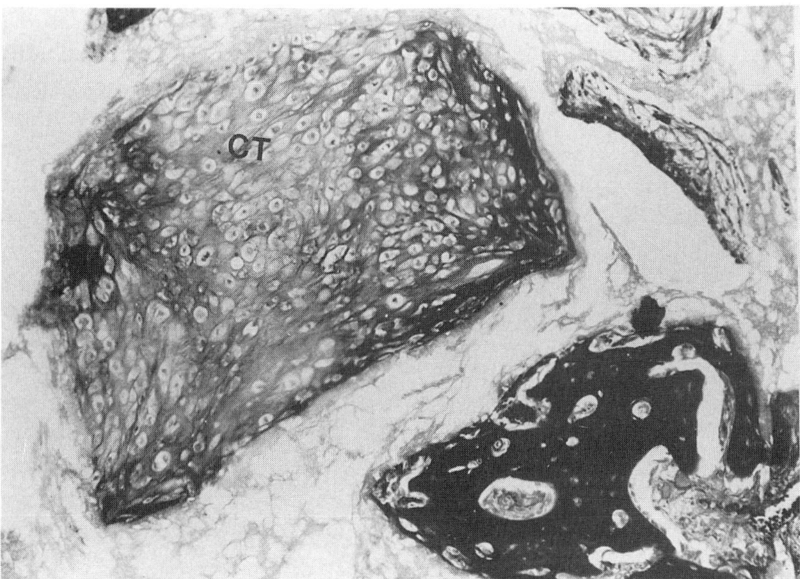

Fig. 2 ***Top,*** *Photomicrograph of histological section of a graft of ceramic and cultured marrow mesenchymal cells harvested three weeks after implantation in nude mouse. Bone matrix (M) is observed lining the walls of the ceramic pores; original magnification, X100.* ***Bottom,*** *Photomicrograph of histological section of a graft of ceramic and cultured marrow mesenchymal cells harvested three weeks after implantation in nude mouse. Cartilage tissue (CT) is observed almost completely filling a pore, with an adjacent pore observed partially filled with bone tissue; original magnification, X100.*

same synegenic young adult or old host rats. The ceramics were allowed to incubate for three, four, or six weeks, after which they were harvested and examined histologically for bone formation.

Results

A comparison of the bone formed by young adult and old donor mesenchymal stem cells implanted into young adult and old hosts is summarized in Table 1. Young donor cells generated consistent bone formation in young host rats (9/9) at all three harvest timepoints, as had been observed in our previous studies. In addition, young donor cells differentiate into bone-forming osteoblasts almost as consistently when implanted into old host rats (11/12). These data demonstrate that in the case of bone differentiation by marrow-derived mesenchymal cells in the ceramic cube assay, the bioactive factors and other inductive agents of the old hosts provide a level of osteoinduction similar to that observed with young host rats.

Also shown in Table 1 is the ability of old donor cells to differentiate when implanted into young and old hosts relative to young adult donor cells. As mentioned above, young adult donor cells generated near equal bone formation after implantation in young and old host rats. Old donor cells also demonstrated similar bone-forming capacity after implantation in young host rats (8/9). After implantation in old host rats, old donor cells failed to differentiate into osteoblasts and make bone at the earliest timepoint (0/3), but demonstrated consistent bone formation in ceramics harvested at the other two timepoints (6/6). These data seem to indicate that old mesenchymal stem cells have potential to differentiate into bone similar to that of young mesenchymal stem cells when provided an optimal osteoinductive environment, which is present in the ceramic cubes implanted into young hosts. The observed delay in differentiation into bone in ceramics containing old donor mesenchymal stem cells implanted into old hosts is unclear, especially since three out of three ceramics containing old donor cells differentiate and form bone at the next timepoint one week later. A larger number of ceramics at each harvest timepoint will be necessary to determine if this difference is significant.

A comparison of the cartilage-forming potential of the young adult and old donor cells implanted into young adult and old host is presented in Table 2. Young donor cells differentiated into chondrocytes in six of nine ceramics implanted into young adult hosts, and seven of 12 ceramics implanted into old hosts. In comparison, old donor cells formed cartilage tissue in three of nine ceramics implanted into young hosts, and four of 12 ceramics implanted into old hosts. At first glance these data appear to indicate that mesenchymal stem cells from young adult donors differentiate more readily

Table 1 Bone formation in ceramic cubes

Sampling Time (weeks)	Cell Density (10^6 cells/ml)	Young Donor Cells		Old Donor Cells	
		Young Host	Old Host	Young Host	Old Host
3	7.5	3/3	2/3	2/3	0/3
4	5.0	-	3/3	-	3/3
6	7.5	3/3	3/3	3/3	3/3
6	5.0	3/3	3/3	3/3	3/3

Table 2 Cartilage formation in ceramic cubes

Sampling Time (weeks)	Cell Density (10^6 cells/ml)	Young Donor Cells		Old Donor Cells	
		Young Host	Old Host	Young Host	Old Host
3	7.5	3/3	1/3	1/3	0/3
4	5.0	-	0/3	-	1/3
6	7.5	2/3	3/3	2/3	3/3
6	5.0	1/3	3/3	0/3	0/3

into chondrocytes than do those from old donors. However, at the higher loading density of 7.5 million cells per ml, young donor cells differentiate into chondrocytes in five of six ceramics, and older cells form cartilage in five of six ceramics. Because chondrogenic differentiation in the ceramic pores is only observed when a bolus of mesenchymal cells completely fills the pore and thereby excludes ingrowth of the host vasculature, the higher loading concentration is a more reliable condition for comparing the chondrogenic potential of the mesenchymal stem cells derived from young adult and old donor rats. In contrast, no significant differences were observed in the ability of young and old hosts to support chondrogenic differentiation from mesenchymal stem cells in the ceramic cubes.

Discussion

The data presented above have allowed us to gain insight into the role of three proposed mechanisms for the role of mesenchymal stem cells in the age-related decline in skeletal tissue formation. First, our direct observations of mesenchymal stem cell cultures derived from young adult and old donor marrow indicate that old rats have significantly fewer colony-forming mesenchymal stem cells in their marrow as compared to young adult rats. The rates of growth of young adult and old rat mesenchymal stem cells do not appear to vary significantly in culture. Current experimentation is focused at providing quantitative values to these observations.

Second, by implanting mesenchymal stem cells from young adult cells into young adult and old rat hosts, the influence of the old host environment was tested, since it has already been firmly established that when young adult cultured mesenchymal stem cells are implanted into young adult host rats, consistent bone and cartilage formation are observed. Here, we observed that the old rat environment supports the differentiation of mesenchymal stem cells into bone and cartilage at a level similar to that observed with young adult hosts.

Third, by implanting mesenchymal stem cells from old rats into young adult hosts along with mesenchymal stem cells from young adult rats, the bone- and cartilage-forming potential of mesenchymal stem cells from old rats was tested in a host environment that has been proven to support mesenchymal stem cell differentiation into bone and cartilage. Here we observed no differences in the ability of mesenchymal stem cells from old rats to differentiate into bone and possibly cartilage, although more samples will be needed to confirm all of these observations.

In summary, the decline in bone synthesis with age appears to occur not because the host's environment is depleted of osteoinductive factors, but rather because fewer stem cells are present in older animals to differentiate into osteoblasts. However, when expanded in culture and implanted in vivo, the rates of bone and cartilage formation from mesenchymal stem cells are relatively the same regardless of the age of the donors or the implanted hosts. These observations may have important significance for generating experimental models to determine the influence of mesenchymal stem cell numbers to the turnover and repair potential of other skeletal tissues. In addition, the ability to isolate and culture mesenchymal stem cells may have important therapeutic potential for augmenting skeletal repair for bone, and for soft skeletal tissues where mesenchymal stem cells reserves may be limited or nonexistent.

Acknowledgement

Supported by grants from the National Institutes of Health.

References

1. Syftestad GT, Urist MR: Bone aging. *Clin Orthop* 1982;162:288-297.
2. Cesnjaj M, Stavljenic A, Vukicevic S: Decreased osteoinductive potential of bone matrix from ovariectomized rats. *Acta Orthop Scand* 1991;62:471-475.
3. Irving JT, LeBolt SA, Schneider EL: Ectopic bone formation and aging. *Clin Orthop* 1981;154:249-253.
4. Nishimoto SK, Chang CH, Gendler E, et al: The effect of aging on bone formation in rats: Biochemical and histological evidence for decreased bone formation capacity. *Calcif Tissue Int* 1985;37:617-624.
5. Nimni ME, Bernick S, Cheung DT, et al: Biochemical differences between dystrophic calcification of cross-linked collagen implants and mineralization during bone induction. *Calcif Tissue Int* 1988;42:313-320.
6. Nimni ME, Bernick S, Ertl D, et al: Ectopic bone formation is enhanced in senescent animals implanted with embryonic cells. *Clin Orthop* 1988;234:255-266.
7. Tsuji T, Hughes FJ, McCulloch CA, et al: Effects of donor age on osteogenic cells of rat bone marrow in vitro. *Mech Ageing Dev* 1990;51:121-132.

8. Goshima J, Goldberg VM, Caplan AI: The osteogenic potential of culture-expanded rat marrow mesenchymal cells assayed in vivo in calcium phosphate ceramic blocks. *Clin Orthop* 1991;262:298-311.
9. Goshima J, Goldberg VM, Caplan AI: The origin of bone formed in composite grafts of porous calcium phosphate ceramic loaded with marrow cells. *Clin Orthop* 1991;269:274-283.
10. Dennis JE, Haynesworth SE, Young RG, et al: Osteogenesis in marrow-derived mesenchymal cell porous ceramic composites transplanted subcutaneously: Effects of fibronectin and laminin on cell retention and rate of osteogenic expression. *Cell Transplant* 1992;1:23.

Future Directions

Use longitudinal research to study factors mediating the impairment and disability relationship.

Longitudinal research is needed to identify the major risk factors and buffers that mediate the effects of musculoskeletal impairments on disability. This research can track the progression of impairments and the subsequent impact on disability in select populations.

Improve measures of musculoskeletal impairments for field research.

To date, epidemiologic research on the relationship between musculoskeletal impairments and resultant disability has used very crude measures of impairment. Future field research on the specific site and type of impairments leading to subsequent disability will require precise measures of impairment that can be used outside of the laboratory.

Develop secondary prevention strategies.

Traditionally, public health researchers have focused on primary prevention strategies in the fight against chronic disabling disease. More research is needed on the development and subsequent testing of promising secondary prevention strategies designed for the early detection of impairments and implementation of interventions to halt, slow, or reverse progression of impairments.

Determine which aspects of mobility task performance in the elderly are critical ones.

Whether a mobility-related task can be performed successfully depends on many component abilities. Age-related changes occur in those component abilities, but those changes may or may not determine whether a task can be performed. Understanding which changes are critical to performance abilities will provide important insights into how to better assess and intervene in mobility impairments.

Devise better assessments of risk for mobility impairments.

Current assessments of mobility are usually qualitative and often are not sensitive enough to detect small changes occurring over time. Quantitative assessment techniques need to be devised that are more sensitive, are capable of early detection of developing impairments, and are simple enough to be used routinely in a clinical setting.

Determine more specifically how changes with age in joint ranges-of-motion, joint torque strengths, and rate-of-strength-development relate to decreasing mobility.

Many mobility task performances do not require large strengths or ranges of motion, but others do. The relation of declines in these factors to declines in mobility has not been well described, and this needs to be addressed. Little is known about age-related declines in rate-of-strength-development, yet strength development rates are clearly important to abilities to perform some tasks.

Provide a better characterization of musculoskeletal soft tissue dysfunction and pathology in the older population, including the identification of presently undefined pathologies.

Presently undefined pathologies are especially likely to be found in those aged 75 and over. Clinicopathologic correlation studies, which include functional, clinical, and biochemical measures, and autopsy data would be especially useful in establishing levels at which various factors contribute to disability or morbidity (morbid ranges).

Compare age-related changes in musculoskeletal soft tissue functions in several age groups.

The distribution of levels of each studied function in each age group would be of great value to basic and clinical researchers in choosing functions for further studies. If a morbid range has not been established for each function measured in the study, case-control or other techniques should be used to determine it, if possible.

Long term longitudinal studies combining clinical, biochemical, genetic, and histopathologic measures are especially needed. Additional noninvasive or minimally traumatic techniques are needed to expand the range of longitudinal measures. Because such longitudinal studies may require many years, stable institutional support through centers or similar organizations is important.

Characterize subjects of studies of age-related changes in musculoskeletal soft-tissue functions with regard to presence and level of variables, such as chronic diseases and physical activity, which could affect the function of interest.

If a large enough sample to permit analyses of the effects of these variables can be obtained, this should be done. If not, less (but still interpretable) information can be gained by narrowing the range of variability (eg, by excluding diseases) to avoid confounding effects. Ignoring these factors will generally result in uninterpretable data.

Seek better animal models for human age-related musculoskeletal soft-tissue pathologies.

Such models should also be as free as possible of pathologies of other systems which contribute to artifactual findings. Paired strains which are as identical as possible, except for the development of the pathology, would be especially valuable. Both classic and molecular genetic techniques should be explored in the development of suitable models.

Determine whether the damage mediated by the Maillard reaction is sufficient to explain the degenerative process in disks and cartilage.

One important question is whether advanced Maillard reaction products predispose the joint to chondrocyte-mediated degradation through an advanced glycosylation end-product receptor pathway. Other questions involve identifying the structure of the major Maillard cross-links and testing the effects of aminoguanidine on inhibition of the advanced Maillard reaction in appropriate models of osteoarthritis. Aminoguanidine is already in clinical trials for its promising effects on diabetic complications. Its ability also to inhibit nitric oxide synthesis suggests that it may have beneficial effects on the immunologic aspects of osteoarthritis.

Determine the effect of human growth hormone on regional distribution of adipose tissue in hyposomatomedinemic older persons.

In hyposomatomedinemic elderly persons who are treated with human growth hormone (hGH), determine how individual organs within the lean body mass change in size, which cell types participate in the response, and the effects on cell size, cell number, and functional capacities of affected organs and tissues.

Determine if low levels of insulin-like growth factor-1 (IGF-1) in older persons are associated with lower lean body mass, increased adiposity, and atrophic changes in other growth hormone target tissues.

Older subjects with low levels should be compared to older subjects with higher growth hormone levels typical of younger persons.

Identify the transcription factors that have important regulatory roles in growth- and differentiation-specific gene expression in the major cell types of tissues or organs that show clear age-related dysfunctions.

These studies should also include identifying the major genes that are the targets of each transcription factor and determining which of these transcription factors show age-related changes in expression or function. If feasible, study the basis for this alteration because, by analogy to what is known in senescent human fibroblasts, such studies may uncover a hierarchy of altered transcriptional control.

Learning the mechanisms by which tissue- or cell type-specific factors are regulated at the levels of both expression and function will assist in development of rational approaches to reversing or intervening in age-related changes.

Firmly establish a link between aging and a decline in number of mesenchymal stem cells in an animal model and in humans.

Additional data is needed to firmly establish the age-related decline of mesenchymal stem cells. Additional colony count analyses should be conducted on an expanded age group of rats; for example, two, six, 12, 18, 24, and 30 months. The data from these analyses will provide a more detailed understanding of the correlation of decline in mesenchymal stem cell number with age.

A similar mesenchymal stem cell colony count analysis data base should be established for humans. Marrow from normal donors in all age groups, as well as patients with skeletal tissue diseases, such as osteoarthritis and osteoporosis, should be included. These data not only will establish the age-related decline of mesenchymal stem cells in humans, but also will provide information relating to any correlation between mesenchymal stem cell number and the development of skeletal disease; for example, do individuals with osteoporosis have fewer mesenchymal stem cells than age-matched normal individuals?

Develop in vivo assays to determine if marrow-derived mesenchymal stem cells can differentiate into soft skeletal cell phenotypes such as cartilage (already established), tendon, and ligament.

The most successful way to develop such assays will probably be to develop surgical models and delivery vehicles for implanting mesenchymal stem cells into defects in the tissues in which the cells should differentiate. This methodology would allow the implanted mesenchymal stem cells to respond to the local environmental influences, including bioactive protein factors, which at present are not understood well enough to be controlled or duplicated in vitro.

Determine if mesenchymal stem cell repositories exist in other skeletal tissues.

Cells from tendon and ligament should be harvested and grown in culture under conditions similar to those used for expanding mesenchymal stem cells from bone marrow. Culture-expanded cells from these tissues could then be tested for mesenchymal stem cell characteristics by placing them into assays that induce their differentiation into various skeletal cell types; for example, culture-expanded cells from tendon and ligament could be tested for bone differentiation potential in the ceramic cube assay. Preliminary data suggest that cells with these characteristics exist in both tendon and ligament. If mesenchymal stem cells can be firmly established to reside in these tissues, the next step would be to determine if their numbers change as a function of age.

Develop animal model assays to induce soft-tissue phenotypic differentiation by mesenchymal stem cells.

In vivo assays need to be developed to determine if marrow-derived mesenchymal stem cells can differentiate into soft skeletal cell phenotypes; cartilage (already established), tendon, and ligament. The most successful logic for developing such assays will likely be to develop surgical models and delivery vehicles to implant mesenchymal stem cells into defects in the tissues in which the mesenchymal stem cells should differentiate. This logic would allow the implanted mesenchymal stem cells to respond to the local environmental influences, including bioactive protein factors, which at present are not understood well enough to be controlled or duplicated in vitro.

Section Two

Articular Cartilage

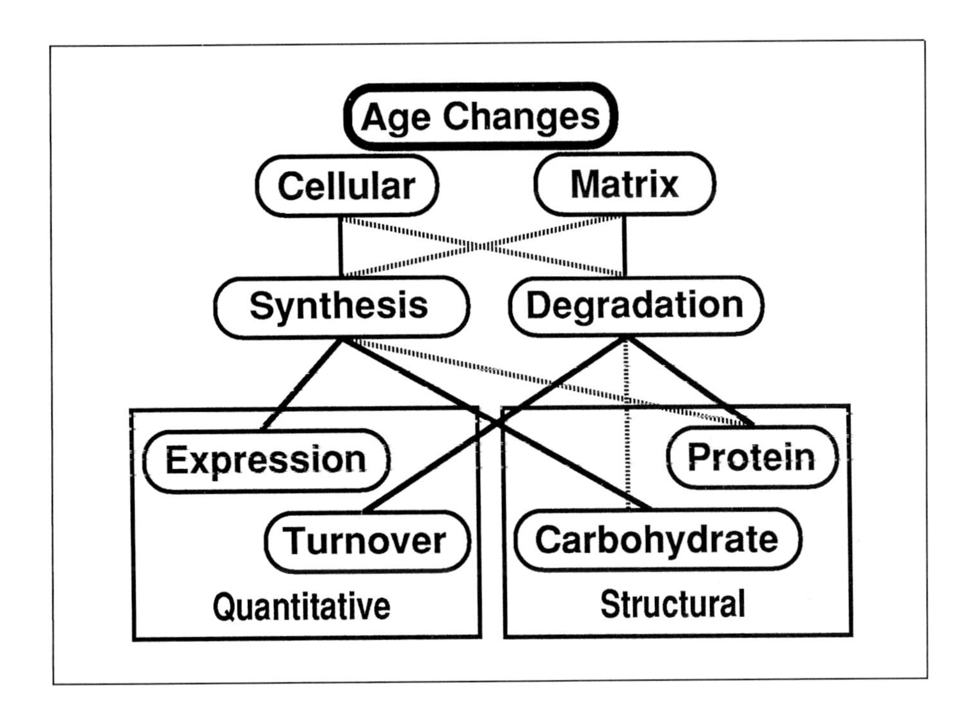

Section Leader

David Eyre, PhD

Section Contributors

Wayne H. Akeson, MD
David Amiel, PhD
Peter G. Bullough, MB, ChB
Alan Grodzinski, ScD
Stefan Lohmander, MD, PhD
Martin Lotz, MD
Peter J. Roughley, PhD
Paul Schreck, MD
Virgil L. Woods, Jr., MD

Overview

Articular cartilage varies greatly in quality at sites within joints and between different joints of the body, with the widest diversity in the elderly. This variability includes functional differences programmed during development but also age-related changes that may represent normal adaptation to load or disuse and reparative and degradative processes. We know very little about the capacity of articular cartilage to respond to altered physical activity; for example, disuse from progressive inactivity and decreased range of motion in the elderly.

Although the articular cartilage section does not deal with the meniscus or other articulating fibrocartilages, it is clear that these structures also alter greatly with age and are integral to the pathology of the aging skeleton. Studies on aging articular and nonarticular fibrocartilages are, therefore, important and represent a general research area to be encouraged in the future. Similarly, neighboring tissues essential for the maintenance of articular cartilage, for example, the synovium and synovial fluid, are not dealt with specifically in this section. It is recognized, however, that studies on the ability of these tissues to continue to support cartilage health with increasing age are critically important.

The five chapters in this section present perspectives from the clinician, pathologist, bioengineer, matrix biochemist, and cell biologist. The clinician who specializes in diagnosing and treating osteoarthritis (OA) needs to know to what degree age-associated changes in human articular cartilage are related to the incidence of clinical joint disease. For example, are such changes actually a result of disease processes, or do some precede and predispose to disease? The pathologist is interested in the changing macro- and microanatomy of joint articular cartilages, including altered cellular morphology and evidence of reparative responses and of tissue remodeling at the bone interface. Bioengineers are becoming increasingly active in developing methods to explore the effect of physical forces on chondrocyte metabolism, in addition to understanding the material properties of cartilage tissue both macro- and microanatomically.

Biochemists, who have observed various changes in the molecular quality of proteoglycans and collagens of articular cartilage with increasing developmental age and, to some degree, with increasing adult age, are now examining the responsible molecular mechanisms. Little is known of age-related changes in the properties of adult articular chondrocytes. Progress in understanding the role of membrane-bound integrin receptors in cell-matrix interactions is beginning to be applied to the study of chondrocytes. The chapter by Woods and associates explores the potential of this approach for probing age-related changes in the responsiveness of chondrocytes to their environment.

The presentations and discussion during the articular cartilage session summarized many advances, but also highlighted our lack of knowledge. For example, there is no hard evidence to link any age-dependent change in articular cartilage to clinical pathology in a causative way. Material property measurements have shown that, rather than getting stiffer in tension with increasing age, as one might expect if collagen becomes increasingly cross-linked, articular cartilage becomes more compliant in tension. This issue needs more study using better micromechanical methods. Finally, there is a priority to understand fully the properties of articular chondrocytes, including their life-cycle in human tissue, their capacity to mount a reparative response, and, in particular, whether reparative properties deteriorate with increasing age. Such deterioration could be an inherent feature of cellular senescence or a result of altered regulatory signals from the extracellular environment.

The breakout session on articular cartilage to formulate future directions was organized as follows: The entire group met for one to two hours to decide on topic subdivisions; five emerged dealing with clinical, morphologic, biomechanical and cellular aspects, and biochemistry of the extracellular matrix. Each subgroup of three to five specialists then met to prepare recommendations on areas of missing knowledge and promising research directions. A common, recurrent problem in each group was the tendency to deal synonymously with OA and aging. Most studies formulated to define pure age-related changes in articular cartilage, as opposed to developmental changes, ran the risk of measuring the consequences of arthritis. All the recommendations, therefore, highlighted this obstacle.

In summary, directions for clinical research included a need to identify aging processes that are risk factors for loss of joint function, more refined diagnostic criteria for joint disease, and universal markers (molecular markers, imaging signals, etc) for pathologic processes that precede and may herald overt joint disease. The morphology of articular cartilage and joint tissues was viewed to be a primary index of joint function, and systematic correlations need to be made between age-associated changes in morphology in each joint at each decade of life as a background for evaluating the morbid anatomy of disease.

Recommendations from the bioengineers included a need for more site-specific and joint-specific analyses of the biomechanical and physical properties of articular cartilage with increasing age. In particular, a need was perceived for micromechanical methods of analysis that could distinguish regional and microscopic sites in cartilage, for example, pericellular from interterritorial matrix and target ultrastructural units, such as the individual cell and its matrix. Experimental systems for evaluating the influence of physical forces on living chondrocytes were emphasized as showing particular promise.

Recommendations were made for an increased understanding of the extracellular matrix through microanatomic studies to define precisely where (relative to the chondrocyte, for example) progressive macromolecular changes occur with increasing age. Structural macromolecules of the matrix, in addition to the major collagen and proteoglycan species, require study and potential molecular markers should be sought that appear or disappear with increasing age in the adult tissue. Mechanisms of proteolytic degradation that modify long-lived matrix proteins in cartilage need to be defined. Processes that lead to altered macromolecular interactions, for example, age-dependent synthetic changes and cumulative protein modifications by nonenzymic glycosylation and oxidative mechanisms, need to be evaluated. The importance of distinguishing normal aging processes in the matrix from arthritic processes was emphasized.

Recommendations for future directions on aging chondrocyte biology were wide ranging and spanned opportunities to characterize better the distinctive phenotypes of articular chondrocytes with site (type of joint, surface or deep zone, etc) and, in particular, how chondrocyte phenotype may change with age. Mechanochemical sensory mechanisms are important to define, as are any changes in the capacity of chondrocytes to proliferate or repair matrix with increasing age. More knowledge is needed of the influence on chondrocytes of systemic factors, including potential secondary effects of such common diseases of the elderly as diabetes and uremia.

Genetic background and the possibility that rare mutations and more common polymorphisms predispose to accelerated cartilage "aging" in certain individuals with no overt inborn disorder of skeletal growth was clearly viewed as an overriding biological concept. The incidence and contribution of genetic variants of matrix macromolecules and chondrocyte regulatory proteins to cartilage longevity requires attention.

Chapter 8

Osteoarthritis: A Major Cause of Disability in the Elderly

Stefan Lohmander, MD, PhD

The title of this workshop—Age-Related Changes in the Musculoskeletal System: A Major Cause of Decreased Mobility With Increasing Age—is to many synonymous with the picture of an elderly lady crippled by painful, arthritic joints. Although this undoubtedly is an oversimplification, the fact remains that arthritis, and in particular the condition we define as osteoarthritis (OA, osteoarthrosis, arthrosis), is a major cause of impairment and disability among the elderly.

OA is a considerable burden both to the individual patient and to society: in 1988 in the United States the total cost of arthritis was estimated to be $54.6 billion, or 43% of the total cost of all musculoskeletal conditions. More than 500,000 arthroplasty procedures are performed in the United States annually. Of these, about 400,000 are performed on the knee and hip. In the knee, 85% of the arthroplasties are done because of OA; the corresponding figure for the hip is 55%.[1] Arthritis, the leading chronic condition reported by the elderly, is reported by almost 50% of persons 65 years of age and older.[1] More than one third of elderly persons who have arthritis are limited in five or more physical activities.[2] The rapidly changing demographic structure in the developed countries will only emphasize the magnitude of this problem. Musculoskeletal pain and arthritis are also major public health problems in the developing countries.[3]

OA presents us with several challenges. The general practitioner or rheumatologist can diminish the pain but cannot modify the progression of the disease. The physiotherapist can lessen some of the consequences of the disease process but, again, cannot modify the progression of it. The orthopaedic surgeon deals only (however successfully) with the end-product of the disease process, not the process itself. The radiologist can identify only the final stages of a disease process that has been persisting in the joint for years and perhaps decades. The epidemiologists have great difficulty trying to measure OA in the population because of a lack of definite and universally applicable criteria for the disease. The biochemistry and the biomechanics scientists observe the progressive breakdown and failure of repair of cartilage in animal models, but do not yet understand the basic mechanisms of tissue breakdown during OA or why repair fails. They are thus unable to tell us how to slow down or reverse the process of cartilage failure in the patient.

Epidemiology

Epidemiologic studies on osteoarthritis are severely hampered by the lack of universally and easily applicable criteria for patient reported symptoms, and for clinical or radiologic diagnosis of OA in different joints. Therefore, a high degree of variability and uncertainty is associated with all reported data.

All reports on OA epidemiology, however, consistently show an almost exponential increase of the prevalence of OA with increasing age. Thus, the prevalence per 100 people of radiologic changes indicative of OA in the knees increased from zero in those aged 25 to 34 to 13.8 in individuals aged 65 to 74.[4] In the hands, the corresponding figures were 3.4 and 84.5, respectively. Changes were more common in women than in men for these two sites. Similarly, radiographic evidence of OA of the knee was present in 27% of subjects aged 63-70; 44% of those aged 80 or older had similar changes.[5] The prevalence of radiologic changes of the hip consistent with primary OA increased from 0.2% at 40 years of age to 11.5% at 85 years.[6] Among individuals 70 years of age and older, more than half report having arthritis, as compared to about one fifth of men and one third of women aged 45 to 64.[6] Of the elderly in this study, more than one third were limited in activities of daily living. The prevalence of self-reported or clinical arthritis of the knee, hand, shoulder, and hip in a general population shows similar age-related increases for all these joints, with knee most common and hip least common.[7,8] Only about half of the subjects aged over 55 years and with symptomatic OA of the knee had radiographic knee OA; about one sixth of asymptomatic subjects of similar age had radiographic OA.[9] These figures demonstrate the lack of correlation between symptoms and clinical signs of OA and the presence of radiographic signs of OA (Fig. 1).

Risk Factors

Osteoarthritis is not a single disease entity but rather a final common end-stage of cartilage and joint failure (Fig. 2). Inherent to this attempt at a general definition is the multifactorial pathogenesis with many recognized risk factors which represent unknown, underlying specific causes (Fig. 3). Major risk factors are both endogenous and exogenous. Of the endogenous factors, age, sex, race, and inherited susceptibility have all been shown to play a role, but with different weights at different joint sites. The rare single base mutation of 519 arg → cys in the human type II procollagen gene, which

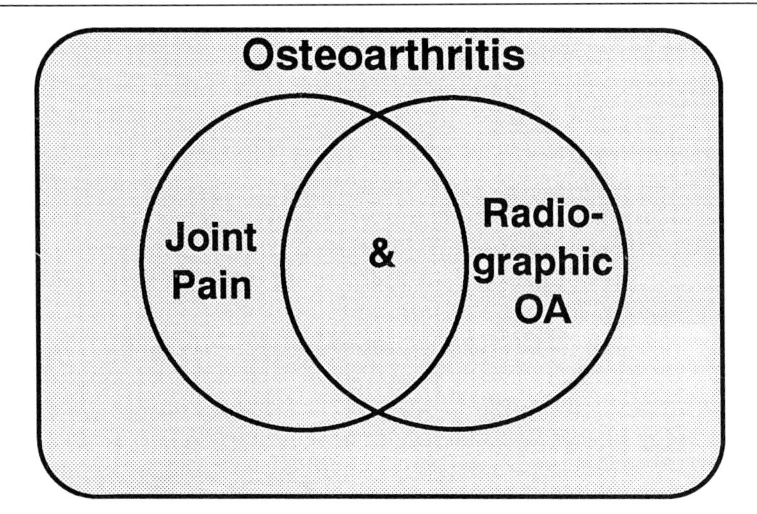

Fig. 1 *Symptoms, clinical signs, and radiographic signs of osteoarthritis show only a limited correlation. Patients with symptoms and clinical signs of OA may have normal radiographs, patients with OA changes on radiographs may be asymptomatic and patients with developing "preOA" or silent OA, may lack all symptoms and signs. Compare figure 2.*

has been shown to cause mild chondrodysplasia and early-onset OA, is at present the only example of a defined inherited susceptibility for OA.[10,11] Exogenous factors such as obesity, use and abuse of joints, and abnormal joint shape are also proven risk factors, again with different weighting for different sites, sex, and age. For example, obesity is strongly correlated with knee and hand OA.[12] Risk factors are probably additive. For example, it was shown that patients with a presumed systemic risk factor shown by the development of distal interphalangeal OA had an increased risk of developing OA in the knee after meniscectomy.[13] There is good evidence for the association of occupational risk factors with OA. For example, farmers and soccer players have an increased relative risk of hip OA,[14-17] and occupations and sports with excessive knee loading are associated with an increased relative risk of knee OA.[14,15,18,19] Although at least part of these increased relative risks may be attributed to cumulative joint injury, the precise connections between these increased risks and the underlying causes are unknown (Fig. 3).

Joint injury is a well-known risk factor for OA of the knee, but it is not known why post-traumatic OA of the often-injured ankle joint is so uncommon. Traumatic injuries to the knee joint are frequent in younger people.[20,21] However, in an academic orthopaedic setting, two thirds of all meniscal injuries are diagnosed in patients older than 30 years of age (unpublished). Patients in this older age group often have advanced cartilage changes at the time of diagnosis, and radiologic changes consistent with OA of the knee are already seen

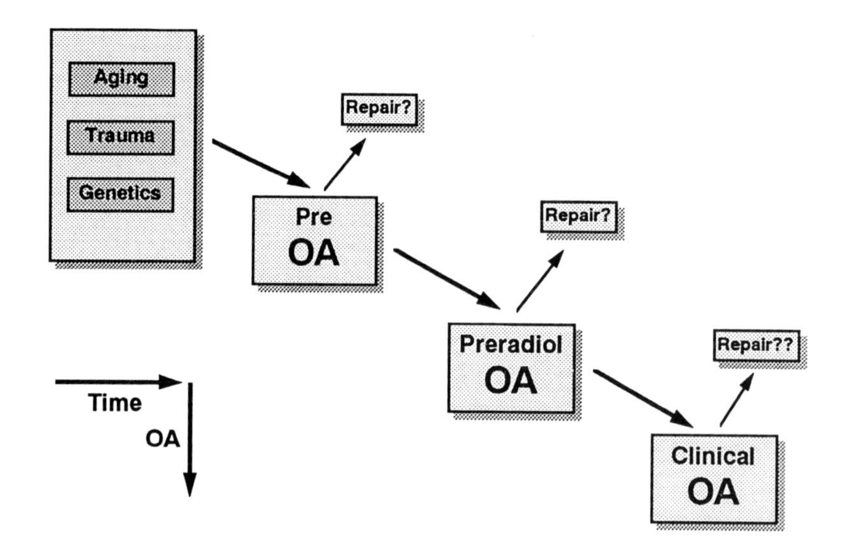

Fig. 2 *The progress of osteoarthritis takes place over a time period of years and sometimes decades. In the population, risk factors such as aging, trauma, genetics may in some individuals initiate a process which leads to changes in the metabolism and composition of the joint cartilage (PreOA). Continued exposure to the same or other risk factors may in some cause a progression of these changes, leading to further abnormalities of the metabolism, composition, biomechanics and morphology of the cartilage (Preradiol OA). Progress may eventually reach the final end-stage (Clinical OA) with classic symptoms, radiologic changes and loss of joint cartilage substance. Progress from one stage to the next is not inevitable, some patients remain at the same level for long periods of time or even show evidence of spontaneous repair. The response of the individual to joint insult is variable.*

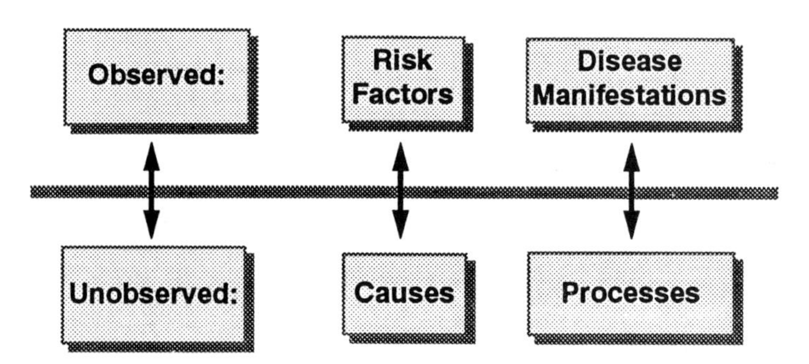

Fig. 3 *The risk factors and disease manifestations we are able to observe in osteoarthritis represent unknown underlying causes and disease processes.*

at an average of 5 years after injury. In comparison, patients in the same study who are younger than 30 years at the time of injury do not show signs of radiologic OA until an average of some 20 to 25 years after injury (Fig. 4). These observations are consistent with several different explanations. For example, it may be that the meniscus rupture in the older joint is intrinsic to an osteoarthritic process which is already well underway at the time of the tear of the meniscus. Alternatively, the joint cartilage in the older patient is less well able to cope with the changed biomechanical loading patterns after meniscus rupture and therefore quickly deteriorates after the injury.

The reasons why age is one of the strongest risk factors for development of OA are not clear. It may be that with increasing senescence the chondrocytes lose the ability to replace and repair joint cartilage matrix lost by injuries or normal wear. Alternatively, the aging cartilage matrix becomes more susceptible to normal cumulated microinjuries and the repair mechanisms of the cells are no longer able to compensate for this increased susceptibility. In both cases there is a discrepancy between the demands of the environment and the capacity of the chondrocytes or the matrix to respond to these demands (Fig. 5). It is also important to note that, although the lag time from the initiating event to the appearance of symptoms or radiographic signs of osteoarthritis varies, it is usually measured in years to decades. This would also tend to bias symptomatic osteoarthritis towards the elderly.

Disease Progression

The rate of progress of OA varies from patient to patient, even within the same age group and with the same type of joint injury.[22] This suggests that such factors as inherent genetic susceptibility, activity level, and differences between joints strongly influence OA development and again emphasizes that the individual response to joint insult is variable. Perhaps this as much reflects a variable capacity for repair as a variable degradative response after joint insult.

Recent data suggest that the progress of knee OA that has been diagnosed by clinical or radiologic signs is not inevitable. In an 11-year follow-up of knee OA diagnosed by symptoms or radiography, only 50% of the knees deteriorated by radiologic criteria and 10% improved, while average visual analog pain scores remained unchanged.[23] In a similar follow-up study, 27% of the patients diagnosed with knee pain and OA were free of pain 20 years after the initial diagnosis.[24] These results contrast with those of a previous longitudinal study of knee OA, where the majority of patients with OA initially diagnosed by radiographic joint space narrowing progressed in their disease.[25] Although lack of universally accepted criteria for OA diagnosis and differences in disease stage at the

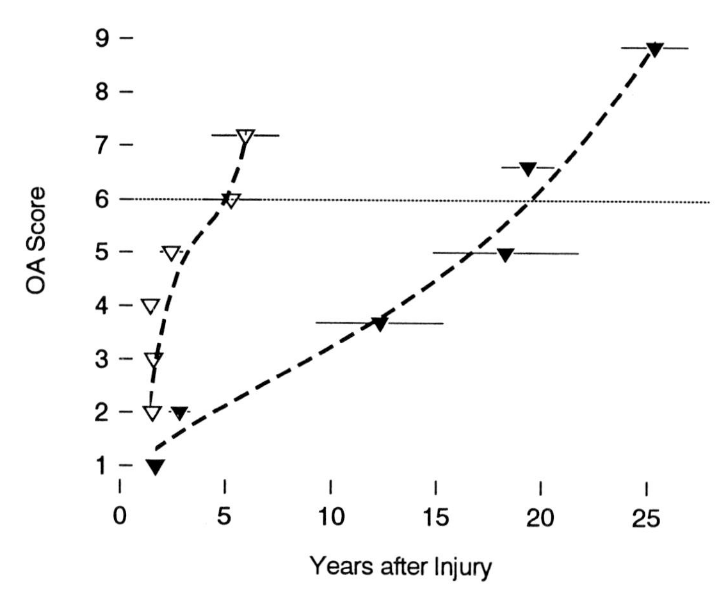

Fig. 4 *Progress of osteoarthritis of the knee after meniscus injury in patients aged under or over 30 years at the time of injury. OA score (77) represents a combined arthroscopic and radiographic grading of the knee joint: 1 = normal cartilage by arthroscopy, 2 = superficial fibrillations in one compartment, 3 = superficial fibrillations in more than one compartment, 4 = deep fissures in one compartment, 5 = deep fissures in more than one compartment, 6 = osteophytes, sclerosis and minor joint space narrowing, 7 = joint space narrowing to less than half normal width (Ahlbäck grade I), 8 = complete disappearance of joint space (Ahlbäck grade II), 9 = bone erosion less than 5 mm (Ahlbäck grade III), 10= bone erosion more than 5 mm (Ahlbäck grade IV). Scores 1-5 all have normal standing X-ray films. Results represent average time after injury and OA score for patients with knee symptoms and arthroscopically verified meniscus injury with the trauma incident before ( ) or after ( ) the age of 30 years. Bars represent SE, lines are 3rd order regression fit. Data represent a total of 732 patients examined by arthroscopy and/or radiography because of acute injury or knee problems caused by previous injury (each patient provided one data point). (Adapted with permission from Dahlberg L, Ryd L, Heinegard D, et al: Proteoglycan fragments in joint fluid: Influence of arthrosis and joint inflammation.* Acta Orthop Scand *1992;63:417-423.)*

initial diagnosis may confound the interpretation of these studies, results such as these merit further investigation. The natural course and prognosis of knee OA at each discrete stage should be the gold standard with which to compare the outcome of surgical or pharmacologic intervention. Furthermore, these observations emphasize the critical need for better methods for early identification of patients in high-risk and poor prognosis groups. Intervention could then focus on these groups. Similarly, clinical trials for OA treatment need to select the patients in the poor prognosis groups and

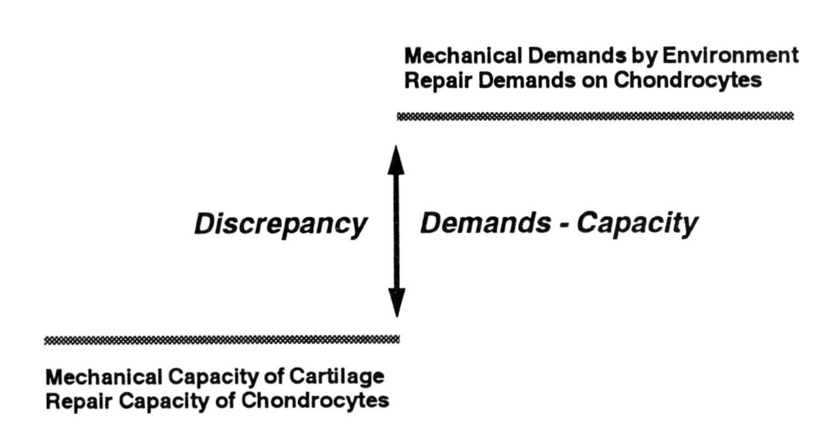

Fig. 5 *Osteoarthritis results from a discrepancy between environmental demands and cartilage capacity. The capacity of the joint cartilage to withstand environmental demands and to repair injury may decrease with age.*

exclude those whose disease will not progress in order not to confound the results of the study.

Pathogenesis

The pathway leading to OA may be viewed as resulting from a discrepancy between the demands of the environment on the joint and the capacity of its cells and matrix to respond to these demands (Fig. 5). Consequently, OA could be the result of excessive demands on a normal tissue or normal stress on a tissue with a lowered capacity to respond to stress. This lowered or defective capacity to respond to stress could be either a failure of the cartilage matrix to withstand mechanical stress or a failure of the chondrocytes to perform the necessary repair and replenishment of matrix components. Perhaps failure of normal repair should receive increased attention as a cause of OA.

The disease mechanisms active at the tissue and cell level in osteoarthritis have not yet been identified. Moreover, their relation to risk factors, symptoms, and clinical signs remains unclear (Fig. 3), and the initiation and progression of OA may well be controlled by different processes.

A number of investigations of human and animal model OA have shown that changes in the metabolism and properties of joint cartilage and loss of molecular fragments from the matrix are early events in the disease process, and that bulk loss of cartilage tissue is a late event.[26-29] It may thus be argued that increased degradation of cartilage matrix is a key event in the development of OA. However, because cartilage matrix is continuously turned over and held in a steady state under physiologic conditions, loss of matrix may also be the result of defective mechanisms for replacement and repair of matrix normally turned over (Fig. 6). The pathology of end-stage OA is thus a result of both excess degradation and aberrant repair.

The physiologic rate of turnover of cartilage matrix is slower in the adult than in the young and is variable for different matrix components.[30,31] Available evidence suggests that the chondrocytes play a major role both in physiologic turnover during development and growth and in the matrix degradation that occurs in osteoarthritis. Cytokines such as interleukin-1 and tumor necrosis factor both stimulate matrix catabolism by chondrocytes and strongly inhibit the synthesis of matrix molecules.[32-34] These cytokines thus have ability both to stimulate cartilage destruction and to inhibit repair activity. However, the significance of this signal pathway in OA has not yet been demonstrated. The complexity of the interaction between the chondrocytes, cytokines, and growth factors is further illustrated by the fact that transforming growth factor-β and insulin-like growth factor-1 are effective antagonists of interleukin-1 action on cartilage.[35,36]

The primary cleavage of cartilage matrix molecules is extracellular and is presumably mediated by proteinases, most of which are probably released by the chondrocytes themselves. An important role in cartilage matrix degradation has been proposed for the metalloproteinase family, the enzymes of which have the capacity to degrade all the components of the cartilage matrix.[37,38] These endopeptidases are secreted in a latent proenzyme form and are activated extracellularly, and the active forms are inhibited by strong binding to tissue inhibitors of metalloproteinases (TIMPs) or to α-2-macroglobulin. An increase in metalloproteinase activity may therefore be caused either by an increase in the amount of activated enzyme or by a decrease in the amount of available inhibitor, or both. Interestingly, such perturbations in the equilibrium between enzyme and inhibitors have been shown to occur both in osteoarthritic cartilage and in joint fluid from osteoarthritic joints.[39,40]

The identity of the protease(s) active in the degradation of cartilage matrix molecules in OA is, however, equivocal. Thus, N-terminal analysis of proteoglycan fragments isolated from human OA joint fluid is consistent with a dominating cleavage of the proteoglycan core protein at the glu 373 - ala 374 bond in the interglobular domain.[41] Interestingly, metalloproteinases have not been shown to generate this cleavage, and the identity of the protease(s) responsible is as yet unclear.

The fragments of degraded cartilage matrix molecules are either taken up by the chondrocytes and further degraded by lysosomal enzymes, or they are lost to the joint fluid by diffusion (Fig. 7). Fragments in the joint fluid may be taken up by and further degraded in the synovial tissue cells, or they may be removed by bulk flow with the synovial fluid to the lymph circulation.[42] A substantial proportion of matrix molecule fragments removed by the lymphatic

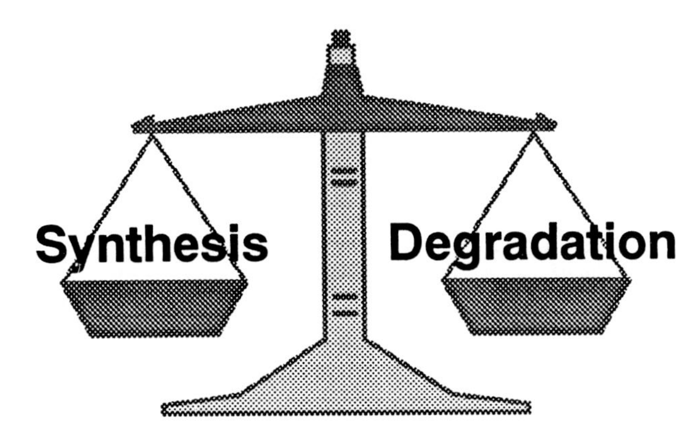

Fig. 6 *Under physiological conditions, the cartilage matrix composition is held in a steady state by a balance between synthesis and degradation. A decrease in cartilage matrix components may arise from a decrease in synthesis or an increase in degradation, or both.*

circulation are eliminated or at least further degraded in the regional lymph nodes.[42,43] The majority of the remaining fragments that reach the blood stream are rapidly (within minutes) removed from circulation, most likely by the liver cells.[44,45] The collagen cross-links, however, apparently survive the circulation and are found enriched in urine.[46,47]

Diagnosis

As noted (Fig. 2), the OA process is a continuum from early initiating events on the cell and tissue level to overt, clinical disease that can be diagnosed by classical criteria. The criteria by which the condition is defined by the cartilage biochemist, general practitioner, orthopaedic surgeon, and epidemiologist will by necessity differ, depending on their different tools, needs, and purposes.

In the clinical setting, we currently diagnose OA by a combination of symptoms, clinical signs, and radiological findings.[48] Because these methods are associated with several uncertainties, ways are being sought by which the diagnosis and assessment of disease progression can be made with greater precision and reproducibility. Among the additional promising techniques currently under evaluation are office-based arthroscopy, magnetic resonance imaging, bone scintigraphy, and body fluid markers of joint cartilage turnover.

Although radiologic changes are usually regarded as the standard in OA diagnosis, it may be questioned whether they represent the gold standard or are fool's gold, to echo the title of Brower's paper

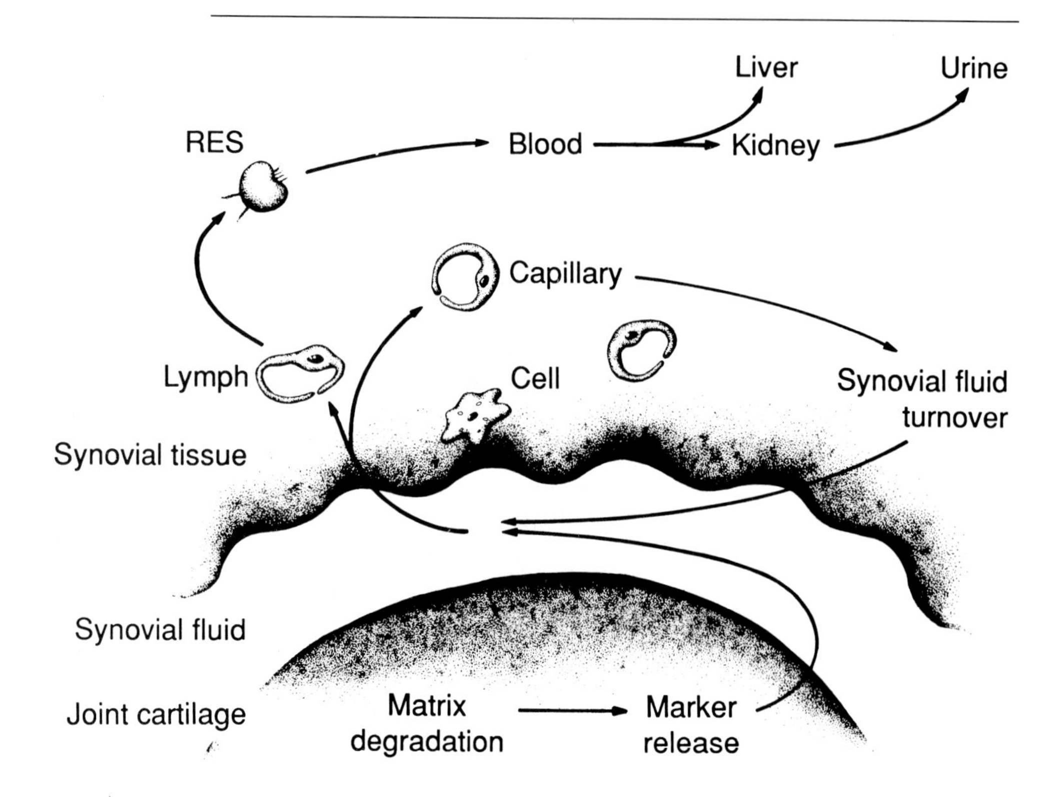

Fig. 7 *Turnover of joint cartilage matrix fragments released into the joint fluid and other body fluid compartments. Fragments released into joint fluid from cartilage are removed mainly by lymphatic flow. Fragments which reach the blood circulation and which are not eliminated in the regional lymph nodes are largely metabolized by the liver. (Reproduced with permission from Lohmander LS: Markers of cartilage metabolism in arthrosis: A review.* Acta Orthop Scand *1991;62:623-632.)*

on the radiographic grading of the course of rheumatoid arthritis.[49] An example are the recently published guidelines for the radiographic assessment of OA of the knee.[50] These propose the use of weight-bearing anteroposterior radiographs, which would not allow the assessment of the commonly occurring OA in the patellofemoral joint. Further, radiologic changes as assessed by the Kellgren and Lawrence[51] score, which takes osteophytes, sclerosis, and joint space narrowing into account, or by the Ahlbäck[52] score, which only grades the degree of joint space narrowing,[52] show only a limited correlation with arthroscopic grading of the joint surfaces.[53,54] Clinical signs, in turn, show only moderate correlation with radiographic changes (Fig. 1).[55] Some recent reports suggest, however, that progressive joint space narrowing in knee or hip OA can be reproducibly measured over periods as short as three years.[56,57] The problem associated with the radiologic grading of OA is, of course, that radiography presents only a record of past destructive events and does not give information on current disease activity.

Bone scintigraphy with the use of ^{99}Technetium-labeled diphosphonates provides information on bone remodeling activity in the joint and may predict surgery or joint space loss over five years in knee OA.[58] Magnetic resonance imaging is more sensitive than either conventional radiography or computed tomography for assessing the extent and severity of osteoarthritic changes.[59] Point scaling systems for the arthroscopic grading of articular cartilage lesions in the hip and knee have been developed.[60,61] Although these techniques are now available for validation in the OA research environment, it will likely take some time before their value has been proven and they are accessible for use in a more routine setting.

The assay of molecular fragments of cartilage matrix molecules released into joint fluid, serum, or urine has been suggested as one way to monitor cartilage turnover in vivo and to serve as a tool for the diagnosis, monitoring, and prognostication of OA and other joint cartilage disease.[22,62] This approach holds great hope for the eventual development of much-needed process markers and for providing a better understanding of the disease process in OA. Several problems remain to be solved, however. For example, if a marker is to serve as a quantitative measure of cartilage turnover, we need to know the metabolism and clearance of this marker in the compartment in question.[63] The determination of ratios between different markers in the same compartment may also yield useful information.[64]

Treatment

The main problems associated with the progression of OA are increasing pain, impairment, and disability. As a consequence of the conditions, a large proportion of both the working-age population and the elderly with OA are limited in their ability to work and in their activities of daily living.[2] Treatment seeks to decrease pain and preserve and restore function. An assortment of beneficial treatment modalities are in use and will vary with the patient, joint site, and the stage of the disease. These include education, walking aids and braces, physiotherapy, drugs, abrasion arthroplasty, osteotomy, and, finally, joint replacement. Irrespective of the method of treatment used and its success in decreasing pain and disability, it may be stated that treatment rarely, if ever, is able to modify the disease process itself, but only the consequences of it. For any of the procedures, efficacy and outcome should be compared with the natural history of a control group with the appropriate age and disease stage. Although this may sound like a cliche, such a comparison is not a trivial task in the light of our limited documentation of the natural history of OA in many sites. The issue of primary prevention is a remote but not completely unreachable goal with our increasing identification of risk factors.[65] Secondary prevention, or modification of the disease process, is not yet possible, but current rapid advancement in the understanding of the basic disease mechanisms makes the future look increasingly hopeful.

The use of non-steroidal anti-inflammatory drugs for the treatment of OA is widespread, and this family of drugs provides well-documented pain relief which, however, occurs at the price of significant side effects, particularly in the elderly.[66] No well-controlled studies have been published that demonstrate any protective effects of NSAIDs on the joint cartilage (chondroprotection). On the contrary, it was suggested that potent inhibitors of prostaglandin synthesis may be inappropriate in the management of OA of the hip, because they can cause the disease to progress more rapidly.[67] These observations have become all the more relevant since paracetamol was proposed to be as effective as NSAIDs in the treatment of knee OA.[68]

The question of disease modification, or chondroprotection, by drug treatment in OA is controversial.[69] Although several interesting studies have reported the protective effects of injections of polysulfated polymers in animal OA models,[70-72] well-controlled studies in the human are still lacking, despite the fact that in several countries these preparations have been in use for a long time.

Arthroscopic abrasion arthroplasty, whereby exposed bone surface on the joint is abraded to produce a vascular response and regrowth of fibrocartilage, has been reported as a good procedure in selected patients.[73] It is advocated that the technique is suitable for patients with moderately advanced unicompartmental knee disease and that three quarters of patients with early OA changes can expect pain relief for as long as five years. In moderately advanced medial compartment knee OA, tibial valgus osteotomy is also a successful technique for patients with an active life-style. With proper patient selection and precise surgery, the results for this procedure are equal to or better than those reported for unicompartmental arthroplasty.[74] The basis for the reported good results of these procedures is not clear. However, both give rise to regrowth of fibrocartilage on the joint surface,[75,76] indicating that even osteoarthritic joint surfaces have a regenerative capacity, which should be further investigated.

The remarkable success story of arthroplasty as a treatment for OA does not need to be retold here. In spite of the overall immense success, some problems that merit attention remain, including costs, implant loosening, and joints for which there are yet no satisfactory implants developed. Arthroplasty will no doubt remain the treatment of choice for advanced OA of the hip and knee, but we should remind ourselves that we treat the consequences of the disease, not the disease process itself. To develop the means to retard cartilage destruction and slow or reverse the progress of osteoarthritis in the patient remains the ultimate target. This goal, which only a few years ago would have seemed a mirage, has grown increasingly accessible as a result of the rapid advances in osteoarthritis research.

Acknowledgmemts

The author's laboratory is supported by research grants from the Swedish Medical Research Council (8713), the King Gustaf V 80-Year Birthday Fund, the Kock, Zoega and Österlund Foundations, the Medical Faculty of Lund University, and Merck Research Laboratories, USA. The inspiration and constructive critique provided by Drs. Paul Dieppe and Harald Roos is gratefully acknowledged.

References

1. Praemer A, Furner S, Rice DP: Musculoskeletal Conditions in the United States. Park Ridge, IL, American Academy of Orthopaedic Surgeons, 1992.
2. Yelin E: Arthritis: The cumulative impact of a common chronic condition. *Arthritis Rheum* 1992;35:489-497.
3. Darmawan J, Valkenburg HA, Muirden KD, et al: Epidemiology of rheumatic diseases in rural and urban populations in Indonesia: A World Health Organization International League Against Rheumatism COPCORD study, stage I, phase 2. *Ann Rheum Dis* 1992;51:525-528.
4. Lawrence RC, Hochberg MC, Kelsey JL, et al: Estimates of the prevalence of selected arthritic and musculoskeletal diseases in the United States. *J Rheumatol* 1989;16:427-441.
5. Felson DT, Naimark A, Anderson J, et al: The prevalence of knee osteoarthritis in the elderly: The Framingham osteoarthritis study. *Arthritis Rheum* 1987;30:914-918.
6. Danielsson L, Lindberg H, Nilsson B: Prevalence of coxarthrosis. *Clin Orthop* 1984;191:110-115.
7. Badley EM, Tennant A: Changing profile of joint disorders with age: Findings from a postal survey of the population of Calderdale, West Yorkshire, United Kingdom. *Ann Rheum Dis* 1992;51:366-371.
8. Steven MM: Prevalence of chronic arthritis in four geographical areas of the Scottish Highlands. *Ann Rheum Dis* 1992;51:186-194.
9. McAlindon TE, Snow S, Cooper C, et al: Radiographic patterns of osteoarthritis of the knee joint in the community: The importance of the patellofemoral joint. *Ann Rheum Dis* 1992;51:844-849.
10. Katzenstein PL, Malemud CJ, Pathria MN, et al: Early-onset primary osteoarthritis and mild chondrodysplasia: Radiographic and pathologic studies with an analysis of cartilage proteoglycans. *Arthritis Rheum* 1990;33:674-684.
11. Ala-Kokko L, Baldwin CT, Moskowitz RW, et al: Single base mutation in the type II procollagen gene (COL2A1) as a cause of primary osteoarthritis associated with a mild chondrodysplasia. *Proc Natl Acad Sci USA* 1990;87:6565-6568.
12. Felson DT, Anderson JJ, Naimark A, et al: Obesity and knee osteoarthritis: The Framingham study. *Ann Intern Med* 1988;109:18-24.
13. Doherty M, Watt I, Dieppe P: Influence of primary generalised osteoarthritis on development of secondary osteoarthritis. *Lancet* 1983;2:8-11.
14. Felson DT: Epidemiology of hip and knee osteoarthritis. *Epidemiol Rev* 1988;10:1-28.
15. Vingård E, Alfredsson L, Goldie I, et al: Occupation and osteoarthrosis of the hip and knee: A register-based cohort study. *Int J Epidemiol* 1991;20:1025-1031.
16. Axmacher B, Lindberg H: Coxarthrosis in farmers. *Clin Orthop* 1993;(Feb.)(287):82-86.

17. Lindberg H, Roos H, Gärdsell P: Prevalence of coxarthrosis in former soccer players: 286 players compared with matched controls. *Acta Orthop Scand* 1993;64:165-167.
18. Roos H, Lindberg H, Gärdsell P, et al: The prevalence of gonarthrosis in former soccer players and its relation to meniscectomy. *Am J Sports Med* 1993, in press.
19. Felson DT, Hannan MT, Naimark A, et al: Occupational physical demands, knee bending, and knee osteoarthritis: Results from the Framingham study. *J Rheumatol* 1991;18:1587-1592.
20. Hede A, Jensen DB, Blyme P, et al: Epidemiology of meniscal lesions in the knee: 1,215 open operations in Copenhagen 1982-84. *Acta Orthop Scand* 1990;6:435-437.
21. Nielsen AB: The epidemiologic aspects of anterior cruciate ligament injuries in athletes. *Acta Orthop Scand* 1991;62(suppl 243):13.
22. Lohmander LS: Markers of cartilage metabolism in arthrosis: A review. *Acta Orthop Scand* 1991;62:623-632.
23. Spector TD, Dacre JE, Harris PA, et al: Radiological progression of osteoarthritis: An 11 year follow up study of the knee. *Ann Rheum Dis* 1992;51:1107-1110.
24. Sahlström A, Johnell O, Redlund-Johnell I: The natural course of arthrosis of the knee. *Acta Orthop Scand* 1992;63(suppl 248):57.
25. Hernborg JS, Nilsson BE: The natural course of untreated osteoarthritis of the knee. *Clin Orthop* 1977;123:130-137.
26. Hoch DH, Grodzinsky AJ, Koob TJ, et al: Early changes in material properties of rabbit articular cartilage after meniscectomy. *J Orthop Res* 1983;1:4-12.
27. Carney SL, Billingham ME, Muir H, et al: Demonstration of increased proteoglycan turnover in cartilage explants from dogs with experimental osteoarthritis. *J Orthop Res* 1984;2:201-206.
28. Carney SL, Billingham ME, Muir H, et al: Structure of newly synthesized (35S)-proteoglycans and (35S)-proteoglycan turnover products of cartilage explant cultures from dogs with experimental osteoarthritis. *J Orthop Res* 1985;3:140-147.
29. Ratcliffe A, Billingham ME, Saed-Nejad F, et al: Increased release of matrix components from articular cartilage in experimental canine osteoarthritis. *J Orthop Res* 1992;10:350-358.
30. Maroudas A: Glycosaminoglycan turn-over in articular cartilage. *Philos Trans Roy Soc Lond* 1975;271:293-313.
31. Lohmander S: Turnover of proteoglycans in guinea pig costal cartilage. *Arch Biochem Biophys* 1977;180:93-101.
32. Saklatvala J: Tumor necrosis factor alpha stimulates resorption and inhibits synthesis of proteoglycan in cartilage. *Nature* 1986;322:547-549.
33. Dingle JT, Tyler JA: Role of intercellular messengers in the control of cartilage matrix dynamics, in Kuettner KE, Schleyerbach R, Hascall VC (eds): *Articular Cartilage Biochemistry*. New York, Raven Press, 1986, pp 181-210.
34. Lefebvre V, Peeters-Joris C, Vaes G: Modulation by interleukin 1 and tumor necrosis factor of production of collagenase, tissue inhibitor of metalloproteinases and collagen types in differentiated and dedifferentiated articular chondrocytes. *Biochim Biophys Acta* 1990;1052:366-378.
35. Fosang AJ, Tyler JA, Hardingham TE: Effect of interleukin-1 and insulin like growth factor-1 on the release of proteoglycan components and hyaluronan from pig articular cartilage in explant culture. *Matrix* 1991;11:17-24.

36. Hardingham TE, Bayliss MT, Rayan V, et al: Effects of growth factors and cytokines on proteoglycan turnover in articular cartilage. *Br J Rheumatol* 1992;31(suppl 1):1-6.
37. Docherty AJ, Murphy G: The tissue metalloproteinase family and the inhibitor TIMP: A study using cDNAs and recombinant proteins. *Ann Rheum Dis* 1990;49:469-479.
38. Woessner JF Jr: Matrix metalloproteinases and their inhibitors in connective tissue remodeling. *FASEB J* 1991;5:2145-2154.
39. Dean DD, Martel-Pelletier J, Pelletier J-P, et al: Evidence for metalloproteinase and metalloproteinase inhibitor imbalance in human osteoarthritic cartilage. *J Clin Invest* 1989;84:678-685.
40. Lohmander LS, Hoerrner LA, Lark MW: Metalloproteinases, tissue inhibitor and proteoglycan fragments in knee synovial fluid in human osteoarthritis. *Arthritis Rheum* 1993;36:181-189.
41. Sandy JD, Flannery CR, Neame PJ, et al: The structure of aggrecan fragments in human synovial fluid: Evidence for the involvement in osteoarthritis of a novel proteinase which cleaves the Glu 373-Ala 374 bond of the interglobular domain. *J Clin Invest* 1992;89:1512-1516.
42. Fraser JR, Kimpton WG, Laurent TC, et al: Uptake and degradation of hyaluronan in lymphatic tissue. *Biochem J* 1988;256:153-158.
43. Tzaicos C, Fraser JR, Tsotsis E, et al: Inhibition of hyaluronan uptake in lymphatic tissue by chondroitin sulphate proteoglycan. *Biochem J* 1989;264:823-828.
44. Engström-Laurent A, Hellström S: The role of liver and kidneys in the removal of circulating hyaluronan: An experimental study in the rat. *Connect Tissue Res* 1990;24:219-224.
45. Smedsrød B, Melkko J, Risteli L, et al: Circulating C-terminal propeptide of type I procollagen is cleared mainly via the mannose receptor in liver endothelial cells. *Biochem J* 1990;271:345-346.
46. Eyre DR, Dickson IR, Van Ness K: Collagen cross-linking in human bone and articular cartilage: Age-related changes in the content of mature hydroxypyridinium residues. *Biochem J* 1988;252:495-500.
47. Seibel MJ, Duncan A, Robins SP: Urinary hydroxy-pyridinium crosslinks provide indices of cartilage and bone involvement in arthritic diseases. *J Rheumatol* 1989;16:964-970.
48. Altman R, Asch E, Bloch D, et al: Development of criteria for the classification and reporting of osteoarthritis: Classification of osteoarthritis of the knee. *Arthritis Rheum* 1986;29:1039-1049.
49. Brower AC: Use of the radiograph to measure the course of rheumatoid arthritis: The gold standard versus fool's gold. *Arthritis Rheum* 1990;33:316-324.
50. Altman RD, Fries JF, Bloch DA, et al: Radiographic assessment of progression in osteoarthritis. *Arthritis Rheum* 1987;30:1214-1225.
51. Kellgren JH, Lawrence JS: Radiological assessment of osteo-arthrosis. *Ann Rheum Dis* 1957;16:494-502.
52. Ahlbäck S: Osteoarthrosis of the knee: A radiographic investigation. *Acta Radiol Suppl (Diagn) (Stockh)* 1968;277:7-72.
53. Lysholm J, Hamberg P, Gillquist J: The correlation between osteoarthrosis as seen on radiographs and on arthroscopy. *Arthroscopy* 1987;3:161-165.
54. Brandt KD, Fife RS, Braunstein EM, et al: Radiographic grading of the severity of knee osteoarthritis: Relation of the Kellgren and Lawrence grade to a grade based on joint space narrowing, and correlation with arthroscopic evidence of articular cartilage degeneration. *Arthritis Rheum* 1991;34:1381-1386.

55. Hart DJ, Spector TD, Brown P, et al: Clinical signs of early osteoarthritis: Reproducibility and relation to x ray changes in 541 women in the general population. *Ann Rheum Dis* 1991;50:467-470.
56. Kirwan JR, Cushnaghan J, Dacre J, et al: Progression of joint space narrowing in knee osteoarthritis. *Arthritis Rheum* 1992;9(suppl):s134
57. Lequesne M, Winkler P, Rodriguez P: Joint space narrowing in primary osteoarthritis of the hip: Results of a three year controlled trial. *Arthritis Rheum* 1992;9(suppl):s135.
58. Dieppe P, Cushnaghan J, Kirwan J, et al: Bone scintigraphy predicts the outcome of knee osteoarthritis. *Arthritis Rheum* 1992;9(suppl):s323.
59. Chan WP, Lang P, Stevens MP, et al: Osteoarthritis of the knee: Comparison of radiography, CT, and MR imaging to assess extent and severity. *Am J Roentgenol* 1991;157:799-806.
60. Holgersson S, Brattström H, Mogensen B, et al: Arthroscopy of the hip in juvenile chronic arthritis. *J Pediatr Orthop* 1981;1:273-278.
61. Noyes FR, Stabler CL: A system for grading articular cartilage lesions at arthroscopy. *Am J Sports Med* 1989;17:505-513.
62. Lohmander LS, Lark MW, Dahlberg L, et al: Cartilage matrix metabolism in osteoarthritis: Markers in synovial fluid, serum and urine. *Clin Biochem* 1992;25:167-174.
63. Levick JR: Synovial fluid. Determinants of volume turnover and material concentration, in Kuettner KE, Schleyerbach R, Peyron J, et al (eds): *Articular Cartilage and Osteoarthritis*. New York, Raven Press, 1992, chap 37, pp 529-541.
64. Saxne T, Heinegård D: Synovial fluid analysis of two groups of proteoglycan epitopes distinguishes early and late cartilage lesions. *Arthritis Rheum* 1992;35:385-390.
65. Felson DT, Zhang Y, Anthony JM, et al: Weight loss reduces the risk for symptomatic knee osteoarthritis in women: The Framingham study. *Ann Intern Med* 1992;116:535-539.
66. Somerville K, Faulkner G, Langman M: Non-steroidal anti-inflammatory drugs and bleeding peptic ulcer. *Lancet* 1986;1:462-464.
67. Rashad S, Revell P, Hemingway A, et al: Effect of non-steroidal anti-inflammatory drugs on the course of osteoarthritis. *Lancet* 1989; 2:519-522.
68. Bradley JD, Brandt KD, Katz, BP, et al: Comparison of an antiinflammatory dose of ibuprofen, an analgesic dose of ibuprofen, and acetaminophen in the treatment of patients with osteoarthritis of the knee. *N Engl J Med* 1991;325:87-91.
69. Burkhardt D, Ghosh P: Laboratory evaluation of antiarthritic drugs as potential chondroprotective agents. *Semin Arthritis Rheum* 1987;17(suppl 1):3-34.
70. Altman RD, Dean DD, Muniz OE, et al: Prophylactic treatment of canine osteoarthritis with glycosaminoglycan polysulfuric acid ester. *Arthritis Rheum* 1989;32:759-766.
71. Dean DD, Muniz OE, Rodriquez I, et al: Amelioration of lapine osteoarthritis by treatment with glycosaminoglycan-peptide association complex (Rumalon). *Arthritis Rheum* 1991;34:304-313.
72. Moskowitz RW, Reese JH, Young RG, et al: The effects of Rumalon, a glycosaminoglycan peptide complex, in a partial meniscectomy model of osteoarthritis in rabbits. *J Rheumatol* 1991;18:205-209.
73. Ewing JW: Arthroscopic treatment of degenerative meniscal lesions and early degenerative arthritis of the knee, in Ewing JW (ed): *Articular Cartilage and Knee Joint Function: Basic Science and Arthroscopy*. New York, Raven Press, 1990, chap 9, pp 137-145.

74. Odenbring S, Egund N, Knutson K, et al: Revision after osteotomy for gonarthrosis: A 10 to 19-year follow-up of 314 cases. *Acta Orthop Scand* 1990;61:128-130.
75. Odenbring S, Egund N, Lindstrand A, et al: Cartilage regeneration after proximal tibial osteotomy for medial gonarthrosis: An arthroscopic, roentgenographic and histologic study. *Clin Orthop* 1992;277:210-216.
76. Bergenudd H, Johnell O, Redlund-Johnell I, et al: The articular cartilage after osteotomy for medial gonarthrosis: Biopsies after 2 years in 19 cases. *Acta Orthop Scand* 1992;63:413-416.

Chapter 9

Age-Related Changes in Articular Cartilage

Peter G. Bullough, MB, ChB
Frank U. Brauer, MD

Osteoarthritis is a disease of the diarthrodial joint, and not merely of articular cartilage. Changes in the synovium, cartilage, and bone are interactive and may be affected by changes in the neuromuscular apparatus, ligaments, and tendons. Changes in each of these structures occur with advancing age. The epidemiologic association between aging and osteoarthritis has led to the common impression that osteoarthritis results from the aging of joint tissues. Because the primary changes in this osteoarthritis are thought to occur in the articular cartilage, it is relevant to consider the evidence that normal articular cartilage "ages."[1]

Aging varies considerably from individual to individual depending on constitutional and occupational factors, conditions of stance and balance, and the continued functional activity of the joint. Pain, stiffness, and limited motion in a joint of an elderly person probably indicate the presence of changes exceeding those of senescence.[2] Although age-related cartilage degeneration presumably occurs in all joints, only a few joints regularly develop arthritis.

The function of a joint depends on many factors, including its shape, the material properties of its various constituent tissues, and neuromuscular control. It is doubtful, therefore, that there is any one etiology for degenerative joint disease. Rather, in any particular instance of osteoarthritis, it is likely that there are several contributing factors. The question to be answered by this symposium thus would seem to be whether senescence is a major contributing factor to osteoarthritis.

In 1934, Keefer and associates[3] pointed out that "knowledge concerning arthritis has been delayed by a lack of detailed information regarding the changes that may be encountered with advancing age." The following questions in that area were posed by Brandt and Fife[1] in 1986: "Do age-related changes arise on a biological basis? Is there a 'running down' of the cellular metabolic machinery with age? Does the chondrocyte become senescent, losing its capacity to maintain a normal extracellular matrix? Or do age-related changes in normal articular cartilage reflect the consequences of microtrauma to which the joint has been subjected during the lifetime of the individual?''

Altered anatomy and physiology can only be comprehended in the light of normal anatomy and physiology. Therefore, we will start with Hunter's[4] 1743 description of what the cartilage does. ''The articulating cartilages are most happily contrived to all purposes of motion in those parts. By their uniform surface, they move upon one another with ease; by their soft, smooth and slippery surface, mutual abrasion is prevented; by their flexibility, the contiguous surfaces are constantly adapted to each other, and the friction diffused equally over the whole; by their elasticity, the violence of any shock, which may happen in running, jumping, etc., is broken and gradually spent, which must have been extremely pernicious, if the hard surfaces of bones had been immediately contiguous.''

How does the cartilage provide for its unique viscoelastic properties, extreme slipperiness, and self-maintenance? In the simplest terms, it does so through the cellular control of its composition (the various collagens, water, proteoglycan, and other noncollagenous proteins) and through the organization (morphology) of these various constituents.

Cartilage Degeneration

Degeneration is a morbid change in the structure of parts, involving disintegration of tissue or the substitution of a lower for a higher form of structure.[5] Thus, degeneration may occur as a result of advancing age itself. This type of degeneration usually is called senescent change, and there is physiologic and anatomic evidence for this change in all tissues of the body.

However, degeneration may also proceed from causes other than senescence; these include disuse (atrophy) and abuse (injury). In describing the various degenerative changes observed, we should also attempt to identify the cause of each change, whether it is age, disuse, or abuse. If the cause is not apparent, which it often is not, that also must be stated.

Senescent Changes

Many authors have reported changes they thought to be senescent. Sokoloff[6] described these changes as ''time-related alterations in cartilaginous tissues that occur independently of structural disintegration.'' Freeman and Meachim[7] said that, ''Ideally the natural history of age-related changes should be observed by 'longitudinal' studies, in which the same individual is compared at different ages. Since this is not at present practicable, it is necessary to accept the limitations of 'transectional' ('transverse') studies, in which different individuals from the same general population are compared in samples taken from various ages." As far as possible, these need to be representative of the population as a whole.

Histologic changes in the joint and in the consistency of articular cartilage in the elderly, which were believed to be the result of senile changes, were first reported by Weichselbaum[8] and Heine.[9] Weichselbaum,[8] in 1877, described both the development of diffuse fibrillation on particular articular surfaces, especially the medial facets of the patellae, and a change in the color of the cartilage to a more yellowish one in older joints. Heine[9] confirmed these observations; however, he stated that the development of fibrillation, although related to old age, may not necessarily be the result of old age, in other words, not necessarily a senescent change. Both authors agreed that fibrillation could be a predispositional factor for the development of degenerative arthritis.

Unfortunately, few authors since Heine have attempted to differentiate degenerative changes associated with advancing age from degenerative changes directly caused by aging. Baseless assumptions have been made, which have resulted in considerable confusion. For example, in a study about changes in the knee joint at various ages, including specimens from the first to the tenth decades of life, Bennett and associates[10] reported alterations "on articular tissues in their transit from apparent normalcy through successive stages of a type-consistent change which evidently occurred in close parallelism to advancing age." They described the early lesions in articular cartilage as consisting of slight irregularities, small elevations, and shallow depressions in the surface from which fibrillated shreds of tissue have been partially detached. They reported that with time these lesions became ulcerated "into clefts and fissures, and large circular erosions with sharp edges approaching the epiphyseal bone." When this happens, the articular surface assumes a velvety appearance as a result of the loss of intercellular substances and the exposure of collagenous fibrils. The authors noted that some degeneration of the knee joint was observable in every subject beyond the age of fifteen. They felt that the earliest detectable lesions were located in areas that received the hardest use, for example, the patella and the exposed weightbearing portions of the femoral and tibial condyles.

In describing the histologic changes observed in the articular cartilage of subjects in the third decade, Bennett and associates[10] stated that the tibial condyles, in the majority of instances, showed somewhat greater abnormalities than were seen in other portions of these joints, and that these changes were confined to the weightbearing parts not covered by the menisci. They went on to say, "the single most constantly observed factor, and one upon which the overwhelming majority of all workers place emphasis, is the association of degenerative joint disease with advancing age." They then noted that because degenerative joint disease may be found in patients of 30 and not be present in a person of 80, it is inappropriate to consider degenerative joint disease merely an ailment of old age. Instead, it is related to the senescence of joint tissues, varying to a certain degree with the hereditary endowment of each individual.

After examination of a random series of synovial joints from 350 necropsies in the city of Liverpool, Freeman and Meachim[7] reported that, "with increasing age, histologically overt fibrillation tends to spread tangentially across the cartilage surface. This spread is mainly by circumferential and radial enlargement from initially peripheral foci, but new central foci can also develop. Susceptibility to tangential spread varies according to anatomical site, and between individuals. Thus, as already mentioned, in older persons much or all of the cartilage surface often becomes affected on sites such as the patella and on the bare area (i.e., not covered by meniscus) of the lateral tibial plateau, while in contrast, sites such as the superior aspect (zenith) of the femoral head quite often remain free from histologically overt fibrillation, even in the elderly . . . Thus, cartilage fibrillation is initially a focal change, so that it seems likely that local factors in the mechanical environment or in the original structure of cartilage are at least partly responsible for its development. Moreover, there is evidence that the local mechanical environment and local character of the cartilage influences the actual morphology of the cartilage lesions."

Sokoloff[11] has stated that age-related changes vary from joint to joint, and that although there is a roughly linear correlation of the degree of joint degeneration with age in the knee, this is not the case in the hip. Sokoloff believes that the findings are consistent with the possibility that, within a given individual, a mechanical factor may operate exponentially while the aging changes progress linearly.

Peyron[12] stressed that degenerative changes appear in the fourth decade, and their prevalence appears to increase slowly, arithmetically until around the sixth decade and rapidly, geometrically thereafter. This skew in the slope probably indicates that several factors could be implicated at that time of life.

In a report on the study of gross pathologic changes in the knee joints of 300 aged individuals, Casscells[13] hypothesized that "although there is nothing in this investigation to establish trauma or mechanical factors as the primary cause of degenerative arthritis, the combination of some mechanical insult to the cartilaginous surface which either releases or permits degradation enzymes to attack the cartilage seems to be the most attractive hypothesis. Certainly, some cartilaginous lesions break down relatively quickly, as evidenced by the fairly marked changes occasionally seen in the patellae of adolescents, while other lesions appear to be relatively stable such as those of Grade I and II in individuals of the seventh and eighth decades of life. The unpredictable manner in which cartilage reacts to injury is still unknown, but the once-gloomy prognosis given articular cartilage in the aged individual is not justified. A careful analysis of the data collected from these 300 specimens fails to reveal any support for the idea that aging of itself is responsible for the appearance of the degenerative lesions nor do they seem to be related to even moderate differences in the comparative heights and shape of the medial and lateral condyles and the depth of the patellar groove on the femur, provided the patella is centered anatomically on the femur.''

In reference to the earlier reports of Heine[9] and Weichselbaum,[8] Casscells[13] goes on, ''The notion that chondromalacia is a disabling, troublesome and almost universal disease affecting mostly older individuals is chiefly derived from clinical impressions of physicians who treat arthritis plus a few reports of European authors published some 50 years ago. The present investigation of 300 cadaver knees, whose average age was 70 years, demonstrates a much lower incidence of degenerative lesions of the articular cartilage than might be expected. There was minimal or no damage to the patella in 62% of the cases and an even lower incidence of 23% in the weight-bearing areas of the joint. Eighty-two percent of the menisci were essentially normal, as were 96% of the cruciates. In patients in the U.S.A., articular cartilage of the knee resists the wear and tear of a normal lifespan remarkably well and infrequently undergoes progressive degradation.''

A contrary view of the age-related changes reported by the authors discussed above, which has been based on studies of the elbow joint, hip joint, and knee joint, has been proposed.[14-16] All areas of cartilage that usually do not articulate with opposed cartilage were reported to show some degree of chondromalacia, characterized by cartilage softening and fibrillation.

How is this paradox to be explained? For many years, a precise fit or congruency was considered to be a normal feature of the joint. Today, it is generally accepted that most joints are, in fact, incongruent. In the knee joint, the gross incongruencies of the opposed surfaces are partially corrected by interposed fibrocartilaginous menisci. Incongruity provides a mechanism for access of the synovial fluid to the cartilage and for controlled loading of the articular surfaces. However, there is a trend toward increased congruency with age, making this protective mechanism less effective.[15-18] Humans cannot use their entire joint surfaces over the day. A person standing for eight hours per day probably spends about six hours (25% of the day) transmitting substantial load through one particular foot. In a sitting position, the value would be much less (5%), equivalent to a time of one to two hours. Some areas of an articular surface may never, in the course of a day, make contact with opposed articular cartilage. In the hip, it is normally the roof of the acetabulum that fails to make contact with the head. Such a noncontact area is typical of young joints. However, if incongruities are lost in the elderly, as seems to be the case, the area that had been a noncontact area would become a habitual contact area, with disastrous consequences for the cartilage on the femoral head with which it would then make constant contact.

Stockwell,[19] after many observations on aging cartilage, noted that specimens that are structurally intact, both macroscopically and histologically, are usually considered suitable for studies of aging as opposed to pathologic investigation. In this regard, Freeman and Meachim[7] noted that observations of articular cartilage in which the surface is intact have revealed only one group of significant age-related changes: the static tensile fracture strength, the static stiffness, and the tensile fatigue strength all decrease. These changes imply a change in the collagen fiber network, but not necessarily in the fibers themselves, because the total amount of collagen in the tissue does not decrease with age. Although the morphology of the collagen in murine and rabbit articular cartilage changes with age, no corresponding data are available for humans, and at present no morphologic nor chemical explanation for these mechanical changes can be given.

Brandt and Fife[1] also suggest that age-related changes in joint biomechanics may be of major importance in the pathogenesis of osteoarthritis. Kempson[20] found that the tensile properties of human articular cartilage from the femoral condyles of the knee deteriorated with increasing age from the middle of the third decade. Weightman[21] showed that the fatigue resistance of femoral head cartilage in tension decreased with age. He showed that the fatigue life of articular cartilage decreases with age to such an extent that fatigue failure in life may be a distinct possibility at physiologically possible stress levels. Much more work is needed in this area.

Another typical finding in old hyaline cartilage is the unmasking of collagen fibers, that is, a focal conversion of the hyaline matrix of cartilage into a characteristic fibrous structure. It has been suggested by Sokoloff[6] that "the nature of the asbestoid fibers is not known, neither their molecular species nor their pathogenesis. It is not known whether they are newly synthesized during the aging of the tissue, or whether they represent alteration of previously existing collagen."

Stockwell[19] found that in human articular cartilage, the cell content remains constant with age. A reduction in the cell density of the superficial zone described by Vignon and associates[22] for the femoral head was seen by Stockwell[19] to be offset by a slight increase in the deeper tissue.

The thickness of articular cartilage does not appear to change with age,[23] and, additionally, the zone of calcified cartilage remains more or less the same thickness throughout life.[24-26] Although this latter finding has been challenged,[27] these findings seem not only to be true within one joint, but for all joints of a single individual.[28,29] Data show a decrease in permeability from 10 years of age until the age of 30 to 40 years.[30] In older adults, nonfibrillated samples from the knee show a variable increase in permeability, which is more marked in the superficial than the deeper layer of the cartilage.[30]

A change in the color of articular cartilage with age has been described by many authors.[8,9,19] Stockwell[19] has noted that "Aged cartilage has a yellow-brown tinge compared with the pallor of the youthful tissue. In human cartilage, the color is prominent in the intervertebral disk, the central zone of costal cartilage, in the menisci and in the basal zone of articular cartilage. In human costal cartilage, the amount of pigment is appreciable from the fourth decade. The brown colorization is, in general, distributed inversely to the degree of tissue basophilia (seen on H&E stained sections) in costal and articular cartilage.''

Hass[31] found that neither lipid nor iron compounds contribute to the pigment. Van der Korst and associates[32] reported that the pigment is not associated with the glycosaminoglycan fraction of the matrix. Instead, it appears to be part of or attached to noncollagenous proteins, which contain large amounts of acidic amino acid residues. Because proteins of this type may be involved in the collagen-proteoglycan interaction, it is possible that pigmentation may reflect an increasing redundancy of this type of protein as proteoglycan concentration falls and thicker collagen fibers form in the matrix.[19,32]

In 1969, Meachim[30] noted that aging of chondrocytes can be considered in terms of the whole cell population, of changes in an individual chondrocyte, and of changes that affect an individual cytoplasmic organelle. He went on to say that ''it is difficult to build up a comprehensive picture of the influence of age on the ultrastructure of adult human articular chondrocytes, but detailed studies of cellular changes with age have been made in mice and rabbits. In rabbits, fine intracytoplasmic filaments are more numerous in older animals, and similar filaments are seen in adult articular cartilage from the human knee. Silberberg and associates[33] depict murine chondrocytes as showing an 'ascending phase' characterized by progressive development of intracytoplasmic structures and a 'descending grade' leading to degeneration of organelles and eventual disintegration of the cell. In rabbit and human articular cartilage, one pattern of cell degeneration is a disintegration of the chondrocyte to form organelle and cytomembrane remnants, rounded membranous bodies, and granular material.''[30]

The presence of empty lacunae in the tangential layer of articular cartilage was thought to be common in the elderly, indicating cell atrophy and finally cell necrosis. However, articular cartilage does not routinely undergo degeneration in the older individual because, as mentioned above, no dramatic change in cell density has been found.

It has been reported that chondrocyte proteoglycan synthesis changes with age. A decline in the ability of older chondrocytes to synthesize and assemble matrix molecules has been suggested,[34] but was refuted by Sokoloff.[35] However, a study of transplanted chondrocytes suggested that older chondrocytes may have an increased catabolic activity.[36]

Some authors have asked whether increasing age may decrease the replicative abilities of older chondrocytes.[37] Contrary to this view, in vitro studies of aged animals revealed that old articular cartilage doesn't lose its regenerative capacity.[38]

In a study of regenerating hyaline cartilage in articular defects of old chickens, Robinson and associates[39] reported that "the utilization of implants containing embryonal chondrocytes to correct defects in articular cartilage is feasible in young animals as well as in aged ones. This view contrasts sharply with other works in which a complete lack of regenerative power in the cartilage of aged animals was found. Furthermore, it appears that the reparative tissue formed by the implants matures at a faster rate in old animals. The newly formed cartilage in the 3-year-old chickens grows faster and occupies larger spaces within the bone than in 4-month-old roosters, and the endochondral ossification in the deep regions occurs earlier and progresses more rapidly."

In the field of proteoglycan biochemistry, some changes in aging cartilage have been reported, which could be relevant to its proneness to develop osteoarthritis. It has been suggested that proteoglycans become smaller with age.[40] Peyron[12] has reported that some small proteoglycans may be not found in cartilage specimens from subjects older than 40 years of age.

Rosenberg[41] has recently published his findings on dermatan sulfate proteoglycans in articular cartilage. These substances probably play an important protective role, preventing the formation of adhesions between the joint surfaces. Rosenberg and associates[42] found that aging bovine articular cartilage contains over ten times more dermatan sulfate proteoglycan, relative to the major, cartilage-specific, chondroitin sulfate proteoglycans, than fetal epiphyseal cartilage. This probably has an important inhibitory effect on repair.

Stockwell[19] reported that both extracellular lipid and the incidence of calcification increase with age. It has been suggested that lipidic material may accumulate as a result of cell necrosis.

The slow progression of calcification in aging cartilage has been associated with remodeling processes taking place at the osteochondral junction.[43] As with modeling at the growth plate in growing individuals, endochondral ossification in the subchondral region occurs through the process of vascular invasion and, in this regard, it is of interest that the vessels in the calcified cartilage have been seen to decrease with age until the seventh decade of life, after which their number increases. More vessels were always described for the more loaded areas of the joints.[44]

Anatomic changes in joint shape with advancing age were first described by Ogston[18] in 1878, as changes at the ends of senile bones, which he suggested were the result of remodeling. He noted that the general surfaces of senile joints are flatter and not so full and plump in their curves; the grooves of articulation with opposing bones are deeper and more marked. This view has been further expanded.[45-47] The remodeling activity is not constant but increases after the sixth decade of life, which may lead to thinning of the zone of calcified cartilage and the gross appearance of bony outgrowths.[44,48] Capillary invasion at the tidemark as well as tidemark duplications are a common histologic finding at the osteochondral junction of elderly joints.

Using white rabbits of different age groups, Walker and associates[49] showed a very rapid movement of the zone of calcified cartilage toward the articular surface in young animals. However, in the older animals they could measure a consistent movement of greater than 2 μm per week. According to Walker and associates,[49] this would represent the transverse of one zone of calcified cartilage thickness/year or a loss of about 8% to 10% of the articular cartilage height in the adult animals and thus, even in the absence of pathology, must be considered a persistent movement of the zone of calcified cartilage. This also has been reported for human articular cartilage, giving rise to the morphologic picture that the shape of the joint in an elderly person differs from that seen in a younger person. The main differences between osteoarthritis and aging are shown in Table 1.

Changes Resulting From Disuse

To date, little attempt has been made to differentiate degenerative changes due to disuse or abuse. The common view has been that cartilage degeneration occurs because of excessive loading. Recently, however, several investigators have taken the view that in some areas that show softening and fibrillation, the changes result from underuse rather than overuse. Just as unused bone and unused muscle atrophy, so may unused cartilage. One way in which cartilage atrophy is manifested may be as a reduction of proteoglycan production by the chondrocyte.

Additionally, hypokinetics, or decreased activity, is a frequent characteristic of the elderly. As Goldstein[51] noted, "Whether originally caused by an injury, a fall, or increased fatigue, an elderly person begins to 'slow down'. A vicious cycle ensues that affects all systems of the body and eventually results in loss of function and possible injury."

Degenerative changes believed to result from disuse in human articular cartilage were first mentioned by Heine[9] in 1926. He described these changes in the cartilage as being well-delineated, round, and often longish groove-like defects in the cartilage covered by a more or less thick, translucent fibrous layer. Heine[9] believed that these changes were the result of resorptive events that may occur independent of age. He suggested that these lesions were independent of the changes seen in degenerative arthritis because they are never observed together.

From his descriptions, it appears that many of the affected subjects in his series were suffering from paraplegia and that, at least in some cases, he believed the resorption to result from overlying synovium.[9] We have observed similar resorptive changes in hammer toes in the area of the superior exposed cartilage of the proximal phalanx where there is overlying synovium.

Table 1 Main differences between osteoarthritis and aging

Osteoarthritis	Aging
Highly anabolic and synthetic process	Normal metabolism
Enzymatic destruction of hard tissue	Normal enzymatic remodeling
Remodeling all tissues about joint, articular and periarticular	Cartilage changes only
Chondrocyte mitosis	No mitosis
Intense increased synthesis, collagen and proteoglycan	Normal rates synthesis, collagen and proteoglycan
Increased water content cartilage	No change
Fibrillation, focal and progressive at weightbearing sites	Fibrillation nonprogressive, nonweightbearing sites
Eburnation, ivory-like	No eburnation
Osteophytes occur with other changes	Osteophytes only with excessive use
No increased collagen X link	Increased collagen X link
Inflammation	No inflammation
No pigment—cartilage	Pigment—cartilage

(Adapted with permission from Bland JH, Cooper SM: Osteoarthritis: A review of the cell biology involved and evidence for reversibility: Management rationally related to known genesis and pathophysiology. *Semin Arthritis Rheum* 1984;14:106-133.)

This suggestion that changes may result from disuse as well as overuse was largely ignored, and all observed degenerative changes continued to be regarded mainly as the result of overuse or abuse.

However, in 1953, Harrison and associates[14] suggested that lack of pressure is an even more compelling cause of cartilage degeneration. They were surprised to find that daily use preserves rather than wears out articular cartilage and that inadequate use is the commonest cause of cartilage degeneration and ensuing vascular invasion. In their study, every femoral head in subjects from 14 years or over showed changes of fibrillation and loss of proteoglycans around the rim or uncovered area of the femoral head. They believed that inadequate use led to vascular invasion through the calcified cartilage reverting the appearance of the osteophytic cartilage to that seen during the growing period. In their view, "the osteoarthritic process thus appears to be an attempt to transform a decaying joint into a youthful one and for this, as in the miraculous rejuvenation depicted in Goethe's Faust, a high price must ultimately be paid."[14]

The findings of Harrison and associates[14] were supported in a 1967 study of the pattern of aging of articular cartilage of the elbow joint.[15] In that study, fibrillation was related primarily to joint mechanics. There were greater changes in multiaxial joints (for example, the radiohumeral joint) than in uniaxial joint surfaces (for example, humeroulnar joint). Moreover, areas of cartilage that do not usually articulate with opposed cartilage always show some degree of chondromalacia, and chondromalacia of the rim of the head of the radius was present even in the youngest of their subjects. In a 1973 study of human hip joints, common areas for fibrillation in all age groups were reported to be "a band of cartilage extending downwards and backwards from the fovea to the inferior articular margin, . . . the periphery of the articular surface (of the femoral head), anteriorly and above, but not posteriorly, . . . and one area, roughly triangular in outline, on the roof of the acetabulum with its base at the superior lip.''[16]

In 1977, a study of load distribution in the knee joint and its relation to observed patterns of degeneration was reported in which the authors described cartilage destruction as a common finding in areas that were highly loaded, as well as in areas that were loaded only slightly or not at all.[52] In their view, the exposed cartilage, that is, the cartilage not covered by the meniscus, degenerated because it was underused, whereas very specifically defined areas under the posterior horn of the lateral menisci degenerate because they are overstressed in certain movements.

A 1985 report of microscopic and biochemical studies of the tibial plateau in the dog suggested that disuse resulted in an irregular surface with rounded cells and an increased concentration of water per unit volume.[53] The collagen in the uncovered area appeared in wavy, aggregated bundles with thinner mean diameters, and binding of proteoglycan was ill-defined. An increased amount of proteoglycans could be extracted, and the tidemark was smooth.[53]

The authors of this study described a different morphologic and biochemical feature for the covered and presumable load-bearing area.[53] There the surface was smooth, with an amorphous electron-dense layer, and cells were flattened. There was an increased intracellular accumulation of lipid in all three layers; there was an increased extracellular accumulation at the surface; and there were increased numbers of extracellular matrix vesicles in the deep zone. Collagen appeared in randomly oriented fibers with thicker mean diameters; there was regular binding of proteoglycan, the concentration of proteoglycan per wet weight was increased, and the tidemark was irregular. The finding that the tidemark was smooth in the exposed area and irregular in the covered area may indicate that the calcification front in the covered area is more active, and, in consequence, there is more active modeling of the subchondral bone. The increased numbers of extracellular lipid-bound matrix vesicles observed in the deep region of the covered area were associated with the presence of calcium-acidic phospholipid-phosphate complexes, and are also an indication of active mineralization and further evidence for increased modeling in this region.

The report of a study on dogs with articular cartilage atrophy, which was induced by the immobilization of the ipsilateral knee in an orthopaedic cast, illustrates the consequences that may arise when previously unloaded cartilage is required to bear weight.[54] The atrophic changes were rapidly and completely reversed when the dogs were permitted to walk in pens following removal of the casts; however, this reversal did not occur when the dogs were subjected to mild treadmill exercise that produced greater compressive stress on the atrophic cartilage. Negative effects of disuse on the tensile modulus of dog articular cartilage also have been demonstrated following four weeks of immobilization.[55] In vitro studies showed that the chondrocyte responds very rapidly to changes in compressive loading by modulating its level of proteoglycan synthesis.[56]

Thus, there is evidence to suggest that there is a window of physiologic stress and that continued optimal functional integrity depends on balanced rates of matrix production and breakdown by the cells. Low levels of mechanical stress (that is, below the physiologic range) are associated with enhanced catabolic activity in cartilage; this, in turn, is associated with increased water content and superficial fibrillation of collagen. Stress within the physiologic range is associated with increased anabolic activity; the articular cartilage is healthy. The result of supraphysiologic stress (that is, the range of stress is exceeded) is cell injury and eventual necrosis (Fig. 1).

Evidence has been provided that the menisci contribute significantly to the transmission of load.[52] On the basis that load is equal to stress times the area of contact, the load-bearing could be considered to follow a pattern in which there is more load on the menisci at first, and then an increasing proportion of the load falls on the exposed cartilage as the load is applied. At a load of twice body weight, the exposed cartilage on the medial side takes about the same amount of load as the meniscus, whereas on the lateral side, most of the load is still carried by the meniscus.[52]

In a finite element analysis, Carter and associates[57] suggested the role that cyclically applied stresses might play in the degenerative process of the hip joint. They showed that areas of high joint contact pressures on the femoral head correspond to high hydrostatic compression in subchondral bone. The magnitude of the subchondral bone compressive hydrostatic stress correlated with cartilage thickness and was highest in the superior femoral head and moderate at the acetabular roof. The seldom contacting surfaces of the medial-inferior and peripheral areas of the femoral head and the roof of the acetabulum had lower hydrostatic compression and significant subchondral bone tensile strains tangential to the joint surface. Initial cartilage fibrillation and osteophyte formation are often found in these areas. These findings led to the suggestion that fluctuating hydrostatic pressure inhibits vascular invasion and the degeneration and ossification of articular cartilage.[57] The generation of tensile strain may promote the degenerative process by direct mechanical mechanisms and fibrillation may occur.

Abuse of the Joint Cartilage

Abuse may affect cartilage either by a short, sudden increase in the load applied to the joint (trauma) or by the continuous process of overloading over a longer period of time (overuse). Several epidemiologic studies have reported an association between musculoskeletal overuse and the prevalence of osteoarthritis. However, as Bagge and associates[58] stated, "most of those studies were on a large scale and often screened unselected populations in an uncontrolled way. Somewhat different results were obtained when more carefully de-

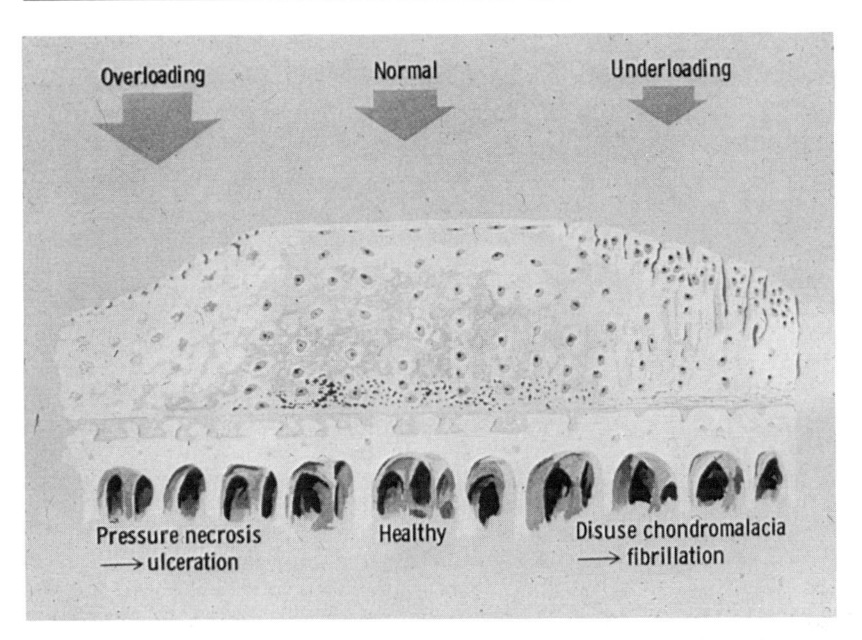

Fig. 1 *Diagram illustrating the effect of load on the health of cartilage. Excess load produces cell death and hence matrix necrosis; too little load results in decreased synthesis of proteoglycans and hence collapse of the matrix. (Reproduced with permission from Teitelbaum SL, Bullough PG: The pathophysiology of bone and joint disease.* Am J Pathol *1979;96:335)*

fined groups were examined to evaluate the consequences of musculoskeletal overuse. No association between heavy loading (occupation and/or physical activity) was then reported. Marathon runners, for instance, did not have an abnormally high prevalence of [osteoarthritis] OA.'' The results of a study of 79-year-old people from Goteborg did not indicate that a heavy occupational history or spare time physical activity preceded the development of osteoarthritis.

Walker and Bernick[59] raised the question of the relationship between natural aging of joint tissues and exercise. They stated, ''Exercise such as jogging, and often aerobic exercise routines, subjects lower limb joints to repetitive rapid loading and unloading. Such exercise is often performed on non-ideal surfaces, such as pavement, non-resilient wooden floors or even concrete floors where minimal shock absorption exists, thus increasing the potential for repetitive microtrauma to joint tissues, in particular the load-bearing surface, articular cartilage. The long-term effects of such exercise await the passage of time to determine whether these effects are beneficial or harmful to joint tissues, if any effect exists . . . Few investigators have examined the interaction between exercise and aging of articular tissues; even fewer researchers have reported studies in which exercise was performed for the majority of the animal's life span. It is not established whether exercise can prevent, retard or reverse any of the progressive age-related changes reported to commence between the second and third decades in humans and different animal species . . . Microtrauma (defined as undetected specific events) during the life span may theoretically result in eventual wear-and-tear changes to articular tissues, and accelerate cartilage destruction. Such microtraumas are most likely to result from high impact loading and unloading, activities such as jogging, running, skipping and the jumping-type exercises which are often part of aerobic exercise programs.''

In 1986, Walker[60] reported that in rats, endurance exercise, in the form of treadmill running, neither appeared to retard the gradual change reported to occur with age in all species, nor, more importantly, retarded the development of defects in the cartilage. In response to the question of Williams and associates[61] as to whether treadmill exercise contributed to the severity of lesions by introducing microtrauma, Walker[60] reported that the more extensive lesions (defects), were significantly more frequent in exercised animals.

Walker[60] concluded that his result supports the conclusion that in rats low intensity treadmill exercise is associated with both the severity and frequency of cartilage lesions. Theoretically, weightbearing exercise may accelerate physiologic changes with age in cartilage by introducing microtrauma, which increases the likelihood of minor defects.

Although the above data appear to implicate exercise as having a harmful effect on cartilage, others have suggested that exercise can have beneficial effects on joint cartilage and tissues. Williams and Brandt[62] have demonstrated that daily treadmill exercise in guinea pigs may have some protective effects on cartilage.

It has been hypothesized that exercise and increased activity levels could minimize the effects of aging, disuse, and disease in the elderly human population.[51] However, continuous heavy use may lead to fatigue failure, for example, stress fractures of the tibia in dancers. Fatigue fractures in cartilage,[63] horizontal splits,[64] minute crevices in the calcified cartilage,[6] and subchondral microfractures[65] have been reported as a common feature in normal, aging joints. Their relationship to the onset of degenerative changes is still unclear. Repetitive impulse loading, leading to an increased number of microfractures in the subchondral bone, could cause stiffening of the subchondral bone followed by damage to the overlying articular cartilage.[66]

A limitation of motion, resulting in excessive and continuous loading of one area, may result in necrosis by interfering with nutrition. We have observed ulcers, which could be attributed to this cause, in the central part of the first metatarsophalangeal joint of patients with hallux rigidus.

In his 1926 study, Heine[9] described morphologic changes in articular cartilage that he thought were caused by overuse and trauma. He described changes caused by overuse as grayish, translucent, circular lesions that are well-defined around the surrounding white or bluish-white cartilage. Such lesions were present in the central part of the acetabulum, in the first phalanx of the first metatarsophalangeal joint, and in the olecranon of the elbow joint. He concluded that these lesions could not be the only and direct cause for osteoarthritis, because they can be seen in early life as well as in old age without any deformities in the affected joint. However, there is no proof that the areas regarded as being overused were in fact overused.[9] For example, the parts of the olecranon and acetabulum which Heine regarded as being altered because of overuse were, in fact, the foci of conjoining of the developmentally separate portions of the articulation.

In 1934, Keefer and associates[3] concluded from their studies that age-related degenerative changes were commonest over the areas of contact subjected to the greatest movement. Bennett and associates[10] believed that the earliest detectable lesions are located in areas that bear the brunt of physiologic use, that is, the patella and the exposed weight-bearing portions of the femoral and tibial condyles. In describing the histologic changes observed in the articular cartilage of subjects in the third decade, they state, "the tibial condyles, in the majority of instances, showed somewhat greater abnormalities than were seen in other portions of these joints . . . . These changes were confined entirely to the weight-bearing parts not covered by the menisci."[10]

However, the authors did not provide information regarding the contact areas on the opposed articular surfaces of the knee under loaded conditions, nor were data available concerning the type, extent, or duration of loading in the different compartments of the joint. Their assumption that the uncovered parts of the tibial plateaus were weightbearing probably was based on the prevailing view that osteoarthritis was a disease of wear and tear. As already discussed in the section on disuse, the changes they described may have resulted from underuse rather than overuse.

Direct injury to the joint or to the cartilage will in most cases result in arthritis. The injury may be mechanical, as in cruciate ligament rupture or transarticular fracture, or it may be a chemical injury. Injury in living beings and tissue results in a process referred to as repair, the object of which is the restoration of a dynamic equilibrium. The process in vascularized tissues is achieved by means of the inflammatory response and by the process of regeneration; in avascular tissues, such as cartilage, it is achieved by regeneration alone.[67]

In osteoarthritis, many of the observed microscopic alterations are the result of reparative phenomena. Heine[9] recognized that small defects in young, healthy cartilage may heal entirely, whereas in old cartilage these defects could lead to severe degenerative changes. Insufficient attention has been given to the processes of cartilage regeneration and, ultimately, to the rebuilding of the joint which, although it is never as efficient as the repair of the axolotl's tail, does occur.

Summary

The changes in articular cartilage in osteoarthritis are markedly different from those of age. However, several important questions[1] remain unanswered:

Do changes of aging, acting additively or synergistically with other etiologic factors, predispose to the development of osteoarthritis? Or, despite the strong association between age and osteoarthritis, does the disease develop independent of aging? Might osteoarthritis occur mainly in older people merely because of a long, latent period between the onset of the disease and the appearance of clinical features? If so, could aging influence the duration of the latent period?

Arthritis cannot be explained without an understanding of joint anatomy and physiology, and until comparatively recently, both of these disciplines were poorly understood. In particular, the maintenance of joint shape and tissue integrity are vital to a functional joint. Neither of these phenomena has received much attention from prior students of arthritis.

The process of degeneration may proceed from senescence, disuse, or abuse. We must learn to recognize the difference.

Finally, when interpreting the histologic changes observed in arthritic joints, we must learn to recognize those that are caused by injury and those that are the result of the reparative process.

References

1. Brandt KD, Fife RS: Ageing in relation to the pathogenesis of osteoarthritis. *Clin Rheum Dis* 1986;12:117-130.
2. Jaffe HL: Degenerative joint diseases, in *Metabolic, Degenerative, and Inflammatory Diseases of Bones and Joints*. Philadelphia, Lea & Febiger, 1972, pp 735-778.
3. Keefer CS, Parker F Jr, Myers WK, et al: Relationship between anatomic changes in knee joint with advancing age and degenerative arthritis. *Arch Intern Med* 1934;53:325-344.
4. Hunter W: Of the structure and diseases of articulating cartilages. *Phil Trans* 1743, pp 267-271.
5. *The Compact Edition of the Oxford English Dictionary*. Oxford, UK, Clarendon Press, 1971.
6. Sokoloff L: Aging and degenerative diseases affecting cartilage, in Hall BK (ed): *Cartilage*. New York, Academic Press, 1983, vol 3, chap 4, pp 109-141.
7. Freeman MAR, Meachim G: Aging and degeneration, in Freeman MAR (ed): *Adult Articular Cartilage*, ed 2. Kent, UK, Pitman Medical, 1979, chap 9, pp 487-543.
8. Weichselbaum A: Die senilen Veränderungen der Gelenke und deren Zusammenhang mit der Arthritis deformans. Sitzungsber d Akad d Wiss Wien, Mathemat. - Naturw. Kl. III, Abt. 4, 75, 1877, pp 193-243.
9. Heine J: Arthritis deformans. *Virchows Arch F Pathol Anat* 1926;260:521-663.
10. Bennett GA, Waine H, Bauer W: Changes in the knee joint at various ages with particular reference to the nature and development of degenerative joint disease. New York. *The Commonwealth Fund*, 1942.

11. Sokoloff L: Loading and motion in relation to ageing and degeneration of joints: Implications for prevention and treatment of osteoarthritis, in Helminen HJ, Kiviranta I, Säämänen AM, et al (eds): *Joint Loading, Biology and Health of Articular Structures*. Kent, UK, Butterworth & Co, 1987, chap 17, pp 412-424.
12. Peyron JG: Risk factors in osteoarthritis: How do they work? *J Rheumatol* 1987;14(Spec. No):1-2.
13. Casscells SW: Gross pathological changes in the knee joint of the aged individual: A study of 300 cases. *Clin Orthop* 1978;132:225-232.
14. Harrison MHM, Schajowicz F, Trueta J: Osteoarthritis of the hip: A study of the nature and evolution of the disease. *J Bone Joint Surg* 1953;35B:598-626.
15. Goodfellow JW, Bullough PG: The pattern of ageing of the articular cartilage of the elbow joint. *J Bone Joint Surg* 1967;49B:175-181.
16. Bullough P, Goodfellow J, O'Connor J: The relationship between degenerative changes and load-bearing in the human hip. *J Bone Joint Surg* 1973;55B:746-758.
17. Ogston A: On articular cartilage. *J Anat Physiol* 1875;10:49-74.
18. Ogston A: On the growth and maintenance of the articular ends of adult bones. *J Anat Physiol* 1878;12:503-517.
19. Stockwell RA: Cartilage degeneration, calcification and chondrocyte death, in Stockwell RA (ed): *Biology of Cartilage Cells*. Cambridge, UK, Cambridge University Press, 1979, chap 8, pp 241-265.
20. Kempson GE: Relationship between the tensile properties of articular cartilage from the human knee and age. *Ann Rheum Dis* 1982; 41:508-511.
21. Weightman B: In vitro fatigue testing of articular cartilage. *Ann Rheum Dis* 1975;34(Suppl 2):108-110.
22. Vignon E, Arlot M, Patricot LM, et al: The cell density of human femoral head cartilage. *Clin Orthop* 1976;121:303-308.
23. Meachim G: Effect of age on the thickness of adult articular cartilage at the shoulder joint. *Ann Rheum Dis* 1971;30:43-46.
24. Green WT Jr, Martin GN, Eanes ED, et al: Microradiographic study of the calcified layer of articular cartilage. *Arch Pathol* 1970;90:151-158.
25. Meachim G, Stockwell RA: The matrix, in Freeman MAR (ed): *Adult Articular Cartilage*. Kent, UK, Pitman Medical, 1979, chap 1, pp 1-68.
26. Haynes DW: The mineralization front of articular cartilage. *Metab Bone Dis Rel Res* 1980;2-S:55-59.
27. Lane LB, Bullough PG: Age-related changes in the thickness of the calcified zone and the number of tidemarks in adult human articular cartilage. *J Bone Joint Surg* 1980;62-B:372-375.
28. Müller-Gerbl M, Schulte E, Putz R: The thickness of the calcified layer of articular cartilage: A function of the load supported? *J Anat* 1987;154:103-111.
29. Müller-Gerbl M, Schulte E, Putz R: The thickness of the calcified layer in different joints of a single individual. *Acta Morphol Neerl Scand* 1987;25:41-49.
30. Meachim G: Age changes in articular cartilage. *Clin Orthop* 1969; 64:33-44.
31. Hass GM: Studies of cartilage: A morphologic and chemical analysis of aging human costal cartilage. *Arch Pathol* 1943;35:275-284.
32. Van der Korst JK, Skoloff L, Miller EJ: Senescent pigmentation of cartilage and degenerative joint disease. *Arch Pathol* 1968;86:40-47.
33. Silberberg R, Silberberg M, Feir D: Life cycle of articular cartilage cells: An electron microscope study of the hip joint of the mouse. *Am J Anat* 1964;114:17-47.

34. Buckwalter JA, Mow VC: Cartilage repair in osteoarthritis, in Moskowitz RW, Howell DS, Goldberg V, et al (eds): *Osteoarthritis, Diagnosis and Medical/Surgical Management.* Philadelphia, WB Saunders, 1992, chap 4, pp 71-107.
35. Sokoloff L: The pathology of osteoarthrosis and the role of aging, in Nuki G (ed): *Aetiopathogenesis of Osteoarthrosis.* Kent, UK, Pitman Medical, 1980, pp 1-15.
36. Evans CH, Georgescu HI, Mazzocchi RA: Does cellular ageing of chondrocytes engender primary osteoarthritis? *Trans Orthop Res Soc* 1981;6:153.
37. Dustmann HO, Puhl W: Altersabhängige Heilungsmöglichkeiten von Knorpelwunden. *Hefte Unfallheilkd* 1977;129:259-264.
38. Hough AJ Jr, Webber RJ: Aging phenomena and osteoarthritis: Cause or coincidence? Claude P. Brown memorial lecture. *Ann Clin Lab Sci* 1986;16:502-510.
39. Robinson D, Halperin N, Nevo Z: Regenerating hyaline cartilage in articular defects of old chickens using implants of embryonal chick chondrocytes embedded in a new natural delivery substance. *Calcif Tissue Int* 1990;46:246-253.
40. Thonar EJ-MA, Bjornsson S, Kuettner KE: Age-related changes in cartilage proteoglycans, in Kuettner KE, Schleyerbach R, Hascall VC (eds): *Articular Cartilage Biochemistry.* New York, Raven Press, 1986, pp 273-287.
41. Rosenberg LC: Structure and function of dermatan sulfate proteoglycans in articular cartilage, in Kuettner KE, Schleyerbach R, Peyron JG, et al (eds): *Articular Cartilage and Osteoarthritis.* New York, Raven Press, 1992, pp 45-62.
42. Rosenberg LG, Choi H, Johnson T, et al: Structural changes in proteoglycans in aging articular cartilages, in Peyron JG (ed): *Osteoarthritis: Current Clinical and Fundamental Problems.* Paris, Ciba-Geigy, 1985, pp 179-191.
43. Bullough PG, Jagannath A: The morphology of the calcification front in articular cartilage: Its significance in joint function. *J Bone Joint Surg* 1983;65A:72-78.
44. Lane LB, Villacin A, Bullough PG: The vascularity and remodelling of subchondral bone and calcified cartilage in adult human femoral and humeral heads: An age- and stress-related phenomenon. *J Bone Joint Surg* 1977;59B:272-278.
45. Johnson LC: Morphologic analysis in pathology: The kinetics of disease and general biology of bone, in Frost HM (ed): *Bone Biodynamics.* Boston, Little, Brown and Company, 1964, pp 543-654.
46. Johnson LC: Structural dynamics of osteoarthritis. *J Rheumatol* 1983;10(suppl 9):22-24.
47. Bullough PG: The geometry of diarthrodial joints, its physiologic maintenance, and the possible significance of age-related changes in geometry-to-load distribution and the development of osteoarthritis. *Clin Orthop* 1981;156:61-66.
48. Nakata K, Bullough PG: The injury and repair of human articular cartilage: A morphological study of 192 cases of coxarthrosis. *J Jpn Orthop Assoc* 1986;60:763-775.
49. Walker G, Carpenter RJ, Oegema TR Jr, et al: Evidence for activity in the tidemark in normal articular cartilage. *Trans Orthop Res Soc* 1990;15:182.
50. Bland JH, Cooper SM: Osteoarthritis: A review of the cell biology involved and evidence for reversibility. Management rationally related to known genesis and pathophysiology. *Semin Arthritis Rheum* 1984;14:106-133.

51. Goldstein TS: The aging human, in Goldstein TS (ed): *Geriatric Orthopaedics: Rehabilitative Management of Common Problems*. Gaithersburg, MD, Aspen Publishing, 1991, chap 1, pp 1-12.
52. Bullough PG, Walker PS: The distribution of load through the knee joint and its possible significance to the observed patterns of articular cartilage breakdown. *Bull Hosp Jt Dis Orthop Inst* 1976;7:110-123.
53. Bullough PG, Yawitz PS, Tafra L, et al: Topographical variations in the morphology and biochemistry of adult canine tibial plateau articular cartilage. *J Orthop Res* 1985;3:1-16.
54. Palmoski MJ, Brandt KD: Running inhibits the reversal of atrophic changes in canine knee cartilage after removal of a leg cast. *Arthritis Rheum* 1981;24:1329-1337.
55. Setton LA, Zimmerman JR, Mow VC, et al: Effects of disuse on the tensile properties and composition of canine knee joint cartilage. *Trans Orthop Res Soc* 1990;15:155.
56. Palmoski MJ, Brandt KD: Effects of static and cyclic compressive loading on articular cartilage plugs in vitro. *Arthritis Rheum* 1984;27:675-681.
57. Carter DR, Rapperport DJ, Fyhrie DP, et al: Relation of coxarthrosis to stresses and morphogenesis, a finite element analysis. *Acta Orthop Scand* 1987;58:611-619.
58. Bagge E, Bjelle A, Eden S, et al: Factors associated with radiographic osteoarthritis: Results from the population study of 70-year-old people in Goteborg. *J Rheumatol* 1991;18:1218-1222.
59. Walker JM, Bernick S: Natural ageing and exercise effects on joints, in Helminen J, Kiviranta I, Säämänen AM, et al (eds): *Joint loading: Biology and Health of Articular Structures*. Kent, UK, Butterworth & Co, 1987, chap 4, pp 89-111.
60. Walker JM: Exercise and its influence on aging in rat knee joints. *J Orthop Sp Phys Ther* 1986;8:310-319.
61. Williams JM, Felten DL, Peterson RG, et al: Effects of surgically induced instability on rat knee articular cartilage. *J Anat* 1982;134(Part 1):103-109.
62. Williams JM, Brandt KD: Exercise increases osteophyte formation and diminishes fibrillation following chemically induced articular cartilage injury. *J Anat* 1984;139(Part 4):599-611.
63. Minns RJ, Steven FS: The collagen fibril organization in human articular cartilage. *J Anat* 1977;123:437-457.
64. Meachim G, Bentley G: Horizontal splitting in patellar articular cartilage. *Arthritis Rheum* 1978;21:669-674.
65. Blackburn J, Hodgskinson R, Currey JD, et al: Mechanical properties of microcallus in human cancellous bone. *J Orthop Res* 1992; 10:237-246.
66. Radin EL, Rose RM: Role of subchondral bone in the initiation and progression of cartilage damage. *Clin Orthop* 1986;213:34-40.
67. Bullough PG: Principles of cell injury, inflammation and repair, in Cruess RL, Williams RRJ (eds): *Adult Orthopaedics*. New York, Churchill Livingston, 1984, vol 1, chap 2, pp 25-46.

Chapter 10

Age-Related Changes in Cartilage: Physical Properties and Cellular Response to Loading

Alan J. Grodzinsky, ScD

Introduction

Articular cartilage functions as a weightbearing wear-resistant material in synovial joints. The ability of cartilage to withstand compressive, tensile, and shear forces during joint loading depends critically on the structural integrity of its extracellular matrix, which is composed of collagens, proteoglycans, and other matrix proteins.[1] While the collagen fibril network is strong in tension, proteoglycans help resist compression due to their bulk compressive stiffness and to electrostatic repulsive interactions between glycosaminoglycan chains.[1-6] The maintenance of adequate concentrations of matrix components and the preservation of a structurally and functionally intact matrix requires a dynamic balance between the synthesis, assembly, degradation, and loss of matrix molecules (Fig. 1). The regulation of these metabolic processes in vivo may involve a combination of cell biological and physical mechanisms. In aging and osteoarthritis, this metabolic balance is modified to varying degrees.

During the past decades, investigators have documented age-related changes in the cellularity and cell morphology of cartilage, the quality and biosynthetic rates of matrix macromolecules, and the biochemical composition of the matrix. Slowly evolving changes in the matrix may also be related to the presence of degradative enzymes derived from chondrocytes or synovium. These modified biosynthetic and degradative processes could lead to irreversible remodeling of the matrix, and concomitant age-related changes in cartilage hydration, fixed charge density, biomechanical and fatigue properties, and in the ability of cartilage to respond to injury.

Although many studies have addressed changes in the physical properties of human and animal cartilages associated with osteoarthritis, relatively fewer studies have focused on changes with age in otherwise normal tissue. Certain age-related changes in the biomechanical, electromechanical, and physiochemical properties of car-

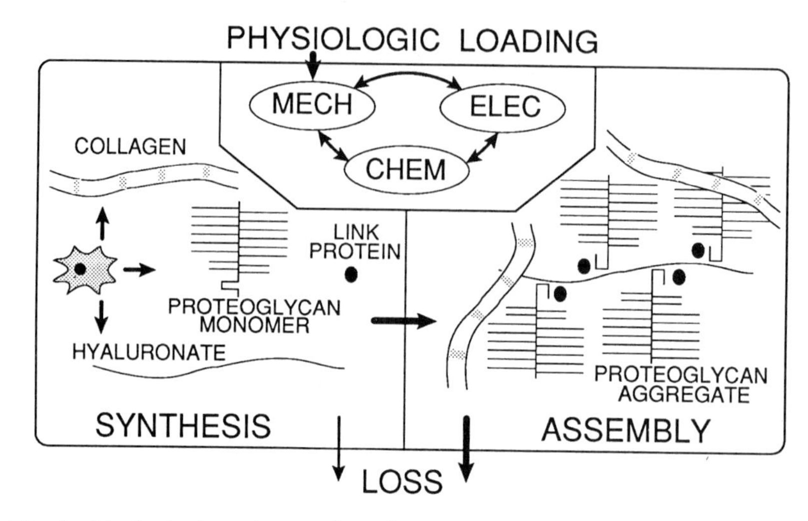

Fig. 1 *Biophysical regulation of cartilage metabolism during mechanical loading. Chondrocytes synthesize matrix components including the structurally important collagens, proteoglycans, hyaluronate, and link protein.These components assemble into a stable extracellular matrix; subsequent catabolic processes cause the release of certain components from the tissue. Alternatively, some biosynthetic products may never become incorporated into the matrix, but instead are immediately lost from the tissue. These metabolic processes may be influenced by mechanical,chemical, or electrical phenomena that are induced by physiological levels of loading. (Reproduced with permission from Sah RL, Grodzinsky AJ, Plaas AHK, et al: Effects of static and dynamic compression on matrix metabolism in cartilage explants, in Kuettner KE, Schleyerbach R, Peyron JG, et al (eds):* Articular Cartilage and Osteoarthritis. *New York, Raven Press, 1992, pp 373-392.)*

tilage have been identified, and are summarized below. These altered physical properties, in turn, may have important consequences on chondrocyte-mediated synthesis, assembly, and degradation of matrix components. Indeed, clinical observations have long suggested that physical forces in the environment of the chondrocytes are important modulators of cartilage matrix metabolism.[7] Recent supporting evidence from experiments with animal models and in vitro systems has demonstrated that chondrocyte biosynthesis and degradation of matrix can be inhibited or stimulated by specific physical forces in the local environment of the cells (Fig. 2).[7,8] Therefore, the ability of cells to respond to normal loading or sustained exercise may change with age as a result of alterations in the physical properties of pericellular, territorial, or interterritorial matrix. Thus, the potential for fatigue failure of cartilage may involve feedback between (a) the micromechanical properties of the matrix and (b) the ability of chondrocytes to modify their matrix metabolic response in order to adapt to changes in joint loading patterns with age.

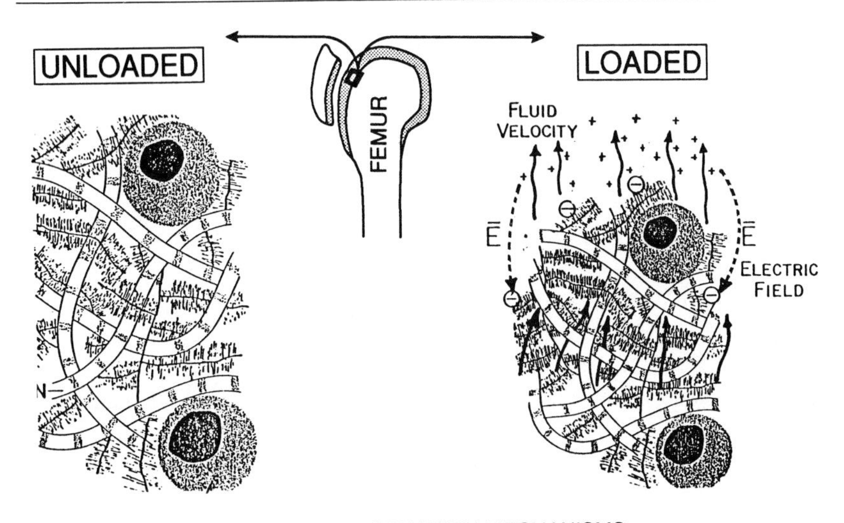

Fig. 2 *Schematic joint loading showing physical phenomena: Chondrocyte deformation, hydrostatic pressurization, fluid flow, electric fields (streaming potentials and currents), matrix consolidation, and physicochemical alterations (altered ion concentrations and osmotic pressure). These forces and flows may modulate chondrocyte metabolism. (Reproduced with permission from Sah RL, Grodzinsky AJ, Plaas AHK, et al: Effects of static and dynamic compression on matrix metabolism in cartilage explants, in Kuettner KE, Schleyerbach R, Peyron JG, et al (eds):* Articular Cartilage and Osteoarthritis. *New York, Raven Press, 1992, pp 373-392. and Gray ML, Pizzanelli AM, Grodzinsky AJ, et al: Mechanical and physicochemical determinants of the chondrocyte biosynthetic response.* J Orthop Res *1988;6:777-792.)*

Physicochemical Properties and Matrix Composition

Using human femoral head cartilage obtained at the time of surgery and at autopsy, investigators have quantified age-related changes in the water content, fixed charge density (associated with matrix glycosaminoglycans), collagen content, thickness and other metabolic parameters in intact tissue.[9-12] While site-specific variations were found, full-thickness cartilage generally exhibited a decrease in water content with age and an increase in fixed charge density,[9,11] provided surface fibrillation was absent.[11] In specimens ranging in age from 3 to 86 years, water content decreased linearly with age;[9] thickness decreased rapidly during the early years, but showed little change after maturity.[9,10] Collagen content on a dry weight basis also decreased significantly with age,[9] mainly as a result of a change in the proportions of other components. Interestingly, changes in the water content and fixed charge density of osteoarthritic cartilage were opposite to those found in normal aging tissue; when equilibrated in saline, fibrillated osteoarthritic cartilage contained much more water and had lower fixed charge density.[11,12]

Tensile Properties

Relatively few studies have focused on age-related changes in the biomechanical properties of human cartilage. Tensile studies have included measurements of tensile stiffness,[2,10,13,14] tensile strength,[2,10,13,14] and fatigue failure.[15] In all these studies, 200 μm-thick dumbbell-shaped specimens oriented along the direction of the superficial collagen fibrils were prepared from various sites along a joint surface and at several depths within the tissue. With specimens subjected to a constant extension rate of 5 mm/min, tensile stiffness was calculated as the slope of stress versus strain curves at specific values of stress[2,13,14] or strain.[10] (Strain was typically measured using photographic or video methods, while stress was calculated as the measured load normalized to the original cross-sectional area;[2,10] tensile strength was reported as the stress at fracture of the specimen.)[2,10]

The tensile strength of knee cartilage from the superficial zone of the patellar groove and condyle was observed to reach a maximum in the third decade and decrease markedly with age thereafter; the strength of specimens from the deep zone decreased continuously with age from 8 to 91 years.[13] Cartilage taken from the femoral head (superficial and middle zones, 19 to 80 year olds) showed a significant decrease in tensile strength with age at both depths;[10] tensile stiffness, however, did not change significantly with age. Kempson[14] recently reported a marked decrease in fracture stress with age in superficial and middle zone specimens from the femoral head, consistent with earlier studies.[10] By comparison, there was no significant change with age in specimens from the talus of the ankle joint.[14] Tensile stiffness at 5 and 10 MPa stress was found to decrease continuously with age in femoral head[14] and knee[2,13] cartilage from middle and deep zones, while little change was found in talar cartilage. Together, these findings were offered as a possible explanation for the observation that osteoarthritis (OA) occurs in the hip and knee at an increasing incidence with age, while OA rarely occurs in the ankle unless there is trauma.[14]

In a study on fatigue failure,[15] similar dumbbell specimens were subjected to repeated applications of a predetermined constant stress until failure. Loading was performed at 3 cycles per minute, each cycle lasting 1 second, and the number of cycles needed to produce failure was recorded;[15] the amplitude of the applied stress was also varied. The fatigue resistance of specimens from the superficial layer of postmortem human femoral heads ranging in age from 20 to 81 years decreased significantly with age.[15] However, fatigue correlated poorly with the collagen and proteoglycan (PG) content of the tissue (with collagen analyzed from half of each tested specimen and PG from immediately adjacent tissue). It was noted that there was a wide variation in fatigue resistance from joint to joint.

In order to interpret the above results on tensile properties, investigators have considered age-related changes in the chemical composition and quality of matrix, and changes in the density and types of crosslinkages in the collagen network at maturity and with increasing age.[2,10,13,14] Although many researchers believe that changes in the collagen mesh make the greatest contribution to the observed changes in tensile strength and stiffness, a definitive understanding of this relationship based on current knowledge of matrix biology and morphology has not yet been achieved. Some age-related changes in animal cartilages have also been studied to gain further insight. For example, the tensile strength of middle and deep zone mature bovine knee cartilage was found to be significantly less than that of immature (open-physis) cartilage, but there was no difference between mature and immature tissue from the superficial zone.[16] In general, more research is needed to understand the relationship between age-related changes in the observed macroscopic tensile properties of cartilage, and changes in the molecular-level structure of the extracellular matrix.

Compressive Properties

As with tensile properties, there have been only a few studies of age-related changes in the compressive properties of human cartilage. These studies have included measurement of (1) the short-term creep indentation behavior of cartilage on the intact femoral head[2,17,18] and of cartilage-bone plugs 10 mm in diameter cored from the femoral head,[10] (2) the equilibrium modulus and hydraulic permeability of lateral facet patellar cartilage using cartilage-bone plugs tested in uniaxial confined compression,[19] and (3) the compliance of cartilage of the femoral head loaded in its natural acetabulum in the intact hip joint.[17,18,20]

From the results of measurements of the creep indentation modulus of femoral head cartilage calculated 2 seconds after application of load, Kempson[2] commented that no significant relationship between compressive properties and age had been found. This is consistent with the data of Roberts and associates,[10] who found no significant change in compressive stiffness (2 second creep modulus) with increasing age, using plugs cored from the zenith and anteroinferior sites on the femoral head of 6 to 90 year olds. Armstrong and Mow[19] used the cartilage biphasic theory to calculate the equilibrium modulus and hydraulic permeability of human knee cartilage from long-term (10,000 seconds) creep compression data. The equilibrium modulus reflects the compressive properties of the extracellular matrix after fluid flow has essentially ceased; in contrast, the 2-second (short-term) modulus can reflect additional stiffness effects as a result of interstitial fluid pressurization and fluid-solid frictional stresses caused by fluid flow after initiation of a creep experiment. There was only a marginal decrease in the equilibrium confined compression modulus with age, and the hydraulic permeability did not change significantly with age.[19] Interestingly, there was a significant decrease in modulus and an increase in permeability with increasing water content.

Armstrong and associates[17,18,20] measured the compressive deformation of cartilage covering the human femoral head when the femoral head was loaded in its natural acetabulum in the intact hip, and when the cartilage itself was compressed by a small indentor. First, the intact joint was subjected to a load five times body weight, and the reduction in cartilage thickness during the period 10 to 45 seconds after load application was assessed radiographically. The joint was then disarticulated, and an indentation test was performed directly on the femoral head cartilage surface. Deformations were recorded 1 second after load application, representing short-term response of the cartilage. The direct indentation response did not change significantly with age, while the deformation of the intact joint was substantially greater in older joints.[18] Armstrong and associates[18] cautioned that the complexities associated with interpreting the intact joint experiment made it difficult to associate the observed age-related changes solely with the compressive stiffness of the cartilage itself.

Thus, the limited evidence currently available suggests that the compressive properties of cartilage do not change significantly with age. This finding is not intuitively consistent with the physicochemical observations of decreased water content and increased charge density found in tissue with increasing age. These latter observations, in isolation, might suggest an increased compressive stiffness associated with the highly charged proteoglycans of the matrix. However, such an argument neglects the contribution of the collagen network in compression and is not consistent with direct measurements that show little or no change in compressive properties with age. It is also interesting to note that investigators have identified significant age-related changes in the quality, structure, synthesis and degradation rates of proteoglycans, collagens, and other matrix macromolecules.[21,22] It is somewhat surprising that such changes do not give rise to associated changes in compressive stiffness. Perhaps the macroscopic compressive properties and methodologies used in the above studies do not fully reflect, or are less sensitive to, the molecular-level changes in matrix constituents that occur with age.

These results suggest the need for further development of micromechanical methods (experimental and theoretical) to quantify matrix physical properties at microstructural levels, ultimately at cellular and molecular length scales. For example, several investigators have begun to correlate the compressive and tensile behavior of thin strips of cartilage with changes in matrix and cell morphology assessed by light and electron microscopy.[23-25] Theoretical micromodels have been developed to account for the effect of fibril orientation on the anisotropy in cartilage equilibrium stiffness,[26] the contribution of electrostatic repulsive interactions between individual glycosaminoglycan (GAG) chains to cartilage swelling pressure and compressive modulus,[27] and the electromechanical transduction behavior of cartilage matrix associated with fluid flow at the level of the GAG chain.[28] More work is needed in this area to better understand age-related changes in matrix structure-function relationships at the molecular level.

Cellular Response to Physical Forces

The physical properties of the extracellular matrix will ultimately determine how joint loading forces are translated into cell-level stresses and strains that may regulate chondrocyte metabolism. In comparing age-related changes found in femoral head versus talar cartilage (described above), Kempson[14] suggested that the mechanical forces experienced by chondrocytes in hip and knee cartilage may change more with age than forces experienced by ankle cartilage. As a result, chondrocyte matrix metabolism in hip and knee cartilage was hypothesized to change more with age, consistent with observed changes in physical properties of the tissue.[14]

In support of this hypothesis, many clinical observations and studies in vivo have shown that joint loading and motion can induce a wide range of metabolic responses in cartilage. Immobilization or reduced loading led to a decrease in proteoglycan synthesis and content.[29,30] Increased dynamic loading led to an increase in proteoglycan synthesis and content.[29,30] More severe static[31] or impact[32] loading caused cartilage deterioration and can lead to osteoarthritic changes.[33] Anterior cruciate ligament transection in dogs led to increases in both synthesis and release of extracellular matrix constituents progressing to osteoarthritis.[34,35] Thus, while some degree of "normal" joint loading appears to promote structural adaptation, "abnormal" mechanical forces predispose cartilage to degeneration. The physical and biologic mechanisms responsible for these alterations are not fully understood and are difficult to identify in vivo. Thus, many recent studies on the metabolic effects of mechanical compression have utilized in vitro cartilage explant organ culture systems.[36-47]

A number of physical phenomena that occur naturally during loading have been identified and quantified (Fig. 2). Compression of cartilage results in deformation of cells and extracellular matrix,[25] hydrostatic pressure gradients, fluid flow, streaming potentials and currents,[48-50] and physicochemical changes including altered matrix water content, fixed charge density, mobile ion concentrations, and osmotic pressure.[4,39,40] A quantitative understanding of the distribution of these forces and flows that exist within the tissue during dynamic compression has evolved over the past decade.[48-53] Any of these mechanical, chemical, or electrical phenomena may modulate matrix metabolism (Fig. 1).

The results of studies on the effects of mechanical compression on matrix metabolism in cartilage explants can be summarized as follows: (1) static compression consistently inhibited synthesis of matrix macromolecules,[36,37,39,40,42] and delayed the rate at which newly synthesized proteoglycans assemble into functional aggregates;[43] (2) above a threshold frequency, a range of dynamic compression regimes could stimulate matrix biosynthesis;[8,42,45,47] (3) under certain conditions, high enough amplitude cyclic compression could lead to accelerated loss of matrix macromolecules and disruption of the collagen network.[44]

Thus, static and dynamic compression has been found to markedly alter chondrocyte and matrix metabolism in cartilage explants. These results mimic many clinical observations and animal studies in which moderate dynamic mechanical loading was found to modulate cartilage matrix metabolism and promote structural adaptation, while "abnormal" mechanical forces predispose cartilage to degeneration, as is seen in osteoarthritis. Physicochemical mechanisms[8,39-41,43] may play an important role in transducing static compression into inhibition of biosynthesis and delayed extracellular processing. Dynamic compression in the 0.01 to 1 cycle/second frequency range may stimulate biosynthesis by generating appropriate hydrostatic pressure,[46] high enough fluid velocities or streaming potentials, or by the modest deformation of cells and matrix that would occur at the low compression amplitudes used.[42] It is interesting that studies of continuous passive motion in vivo[54] have used a frequency of 0.025 cycle/second, which is within the range of biosynthetic stimulation observed in vitro.[42]

The regulation of chondrocyte metabolic response by physical forces may be critically important to understanding the origin of certain age-related changes in cartilage.[14] However, to our knowledge, there have been no studies on the effects of physical forces on the metabolism of human cartilage as a function of age. From studies published thus far, it is clear that the inhibitory effects of static compression on biosynthesis have been observed in immature (eg, newborn calf)[40,42] as well as mature cartilage (adult dog;[37] adult human).[39] Similarly, the stimulatory effects of dynamic loading have been observed in immature[42] and mature[47] bovine cartilage. The effects of compression on matrix assembly have thus far only been examined in immature tissue.[43]

Finally, the role of cell-matrix interactions in mediating the effects of compression on chondrocyte metabolism has recently been studied using a chondrocyte/agarose gel culture system.[55,56] Buschmann and associates[55] observed that bovine chondrocytes in agarose culture synthesized and deposited a mechanically functional matrix with time in culture. At later times in culture, when significant matrix had accumulated (Fig. 3, *top*), static compression of chondrocyte/agarose disk specimens caused an inhibition of proteoglycan synthesis (Fig. 3, *bottom*), which was qualitatively similar to that seen in intact cartilage disks.[42] However, no significant

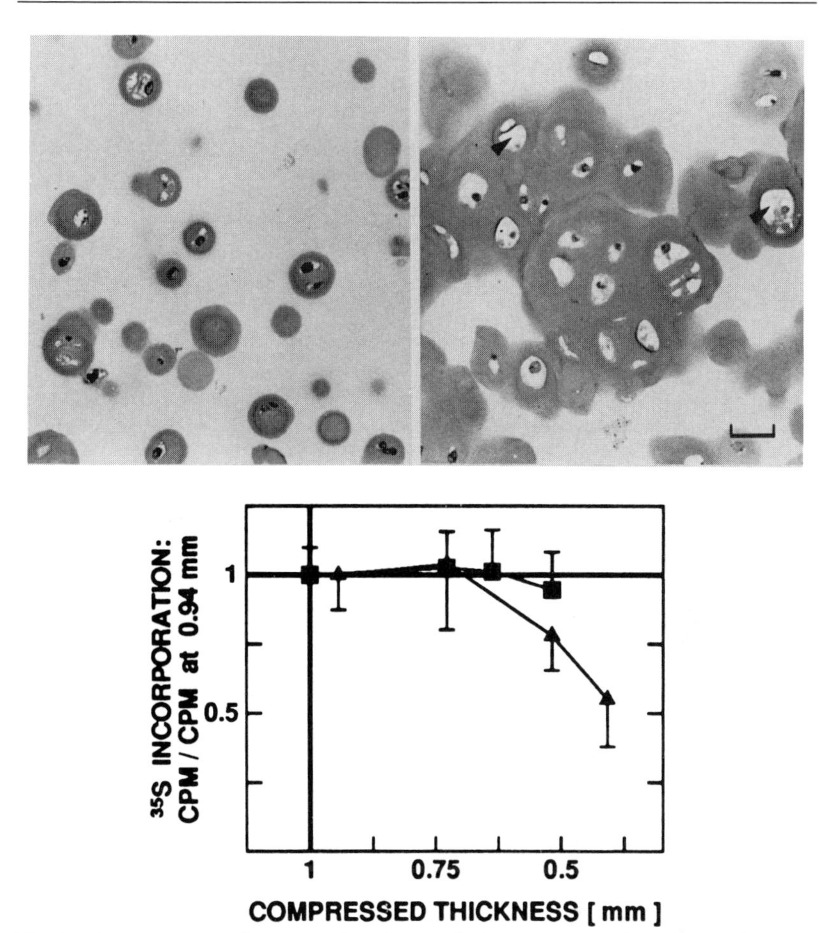

Fig. 3 *Matrix accumulation in chondrocyte-laden agarose gels and correlation with effect of static compression on biosynthesis.* ***Top,*** *Light micrographs of chondrocytes cultured in 2% (w/v) agarose for nine (left) and 36 (right) days. Thick (1 μm) section stained with Toluidine Blue O.* ***Bottom,*** *Effect of static 12-h compression on proteoglycan synthesis in agarose cultures at day two (■) and day 41 (▲). In the former, there was no significant difference in ^{35}S-sulfate incorporation between control and compressed chondrocyte/agarose disks, whereas in the latter, static compression caused a significant inhibition of ^{35}S-sulfate incorporation similar to that observed in cartilage explants. (Reproduced with permission from Buschmann MD, Gluzband YA, Grodzinsky AJ, et al: Chondrocytes in agarose gel synthesize a mechanically functional extracellular matrix.* J Orthop Res *1992;10:745-758, and Buschmann MD, Gluzband YA, Grodzinsky AJ, et al: Mechanical compression modulates matrix biosynthesis in chondrocyte/agarose gel culture.* Trans Orthop Res Soc *1991;1:75.)*

change in biosynthesis was seen when the same static compression was applied at early times in culture before matrix had been deposited (Fig. 3, *bottom*), even though the cells were observed to be similarly deformed.[56] This observation is an example of the hypothesized relationship between the physical properties of the matrix, the physical microenvironment of the cells, and the transduction of applied forces into cellular responses. It is also clear from Fig. 3 that the matrix distribution even at later times is very nonuniform. Thus, micromechanical models and measurements would be very useful in further quantifying the mechanisms underlying this response.

Acknowledgements

Supported by NIH Grant AR33236, Veterans Administration Grant V525P-1742, and NSF Grant BCS-9111401.

References

1. Buckwalter J, Rosenberg L, Coutts R, et al: Articular cartilage: Injury and repair, in Woo SLY, Buckwalter JA (eds): *Injury and Repair of the Musculoskeletal Soft Tissues*. Park Ridge, IL, American Academy of Orthopaedic Surgeons, 1988, pp 465-482.
2. Kempson GE: The mechanical properties of articular cartilage, in Sokoloff L (ed): *The Joints and Synovial Fluid*. New York, Academic Press, 1980, vol 2, pp 177-238.
3. Maroudas A: Physico-chemical properties of articular cartilage, in Freeman MAR (ed): *Adult Articular Cartilage*, ed 2. Kent, England, Pitman Medical, 1979, pp 215-290.
4. Grodzinsky AJ: Electromechanical and physicochemical properties of connective tissue. *Crit Rev Biomed Eng* 1983;9:133-199.
5. Broom ND, Silyn-Roberts H: Collagen-collagen versus collagen-proteoglycan interactions in the determination of cartilage strength. *Arthritis Rheum* 1990;33:1512-1517.
6. Mow VC, Ratcliffe A, Poole AR: Cartilage and diarthrodial joints as paradigms for hierarchical materials and structures. *Biomaterials* 1992;13:67-97.
7. Helminen HJ, Kiviranta I, Tammi M, et al (eds): *Joint Loading: Biology and Health of Articular Structures*. Bristol, John Wright, 1987.
8. Sah RL-Y, Grodzinsky AJ, Plaas AHK, et al: Effects of static and dynamic compression on matrix metabolism in cartilage explants, in Kuettner KE, Schleyerbach R, Peyron JG, et al (eds): *Articular Cartilage and Osteoarthritis*. New York, Raven Press, 1992, pp 373-392.
9. Venn MF: Variation of chemical composition with age in human femoral head cartilage. *Ann Rheum Dis* 1978;37:168-174.
10. Roberts S, Weightman B, Urban J, et al: Mechanical and biochemical properties of human articular cartilage in osteoarthritic femoral heads and in autopsy specimens. *J Bone Joint Surg* 1986;68B:278-288.
11. Grushko G, Schneiderman R, Maroudas A: Some biochemical and biophysical parameters for the study of the pathogenesis of osteoarthritis: A comparison between the processes of ageing and degeneration in human hip cartilage. *Connect Tissue Res* 1989; 19:149-176.
12. Maroudas A, Schneiderman R, Weinberg C, et al: Choice of specimens in comparative studies involving human femoral head cartilage, in Maroudas A, Kuettner KE (eds): *Methods in Cartilage Research*. New York, Academic Press, 1990, pp 9-17.
13. Kempson GE: Relationship between the tensile properties of articular cartilage from the human knee and age. *Ann Rheum Dis* 1982; 41:509-511.

14. Kempson GE: Age-related changes in the tensile properties of human articular cartilage: A comparative study between the femoral head of the hip joint and the talus of the ankle joint. *Biochim Biophys Acta* 1991;1075:223-230.
15. Weightman B: Tensile fatigue of human articular cartilage. *J Biomech* 1976;9:193-200.
16. Roth V, Mow VC: The intrinsic tensile behavior of the matrix of bovine articular cartilage and its variation with age. *J Bone Joint Surg* 1980;62A:1102-1117.
17. Gardner DL, Elliott RJ, Armstrong CG, et al: The relationship between age, thickness, surface structure, compliance and composition of human femoral head articular cartilage, in Nuki G (ed): *The Aetiopathogenesis of Osteoarthrosis*. Tunbridge Wells, UK, Pitman Medical Publishing, 1980, pp 65-83.
18. Armstrong CG, Bahrani AS, Gardner DL: Changes in the deformational behavior of human hip cartilage with age. *J Biomech Eng* 1980; 102:214-220.
19. Armstrong CG, Mow VC: Variations in the intrinsic mechanical properties of human articular cartilage with age, degeneration and water content. *J Bone Joint Surg* 1982;64A:88-94.
20. Armstrong CG, Bahrani AS, Gardner DL: In vitro measurement of articular cartilage deformations in the intact human hip joint under load. *J Bone Joint Surg* 1979;61A:744-755.
21. Hardingham T, Bayliss M: Proteoglycans of articular cartilage: Changes in aging and in joint disease. *Semin Arthritis Rheum* 1990;20S:12-33.
22. Eyre DR, Dickson IR, Van Ness K: Collagen cross-linking in human bone and articular cartilage: Age-related changes in the content of mature hydroxypyridinium residues. *Biochem J* 1988;252:495-500.
23. Broom ND, Myers DB: A study of the structural response of wet hyaline cartilage to various loading situations. *Connect Tissue Res* 1980;7:227-237.
24. O'Connor P, Orford CR, Gardner DL: Differential response to compressive loads of zones of canine hyaline articular cartilage: Micromechanical light and electron microscopic studies. *Ann Rheum Dis* 1988;47:414-420.
25. Poole CA, Flint MHY, Beaumont BW: Morphological and functional interrelationships of articular cartilage matrices. *J Anat* 1984; 138:113-138.
26. Farquhar T, Dawson PR, Torzilli PA: A microstructural model for the anisotropic drained stiffness of articular cartilage. *Biomech Eng* 1990;113:414-425.
27. Buschmann MD, Gluzband YA, Grodzinsky AJ, et al: Proteoglycan associated electrostatic forces and the development of functional mechanical properties in chondrocyte/agarose gel cultures. *Trans Orthop Res Soc* 1993;18:74.
28. Eisenberg SR, Grodzinsky AJ: Electrokinetic micromodel of extracellular matrix and other polyelectrolyte networks. *Physiocochem Hydrodynamics* 1988;10:517-539.
29. Caterson B, Lowther DA: Changes in the metabolism of the proteoglycans from sheep articular cartilage in response to mechanical stress. *Biochim Bioiphys Acta* 1978;540:412-422.
30. Kiviranta I, Jurvfelin J, Tammi M, et al: Weight bearing controls glycosaminoglycan concentration and articular cartilage thickness in the knee joints of young beagle dogs. *Arthritis Rheum* 1987;30:801-809.

31. Gritzka TL, Fry LR, Cheesman RL, et al: Deterioration of articular cartilage caused by continuous compression in a moving rabbit joint: A light and electron microscopic study. *J Bone Joint Surg* 19713;55A:1698-1720.
32. Radin EL, Martin RB, Burr DB, et al: Effects of mechanical loading on the tissues of the rabbit knee. *J Orthop Res* 1984;2:221-234.
33. Thompson RC Jr, Oegema TR Jr, Lewis JL, et al: Osteoarthrotic changes after acute transarticular load: An animal model. *J Bone Joint Surg* 1991;73A:990-1001.
34. Brandt KD: Transection of the anterior cruciate ligament in the dog: A model of osteoarthritis. *Semin Arthritis Rheum* 1991;21:22-32.
35. Muir IHM: Current and future trends in articular cartilage research and osteoarthritis, in Kuettner K, Schleyerbach R, Hascal VC (eds): *Articular Cartilage Biochemistry*. New York, Raven Press, 1986, pp 423-440.
36. Jones IL, Klamfeldt DDS, Sandstrom T: The effect of continuous mechanical pressure upon the turnover of articular cartilage proteoglycans in vitro. *Clin Orthop* 1982;165:283-289.
37. Palmoski MJ, Brandt KD: Effects of static and cyclic compressive loading on articular cartilage plugs in vitro. *Arthritis Rheum* 1984;27:675-681.
38. van Kampen GP, Veldhuijzen JP, Kuijer R, et al: Cartilage response to mechanical force in high-density chondrocyte cultures. *Arthritis Rheum* 1985;28:419-424.
39. Schneiderman R, Keret D, Maroudas A: Effects of mechanical and osmotic pressure on the rate of glycosaminoglycan synthesis in the human adult femoral head cartilage: An in vitro study. *J Orthop Res* 1986;4:393-408.
40. Gray ML, Pizzanelli AM, Grodzinsky AJ, et al: Mechanical and physiochemical determinants of the chondrocyte biosynthetic response. *J Orthop Res* 1988;6:777-792.
41. Urban JP, Bayliss MT: Regulation of proteoglycan synthesis rate in cartilage in vitro: Influence of extracellular ionic composition. *Biochim Biophys Acta* 1989;992:59-65.
42. Sah RL, Kim YJ, Doong JY, et al: Biosynthetic response of cartilage explants to dynamic compression. *J Orthop Res* 1989;7:619-636.
43. Sah RL, Grodzinsky AJ, Plaas AH, et al: Effects of tissue compression on the hyaluronate-binding properties of newly synthesized proteoglycans in cartilage explants. *Biochem J* 1990;267:803-808.
44. Sah RL, Doong JY, Grodzinsky AJ, et al: Effects of compression on the loss of newly synthesized proteoglycans and proteins from cartilage explants. *Arch Biochem Biophys* 1991;286:20-29.
45. Larsson T, Aspden RM, Heinegard D: Effects of mechanical load on cartilage matrix biosynthesis in vitro. *Matrix* 1991;11:388-394.
46. Hall AC, Urban JP, Gehl KA: The effects of hydrostatic pressure on matrix synthesis in articular cartilage. *J Orthop Res* 1991;9:1-10.
47. Parkkinen JJ, Lammi MJ, Helminen HJ, et al: Local stimulation of proteoglycan synthesis in articular cartilage explants by dynamic compression in vitro. *J Orthop Res* 1992;10:610-620.
48. Mow VC, Holmes MH, Lai WM: Fluid transport and mechanical properties of articular cartilage: A review. *J Biomech* 1984;17:377-394.
49. Mak AF: Unconfined compression of hydrated viscoelastic tissues: A biophasic poroviscoelastic analysis. *Biorheology* 1986;23:371-383.
50. Frank EH, Grodzinsky AJ: Cartilage electromechanics-I: Electrokinetic transduction and the effects of electrolyte pH and ionic strength. *J Biomech* 1987;20:615-627.

51. Frank EH, Grodzinsky AJ: Cartilage electromechanics-II: A continuum model of cartilage electrokinetic and correlation with experiments. *J Biomech* 1987;20:629-639.
52. Eisenberg SR, Grodzinsky AJ: The kinetics of chemically induced nonequilibrium swelling of articular cartilage and corneal stroma. *J Biomech Eng* 1987;109:79-89.
53. Lai WM, Hou JS, Mow VC: A triphasic theory for the swelling and deformation behaviors of articular cartilage. *J Biomech Eng* 1991;113:245-258.
54. Salter RB, Simmonds DF, Malcolm BW, et al: The biological effect of continuous passive motion on the healing of full-thickness defects in articular cartilage: An experimental investigation in the rabbit. *J Bone Joint Surg* 1980;62A:1232-1251.
55. Buschmann MD, Gluzband YA, Grodzinsky AJ, et al: Chondrocytes in agarose culture synthesize a mechanically functional extracellular matrix. *J Orthop Res* 1992;10:745-758.
56. Buschmann MD, Gluzband YA, Grodzinsky AJ, et al: Mechanical compression modulates matrix biosynthesis in chondrocyte/agarose gel culture. *Trans Comb Orthop Res Soc USA, Japan, Canada* 1991;1:75.

Chapter 11

Articular Cartilage Matrix Changes With Aging

Peter J. Roughley, PhD

The structure of the articular cartilage extracellular matrix changes considerably with age, from its development in the fetus to its final form in the mature adult. These changes not only reflect differences in the amount of the various macromolecular components, but also changes in the actual structure of the individual components and the manner in which they interact with one another. The changes may originate within the chondrocytes or within the extracellular matrix (Fig. 1). The intracellular changes can be considered to be principally of a synthetic origin. They might, for example, affect the abundance of a given molecule by altering the rate of gene transcription or messenger RNA (mRNA) translation, or they can affect the molecular structure by changing post-translational processing. In some cases, newly synthesized macromolecules may undergo intracellular degradation, thereby limiting their expression. The extracellular changes can be of a synthetic nature as well, particularly with respect to covalent cross-link formation in collagens. Extracellular change, however, more commonly reflects degradative events caused by the action of proteolytic agents. Such degradative processing is most marked in the cartilage proteoglycans. Degradative events may affect the abundance of a given molecule in the matrix, although they generally tend to give rise to a modified protein structure. Age-related tissue changes may be preprogrammed with respect to chondrocyte maturation and senescence. In any given individual, however, a specific change may be influenced by a variety of external factors that can affect both gene expression and protein or carbohydrate processing. Such factors might include nutrition, mechanical stress, and growth factor/cytokine stimulation, and potentially would be operative intermittently throughout life. In contrast, the preprogrammed changes may be operative only during a particular phase of life, such as fetal development, juvenile growth, or adult aging. In many cases, it remains a moot point as to whether the age-related changes in matrix structure are beneficial or detrimental to cartilage function.

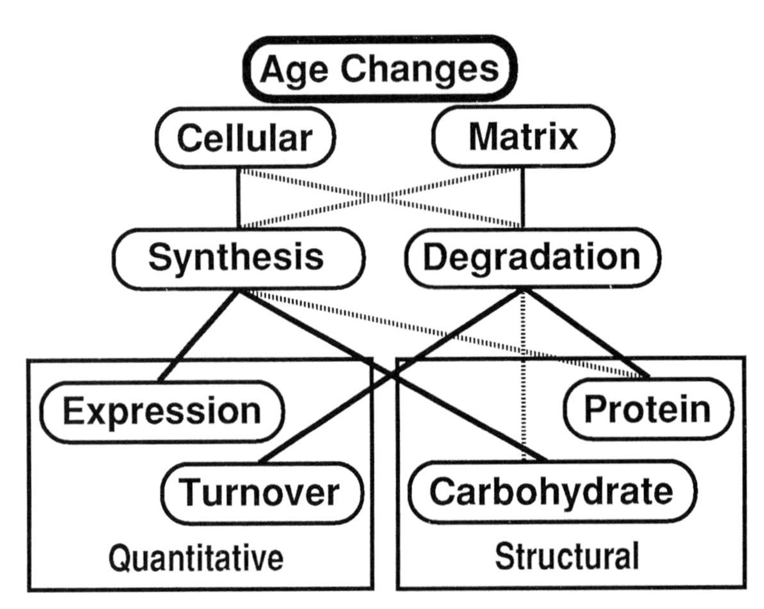

Fig. 1 *Diagrammatic relationship between age-related changes in matrix macromolecules and their origins. Age-related changes can arise either within the cell or within the matrix, and in either site may be either of a synthetic or a degradative origin. The two types of change may affect the macromolecules either quantitatively or structurally. The most common relationships are indicated by the solid lines, whereas less common relationships are indicated by the dashed lines.*

Aggrecan

More data are available on age-related changes in the proteoglycan aggrecan than in any other matrix macromolecule, probably because aggrecan is abundant, readily extractable, and relatively easy to purify. The proteoglycan derives its name from its ability to form large molecular aggregates in association with hyaluronic acid in the cartilage matrix,[1] a property retained throughout life even though the structure of the proteoglycan varies considerably.[2] In humans, aggrecan synthesis at the mRNA level is most prevalent in the young, although the concentration of the molecule in the matrix remains high at all ages. The most remarkable changes in aggrecan are structural rather than quantitative in nature, and may arise by either synthetic or degradative routes (Fig. 2).

Intracellular synthetic events result principally in changes in the relative abundance and the structure of glycosaminoglycans. The aggrecan molecule is substituted with numerous chondroitin sulfate and keratan sulfate chains.[3] In the fetus and at birth, chondroitin sulfate is most abundant,[4] whereas during juvenile growth the relative abundance of keratan sulfate increases.[5,6] The decrease in the

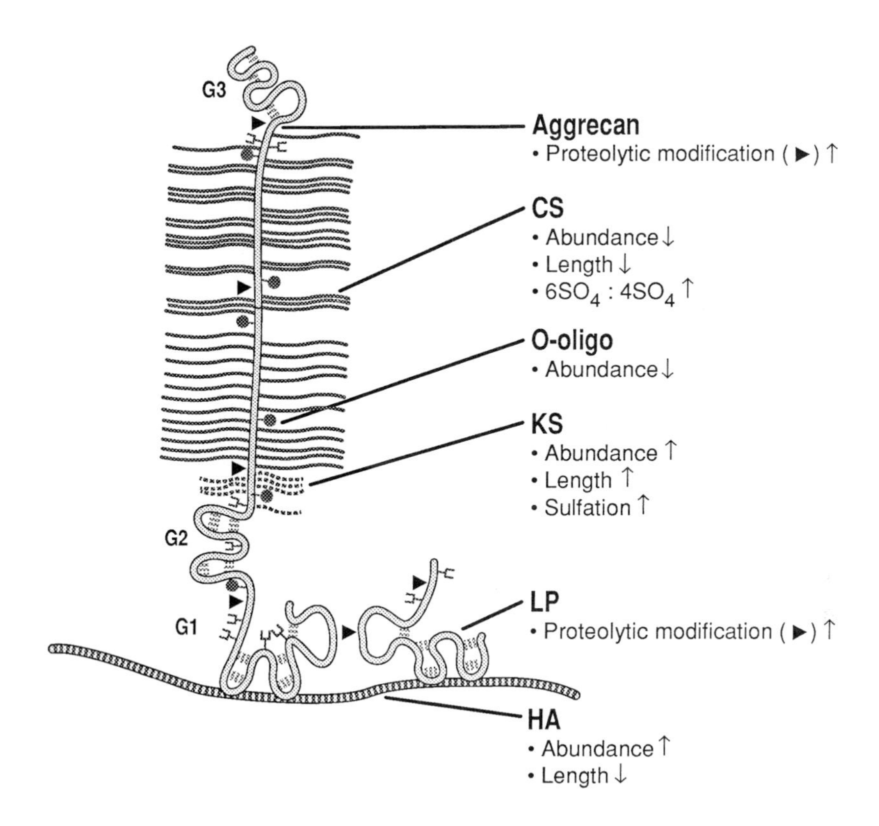

Fig. 2 *Age-related changes in the structure of proteoglycan aggregates. The three components of the proteoglycan aggregate—aggrecan, link protein (LP), and hyaluronic acid (HA)—are depicted together with the component chondroitin sulfate (CS), O-linked oligosaccharide (O-oligo), and keratan sulfate (KS) chains of the aggrecan molecule. The globular regions (G1, G2, and G3) of the aggrecan molecule are also indicated. Changes are indicated as either increases (↑) or decreases (↓) with age.*

amount of chondroitin sulfate may be due in part to a decrease in the activity of the glycosyl transferases that are responsible for linkage region synthesis.[7] The origin of the increase in keratan sulfate is less clear, although it appears to occur at the expense of the O-linked oligosaccharides, which are structurally related to the keratan sulfate linkage region.[8-10] It has been postulated that this change in the relative abundance of the two glycosaminoglycans might be caused by a change in the availability of oxygen in the avascular articular cartilage of the adult compared to the somewhat vascularized cartilage of the epiphysis in the young person.[11] The oxygenated environment of the young cartilage may favor the formation of glucuronic acid and, hence, chondroitin sulfate.

In terms of structure, the chondroitin sulfate chains become shorter with age, and their N-acetylgalactosamine residues become predominantly sulfated at the six position rather than the four position.[5,9] In contrast, the keratan sulfate chains increase in length[8,10] and become more sulfated, presumably because of the increased substitution on the galactose residues.[12] The activity of the sulfotransferases is likely to exhibit an age-related variation in expression. These changes in glycosaminoglycan structure should not be viewed as true aging changes, however, because they occur predominantly during fetal and juvenile development, rather than in the adult.

Changes in the core protein can occur as a result of synthetic events, because alternative splicing occurs within two domains of the G3 region.[3,13] There is no evidence that this is an age-dependent event, however. In contrast, proteolytic processing of the core protein occurs throughout life, and its effect on the aggrecan molecule increases with age. Three principal regions of modification have been described: adjacent to the G3 region,[14] within the glycosaminoglycan attachment region,[14] and between the G1 and G2 regions.[15] The former two regions account for most of the observed decrease in aggrecan size with age,[5,16] and the latter region gives rise to an age-related increase in the abundance of free G1 regions present in the proteoglycan aggregates.[17] These modifications are expected to fundamentally alter the hydrodynamic properties of the proteoglycans. The proteolytic agents responsible for the changes are not fully elucidated, although proteinases are thought to play a major role. The metalloproteinase stromelysin may be particularly important in the cleavage between the G1 and G2 regions.[18] To date, there is little evidence to support a role for nonenzymic mechanisms accounting for proteolytic processing, although the action of free radicals cannot be discounted completely.[19]

Link Protein

Link protein is a glycoprotein that is involved in stabilizing the proteoglycan aggregates[1] through interactions with both the aggrecan and hyaluronate molecules.[20] It shares structural homology with the G1 region of aggrecan.[21] There is little information concerning age-related changes in link protein synthesis or abundance within the matrix. Link protein appears to be present throughout life, and its concentration may remain fairly constant or may decrease slightly with age.[22]

Proteolytic processing of the link protein changes considerably with age, however.[23] Processing occurs in two principal sites: near the amino terminus of the molecules and within the adjacent disulfide-bonded loop (Fig. 2). In vitro, the former region appears to be accessible to the majority of proteolytic agents, whereas access to the latter region is restricted.[24] To date, the only agents known to cleave within this disulfide-bonded loop are cathepsin L and hydroxyl radicals, although neither agent by itself can mimic the natural cleavage sites. Two of the natural cleavage sites near the amino terminus of the molecule can be mimicked in vitro by proteinases. One represents the cleavage site of stromelysin, and the other could be due to cathepsin B or cathepsin G. The role of stromelysin, however, appears to be greatest in the juvenile rather than in the adult.[25] While hydroxyl radicals will cleave within this region in vitro,[19] there is no evidence for their participation during aging.[24] The reason for the progressive accumulation of proteolytically modified link proteins throughout life is probably their continued interaction within the proteoglycan aggregates, suggesting that processing may have little effect on their functional role.

Hyaluronate

Because hyaluronic acid is essential for proteoglycan aggregate formation,[1] its presence within the tissue is critical for normal cartilage function. As human cartilage ages, the tissue concentration of hyaluronate increases even as its molecular size decreases (Fig. 2).[26,27] The decrease in the molecular size of hyaluronate appears to be caused by degradation rather than by synthesis, because at all ages the newly synthesized molecules are of similar size,[26] at least in vitro, although this may not be the case in vivo. Hyaluronate synthesis involves no protein or lipid primer; synthesis occurs at the plasma membrane directly from UDP-glucuronic acid and UDP-N-acetylglucosamine through the action of hyaluronate synthetase.[28]

The in vivo availability of the precursors and the abundance of the synthetase could influence both the quantity and size of the hyaluronate. If degradation is the origin of the size change, then the degradation is more likely to be the result of nonenzymic rather than enzymic mechanisms, because hyaluronidase activity has not been described in the cartilage matrix. Indeed, hyaluronate is very susceptible to depolymerization by free radicals,[29] although this degradative process is limited when the molecule is surrounded by aggrecan and link protein in a proteoglycan aggregate.[19] Furthermore, it is not clear whether the increase in hyaluronate concentration with age represents an age-related increase in synthesis or merely a time-related accumulation.

Other Matrix Proteoglycans

In addition to aggrecan, the articular cartilage matrix contains at least three other proteoglycans[30]: decorin, biglycan, and fibromodulin (Fig. 3). These three proteoglycans are related structurally [31] and collectively have been called nonaggregating proteoglycans because of their apparent inability to interact with hyaluronic acid. While decorin and fibromodulin interact with collagen fibrils,[32] the matrix state of biglycan is less clear, although it appears to be concentrated in the vicinity of the cells.[33] There are also differences in the glycosaminoglycan substitution of the three proteoglycans, with decorin and biglycan having one and two dermatan sulfate chains, respec-

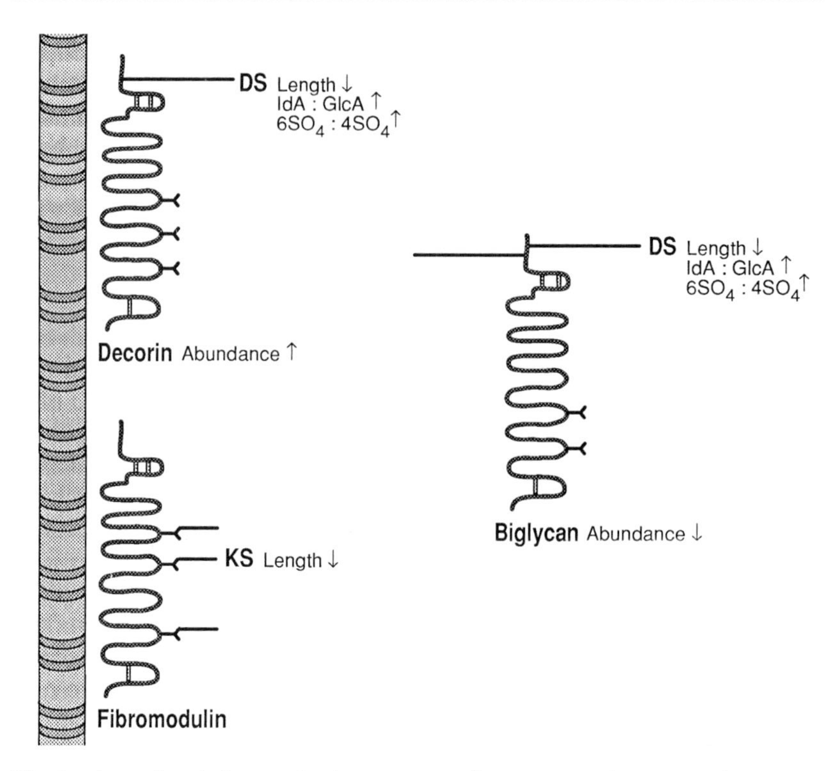

Fig. 3 *Age-related changes in the structure of nonaggregating proteoglycans. The proteoglycans decorin and fibromodulin are depicted in association with a collagen fibril, whereas the proteoglycan biglycan is more remote. Changes in proteoglycan abundance, or the structure of their component dermatan sulfate (DS) or keratan sulfate (KS) chains, are depicted as increases (↑) or decreases (↓) with age.*

tively, in their amino terminal regions,[34] whereas fibromodulin has several keratan sulfate chains in its central region.[35] This region in decorin and biglycan is occupied by N-linked oligosaccharides.

The tissue concentration of decorin increases with age, whereas that of biglycan remains relatively constant.[36] As a result, decorin is about half as abundant as biglycan in the newborn and about twice as abundant as biglycan in the adult. The increase in decorin concentration, however, is not linear with age and appears to be maximal at between 15 and 25 years of age.[37] Changes in the level of synthesis by the chondrocytes can account for much of this variation in matrix abundance, because message levels for biglycan decrease with age, whereas those for decorin increase. This also correlates at the level of proteoglycan secretion, at least in vitro, where the production of decorin is greater than that of biglycan in the adult but not in the newborn.[38]

In terms of structural changes, the dermatan sulfate chains of both decorin and biglycan are shorter in the adult than in the newborn, and the adult has a greater proportion of glucuronic acid epimerization to iduronic acid.[38] The level of N-acetylgalactosamine-6-sulfation increases somewhat with age, although in contrast to the chondroitin sulfate chains of aggrecan, 4-sulfation is always predominant. This may relate to a preference for 4-sulfation adjacent to iduronate residues. These changes in glycosaminoglycan structure undoubtedly are due to variation at the level of intracellular synthesis. In comparison, evidence for changes at the level of matrix degradation are scant for both of these proteoglycans, although it has been suggested that removal of the amino terminal region bearing the dermatan sulfate chains may occur.[39] Certainly there is no evidence to date for the type of age-related increase in proteolytic processing that is evident for aggrecan and link protein, although such a situation might be expected.

Age-related trends in fibromodulin are much less clear than for decorin and biglycan. Indeed, the tissue form of fibromodulin has yet to be defined in human cartilage. Fibromodulin may exist in some tissues as a glycoprotein rather than as a proteoglycan, with the proteoglycan form being derived by extension of the N-linked oligosaccharides of the glycoprotein form with sulfated polylactosamine moieties. A similar process in relation to O-linked oligosaccharides has been postulated to account for some of the age-related increase in the keratan sulfate content of aggrecan. Such a change at the level of synthesis may account for the shorter keratan sulfate chains observed in fibromodulin from mature versus immature bovine cartilage.[40] This decrease in fibromodulin keratan sulfate chain length with age is in contrast to the increase in length with age observed for the keratan sulfate chains of aggrecan.

Collagens

Articular cartilage contains at least five different collagen types within its matrix: types II, V, VI, IX, and XI. Type II collagen is by far the major component at all ages, representing from 80% to 95% of the total collagen content of the tissue.[41,42] The other components generally represent only 1% to 2% in the adult, although their tissue content is higher in growing cartilage.[43] Types II, V, and XI collagens form the structural fibrils of the matrix, and probably are found in heterotypic fibrils.[44] The association of type V or XI collagen with a type II collagen fibril is thought to limit fibril diameter, possibly because of the continued presence of relatively bulky nonhelical terminal propeptides on these molecules. Type IX collagen does not form fibrils by itself, but is associated with fibrils, being present at the fibril surface (Fig. 4).[45] It consists of three collagenous domains separating nonhelical regions. Interaction with the collagen fibrils appears to involve the C-terminal and central collagenous domains. The N-terminal collagenous domain projects away from the collagen fibrils and is terminated by a basic globular domain, which has been hypothesized as forming a linkage site be-

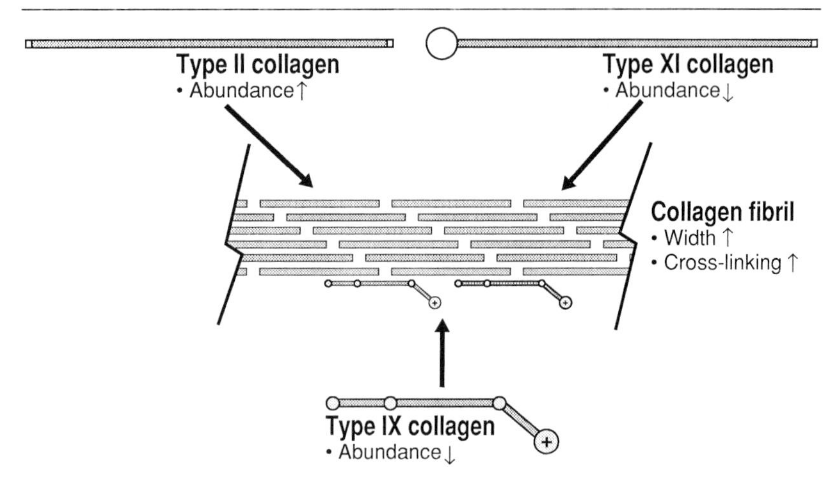

Fig. 4 *Age-related changes in the structure of collagen fibrils. A collagen fibril is depicted as quarter-staggered associations of type II and type XI collagen molecules, with type IX collagen molecules aligned on the outer surface of the fibril. Changes in abundance of the collagen types and in the width and degree of cross-linking of the collagen fibril are indicated as increases (↑) or decreases (↓) with age.*

tween the fibrillar collagen network and the surrounding acidic proteoglycans.[46] Changing the relative amounts of these collagen types could greatly influence tissue properties. In contrast to the above collagen types, type VI collagen is concentrated in the pericellular environment and its specific molecular interactions are not clear.[47] Type IX collagen also has a unique feature, in that it possesses a glycosaminoglycan chain attached to the nonhelical region that separates the central and N-terminal collagenous domains; thus, type IX collagen may be considered to be a proteoglycan. It is not clear, however, whether the glycosaminoglycan chain is present on all molecules within the matrix or whether it confers any unique properties to the molecule.

In addition to altering the quantity of each collagen type, changes at the level of synthesis also may produce structural variation with age. Intracellularly, this is reflected in the type of glycosylation present on hydroxylysine residues, with the proportion of glucose-galactose disaccharides relative to single galactose units decreasing with age.[48] As with glycosaminoglycan synthesis in the proteoglycans, the change probably reflects an age-related variation in the activity of the relevant glycosyl transferases. The content of hydroxylysine relative to hydroxyproline also decreases with age. This may not reflect age-related changes in the activity of the relevant hydroxylases, but rather an increase with age in the utilization of hydroxylysine for the formation of cross-links between the collagen molecules.[49,50] This process, which results ultimately in the formation of pyridinoline-type cross-links, is mediated enzymically in its initial step.[51] It may be considered to be one of the few age-related changes in matrix structure that is mediated by a synthetic event occurring within the matrix rather than within the cell. The pyridinoline-type cross-links not only link fibrillar collagen molecules together, but also link the fibrillar collagen molecules to the associated type IX collagen molecules.[42] The increase in pyridinoline-type cross-links may be associated with the growth period of the individual rather than aging in the adult,[50] as are many of the age-related trends in proteoglycan described previously. There are, however, other types of protein cross-linking that increase as a function of age in the adult, such as the pentosidine residues formed as a non-enzymic glycosylation product.[52,53]

The overall effect of increased collagen cross-linking is to produce a fibrillar framework that is more resistant to mechanical, thermal, and enzymic disruption. The relative resistance of the mature collagen fibrils to proteolytic damage probably accounts for the lack of major degradative changes being observed in their structure during normal aging, in contrast to the vast changes observed in the aggrecan molecules. A further contributing factor is the resistance of the triple helical region of fibrillar collagen molecules to attack by all proteinases, except specific collagenases. The nonhelical regions, however, are susceptible to more general proteolytic cleavage, and modification in these regions might occur to some degree from proteinases involved in proteoglycan degradation, such as stromelysin.[54]

Molecular Function

Although the age-related changes in many macromolecules have been well documented, it is less clear in most cases whether these changes are beneficial or detrimental to the function of the molecule within the tissue. In general terms, the degradative events are commonly viewed as being detrimental, whereas the synthetic events are viewed as beneficial, although exceptions can be made. When one is trying to determine such effects, it is important to distinguish between those changes that are characteristic of growth and development and those that are true aging changes occurring throughout life. Again, the former often can be viewed as beneficial, whereas the latter commonly are seen as being detrimental.

Aggrecan function is thought to depend on three parameters: the anionic charge contributed by each glycosaminoglycan chain, the size of the core protein that determines the number of glycosaminoglycan chains present on a given molecule, and the continued ability to interact with hyaluronic acid. During fetal development, chondroitin sulfate is the predominant glycosaminoglycan and its sulfation increases with age, which presumably benefits the molecule. During juvenile growth, the size of the chondroitin sulfate chains decreases, which reduces their net change; this may be viewed as detrimental. At the same time, however, the content of keratan sulfate increases in a compensatory mechanism, so that it is difficult to know whether the overall function of the molecule is impaired. In a similar manner, it is difficult to know whether the change in sulfation position on the chondroitin sulfate is of any functional consequence. These changes are essentially complete by the time growth is completed. It has been postulated that the changes do not truly reflect changes in articular cartilage, or at least that cartilage at the articular surface of the joint,[55] because in the young juvenile there is still a considerable amount of cartilage in the epiphyses that is destined to be replaced by bone as the secondary centers of ossification grow. The changes described above, therefore, may reflect the varying proportions of cartilage types with growth.

The degradative changes in aggrecan structure occur throughout life, and would appear to be less ambiguous in their interpretation, because increased proteolysis produces molecules of a smaller size and possessing less glycosaminoglycan chains. This decrease in proteoglycan size may be compensated for to some extent by laying down new proteoglycan aggregates, thereby maintaining the total charge density of the tissue. The increase in hyaluronate content with age would support this concept. With time, these, too, would degrade, however. Ultimately, this results in the accumulation of molecules representing only the G1 region of aggrecan, which are of little functional relevance because they are deficient in glycosaminoglycans. Furthermore, they may be considered a hindrance to tissue function by occupying valuable matrix space and sites on the hyaluronate molecules. This accumulation is presumably aided by the interaction of the link proteins, which, although undergoing extensive proteolytic modification throughout life, do not appear to lose their functional ability to stabilize the binding of the G1 region to hyaluronate. Thus, the end result with respect to proteoglycan aggregates is probably a less functional matrix with age.

The effect of the other matrix proteoglycans on function is less clear. Because the function of the dermatan sulfate chains on decorin and biglycan is not known, it is not possible to determine whether structural changes are beneficial or detrimental. However, some dermatan sulfate chains have the ability to self-associate,[56] a property that can be influenced by the iduronate content of the molecule. The ability of decorin to bind the collagen fibrils via its protein core and to self-associate via its dermatan sulfate chains could provide a mechanism for stabilizing the interaction of collagen fibrils. The increased decorin production in the adult could be viewed as a means of maintaining such interactions once tissue remodelling associated with growth has ceased. An opposite role could be postulated for fibromodulin, in which keratan sulfate chains decrease in size with age. In the young person, it is possible that the increased hydration properties of the keratan sulfate chains may limit collagen fibril associations.

The diameter of the collagen fibrils tends to increase with age, a property thought to be related, at least in part, to the decreased content of type XI collagen relative to type II collagen within a given fibril. Wider collagen fibrils are expected to be less flexible and give rise to a more rigid cartilage matrix, which would limit the hydration properties of the proteoglycans. This property would be enhanced by the increased cross-linking between collagen molecules within a fibril. Thus, while the changes in collagen fibril diameter and cross-linking may give rise to a mechanically stronger structure, it is less clear whether such a structure is functionally superior in the tissue with respect to load transmission. Certainly, the increased stability of the cross-linked collagen fibrils can be viewed as beneficial, because they allow the integrity of the matrix architecture to be maintained under proteolytic conditions in which proteoglycans are being degraded. Other changes in collagen molecules are less clear in their potential functional significance, such as the type of glycosylation present on hydroxylysine residues in the triple helix. Furthermore, it is not clear how the decrease in type IX collagen with age may affect tissue function, although if the molecule serves as a source of interaction between the fibrillar collagen network and the proteoglycans, then changes in its structure or abundance will be expected to be of functional consequence.

In some respects the more rigid, less hydrated articular cartilage of the adult can be viewed as being functionally inferior to that of the juvenile, although this mechanical strengthening may be the only mechanism whereby a tissue with a poor reparative potential can withstand the rigors of life for 70 years or longer. Such considerations are important when the repair of cartilage defects is being attempted. Frequently, chondrocytes used in repair give rise to an immature matrix composition, which might be inappropriate in the adult. Ideally, one might wish to have the stable collagen structure of the adult coupled with the more intact proteoglycan structure of the infant. The use of selective growth factors might permit such a structure to some degree by controlling the synthesis of the different molecular components.

It has been argued for some time whether the age-related changes in cartilage matrix organization could be a predisposing factor to osteoarthritic degeneration of the tissue. Certainly, those individuals in whom the degradative changes are accelerated could be viewed as being of increased susceptibility to degenerative tissue changes. It is likely that such an acceleration is outside the direct control of the chondrocytes and would be affected by many external parameters that influence proteinase production and activity.

References

1. Hascall VC: Proteoglycans: The chondroitin sulfate/keratan sulfate proteoglycan of cartilage. *ISI Atlas Sci Biochem* 1988;1:189-198.

2. Roughley PJ, Mort JS: Ageing and the aggregating proteoglycans of human articular cartilage. *Clin Sci* 1986;71:337-344.

3. Doege KJ, Sasaki M, Kimura T, et al: Complete coding sequence and deduced primary structure of the human cartilage large aggregating proteoglycan, aggrecan: Human specific repeats, and additional alternatively spliced forms. *J Biol Chem* 1991;266:894-902.
4. Roughley PJ, White RJ, Glant TT: The structure and abundance of cartilage proteoglycan during early development of the human fetus. *Pediatr Res* 1987;22:409-413.
5. Roughley PJ, White RJ: Age-related changes in the structure of the proteoglycan subunits from human articular cartilage. *J Biol Chem* 1980;255:217-224.
6. Elliott RJ, Gardner DL: Changes with age in the glycosaminoglycans of human articular cartilage. *Ann Rheum Dis* 1979;38:371-377.
7. Wolf B, Steiner E, Keller R, et al: Determination of UDP-xylose-core protein xylosyl transferase in human serum and age-dependent decrease of activity in cartilage. *Fresenius Z Anal Chem* 1982;311:433-434.
8. Santer V, White RJ, Roughley PJ: O-linked oligosaccharides of human articular cartilage proteoglycan. *Biochim Biophys Acta* 1982; 716:277-282.
9. Inerot S, Heinegård D: Bovine tracheal cartilage proteoglycans: Variations in structure and composition with age. *Collagen Rel Res* 1983;3:245-262.
10. Sweet MB, Thonar EJ, Marsh J: Age-related changes in proteoglycan structure. *Arch Biochem Biophys* 1979;198:439-448.
11. Scott JE, Stockwell RA, Balduini C, et al: Keratan sulphate: A functional substitute for chondroitin sulphate in O_2 deficient tissues? *Pathol Biol* 1989;37:742-745.
12. Choi HU, Meyer K: The structure of keratan sulphates from various sources. *Biochem J* 1975;151:543-553.
13. Baldwin CT, Reginato AM, Prockop DJ: A new epidermal growth factor-like domain in the human core protein for the large cartilage-specific proteoglycan: Evidence for alternative splicing of the domain. *J Biol Chem* 1989;264:15747-15750.
14. Paulsson M, Mörgelin M, Wiedemann H, et al: Extended and globular protein domains in cartilage proteoglycans. *Biochem J* 1987; 245:763-772.
15. Sandy JD, Boynton RE, Flannery CR: Analysis of the catabolism of aggrecan in cartilage explants by quantitation of peptides from the three globular domains. *J Biol Chem* 1991;266:8198-8205.
16. Buckwalter JA, Rosenberg L: Structural changes during development in bovine fetal epiphyseal cartilage. *Collagen Rel Res* 1983;3:489-504.
17. Roughley PJ, White RJ, Poole AR: Identification of a hyaluronic acid-binding protein that interferes with the preparation of high buoyant-density proteoglycan aggregates from adult human articular cartilage. *Biochem J* 1985; 231:129-138.
18. Flannery CR, Lark MW, Sandy JD: Identification of a stromelysin cleavage site within the interglobular domain of human aggrecan: Evidence for proteolysis at this site in vivo in human articular cartilage. *J Biol Chem* 1992;267:1008-1014.
19. Roberts CR, Roughley PJ, Mort JS: Degradation of human proteoglycan aggregate induced by hydrogen peroxide: Protein fragmentation, amino acid modification and hyaluronic acid cleavage. *Biochem J* 1989;259:805-811.
20. Périn J-P, Bonnet F, Thurieau C, et al: Link protein interactions with hyaluronate and proteoglycans: Characterization of two distinct domains in bovine cartilage link proteins. *J Biol Chem* 1987;262: 13269-13272.

21. Neame PJ, Christner JE, Baker JR: Cartilage proteoglycan aggregates: The link protein and proteoglycan amino-terminal globular domains have similar structures. *J Biol Chem* 1987;262:17768-17778.
22. Plaas AH, Sandy JD, Kimura JH: Biosynthesis of cartilage proteoglycan and link protein by articular chondrocytes from immature and mature rabbits. *J Biol Chem* 1988;263:7560-7566.
23. Mort JS, Poole AR, Roughley PJ: Age-related changes in the structure of proteoglycan link proteins present in normal human articular cartilage. *Biochem J* 1983;214:269-272.
24. Nguyen Q, Liu J, Roughley PJ, et al: Link protein as a monitor in situ of endogenous proteolysis in adult human articular cartilage. *Biochem J* 1991;278:143-147.
25. Hughes CE, Caterson B, White RJ, et al: Monoclonal antibodies recognizing protease-generated neoepitopes from cartilage proteoglycan degradation: Application to studies of human link protein cleavage by stromelysin. *J Biol Chem* 1992;267:16011-16014.
26. Holmes MW, Bayliss MT, Muir H: Hyaluronic acid in human articular cartilage: Age-related changes in content and size. *Biochem J* 1988;250:435-441.
27. Thonar EJ, Sweet MB, Immelman AR, et al: Hyaluronate in articular cartilage: Age-related changes. *Calcif Tissue Res* 1978;26:19-21.
28. Prehm P: Biosynthesis of hyaluronate. *Agents Actions* 1988;23:36-37.
29. McNeil JD, Wiebkin OW, Betts WH, et al: Depolymerisation products of hyaluronic acid after exposure to oxygen-derived free radicals. *Ann Rheum Dis* 1985;44:780-789.
30. Hardingham TE, Fosang AJ: Proteoglycans: Many forms and many functions. *FASEB J* 1992;6:861-870.
31. Oldberg Å, Antonsson P, Lindblom K, et al: A collagen-binding 59-kd protein (fibromodulin) is structurally related to the small interstitial proteoglycans PG-S1 and PG-S2 (decorin). *EMBO J* 1989;8:2601-2604.
32. Hedbom E, Heinegård D: Interaction of a 59-kDa connective tissue matrix protein with collagen I and collagen II. *J Biol Chem* 1989;264:6898-6905.
33. Bianco P, Fisher LW, Young MF, et al: Expression and localization of the two small proteoglycans biglycan and decorin in developing human skeletal and non-skeletal tissues. *J Histochem Cytochem* 1990;38: 1549-1563.
34. Neame PJ, Choi HU, Rosenberg LC: The primary structure of the core protein of the small, leucine-rich proteoglycan (PG1) from bovine articular cartilage. *J Biol Chem* 1989;264:8653-8661.
35. Plaas AH, Neame PJ, Nivens CM, et al: Identification of the keratan sulfate attachment sites on bovine fibromodulin. *J Biol Chem* 1990;265:20634-20640.
36. Poole AR, Reiner A, Ionescu M, et al: Immunochemical analyses of the small proteoglycans decorin and biglycan in normal and osteoarthritic human articular cartilages. *Trans Combined Orthop Res Soc* USA, Japan, Canada 1991;80.
37. Sampaio Lde O, Bayliss MT, Hardingham TE, et al: Dermatan sulphate proteoglycan from human articular cartilage: Variation in its content with age and its structural comparison with a small chondroitin sulphate proteoglycan from pig laryngeal cartilage. *Biochem J* 1988;254:757-764.
38. Melching LI, Roughley PJ: The synthesis of dermatan sulphate proteoglycans by fetal and adult human articular cartilage. *Biochem J* 1989;261:501-508.
39. Vogel KG, Koob TJ, Fisher LW: Characterization and interactions of a fragment of the core protein of the small proteoglycan (PGII) from bovine tendon. *Biochem Biophys Res Commun* 1987;148:658-663.

40. Plaas AHK, Barry FP, Wong-Palms S: Keratan sulfate substitution on cartilage matrix molecules, in Kuettner KE, Schleyerbach R, Peyron JG, et al (eds): *Articular Cartilage and Osteoarthritis*. New York, Raven Press, 1992, chap 5, pp 69-79.
41. Mayne R: Cartilage collagens: What is their function, and are they involved in articular disease? *Arthritis Rheum* 1989;32:241-246.
42. Eyre DR, Wu JJ, Woods P: Cartilage-specific collagens: Structural studies, in Kuettner KE, Schleyerbach R, Peyron JG, et al (eds): *Articular Cartilage and Osteoarthritis*. New York, Raven Press, 1992, chap 9, pp 119-131.
43. Németh-Csóka M, Mészáros T: Minor collagens in arthrotic human cartilage: Change in content of 1α, 2α, 3α, and M-collagen with age and in osteoarthrosis. *Acta Orthop Scand* 1983;54:613-619.
44. Burgeson RE: New collagens, new concepts. *Annu Rev Cell Biol* 1988;4:551-577.
45. Shaw LM, Olsen BR: FACIT collagens: Diverse molecular bridges in extracellular matrices. *Trends Biochem Sci* 1991;16:191-194.
46. Smith GN Jr, Brandt KD: Hypothesis: Can type IX collagen "glue" together intersecting type II fibers in articular cartilage matrix? A proposed mechanism. *J Rheumatol* 1992;19:14-17.
47. Ronzière MC, Ricard-Blum S, Tiollier J, et al: Comparative analysis of collagens solubilized from human foetal, and normal and osteoarthritic adult articular cartilage, with emphasis on type VI collagen. *Biochim Biophys Acta* 1990;1038:222-230.
48. Slavin S, Shurlan V, Eyre DR, et al: Age and site-related changes in the composition of type II collagen in mammalian articular cartilage. *Trans Orthop Res Soc* 1979;4:178.
49. Nakano T, Thomson JR, Aherne FX: Growth dependent changes in the content of the collagen crosslink, pyridinoline, in joint cartilage from pigs. *Can J Animal Sci* 1983;63:677-681.
50. Eyre DR, Dickson IR, Van Ness K: Collagen cross-linking in human bone and articular cartilage: Age-related changes in the content of mature hydroxypyridinium residues. *Biochem J* 1988;252:495-500.
51. Ricard-Blum S, Ville G: Collagen cross-linking. *Cell Mol Biol* 1988;34:581-590.
52. Uchiyama A, Ohishi T, Takahashi M, et al: Fluorophores from aging human articular cartilage. *J Biochem* 1991;110:714-718.
53. Reiser K, McCormick RJ, Rucker RB: Enzymatic and nonenzymatic cross-linking of collagen and elastin. *FASEB J* 1992;6:2439-2449.
54. Wu JJ, Lark MW, Chun LE, et al: Sites of stromelysin cleavage in collagen types II, IX, X, and XI of cartilage. *J Biol Chem* 1991;266:5625-5628.
55. Harab RC, Mouräo PA: Increase of chondroitin 4-sulfate concentration in the endochondral ossification cartilage of normal dogs. *Biochim Biophys Acta* 1989;992:237-240.
56. Fransson LA, Cöster L: Interaction between dermatan sulphate chains: II. Structural studies on aggregating glycan chains and oligosaccharides with affinity for dermatan sulphate-substituted agarose. *Biochim Biophys Acta* 1979;582:132-144.

Chapter 12

The Role of Integrin Adhesion Receptors in Aging of Cartilage and Osteoarthritis

Virgil L. Woods, Jr, MD
Paul Schreck, MD
David Amiel, PhD
Wayne H. Akeson, MD
Martin Lotz, MD

Age-related changes are known to occur in the composition and structure of human articular cartilage, and it is likely that many of these are the result of altered chondrocyte function.[1-9] Age-related alterations in chondrocyte function have been proposed to be the result of altered production of extracellular matrix (ECM) components and proteases, which in turn may result from altered responsiveness of aged chondrocytes to monokines and/or growth factors, including interleukin-1 (IL-1), tumor necrosis factor (TNF), transforming growth factor β(TGF-β), and interleukin-6 (IL-6).[10-17] The strong association between aging and the occurrence of osteoarthritis argues that age-related alterations in cartilage may produce a fertile substrate for the action of additional factors, resulting in the osteoarthritic process. In contrast to aging chondrocytes, cells from osteoarthritic cartilage synthesize increased amounts of a variety of matrix components and are generally characterized by a hypermetabolic state.[9,18-20] This chondrocyte hyperactivity has been proposed to be an unsuccessful attempt to repair degenerated cartilage. It is likely that both the degenerative and reparative components of the osteoarthritic process may be the result of altered chondrocyte function. The factors that initiate and sustain altered chondrocyte function in osteoarthritis are poorly characterized.

Integrin Receptors

The integrins are a family of adhesion-mediating cell-surface receptors present in one form or another on the cells of virtually every tissue of the body. They mediate a wide variety of events central to tissue morphogenesis, hemostasis, and tumor metastasis.[21] Considerable evidence indicates that they also play critical roles in the various cellular processes required for tissue remodeling and wound repair. All integrins are heterodimers composed of two transmembrane polypeptide chains, (α and β), each having large extracellular domains that join together to form binding sites for specific ECM ligands, and a small cytoplasmic domain that has specific binding sites for elements of the cytoskeleton.[21-25] Integrins have been conveniently considered as three subfamilies based on the particular β chain (β_1-β_3) present in the heterodimer. At least 19 distinct structural forms of integrin heterodimers have been identified to date. The β_1-containing subfamily of integrins is termed the very late activation (VLA) group.[26] There are at least seven members in this subfamily, VLA-1 through VLA-7; each member is formed by the heterodimeric association of a β_1 chain with the appropriate α chain (α_1 through α_7). They have the widest tissue distribution of all of the subfamilies, are present on every connective tissue so far studied, and can bind specifically to ECM proteins including fibronectin, collagen, and laminin. The β_2-containing subfamily of integrins is restricted in distribution to leukocytes and reacts primarily with parenchymal cell ligands including intracellular adhesion molecule-1 (ICAM-1) and ICAM-2.

The β_3 subunit-containing integrins include the fibrinogen receptor of platelets (GPIIb/IIIa) and the vitronectin receptor. While GPIIb/IIIa is restricted in distribution to platelets, the vitronectin receptor (v) has widespread tissue distribution. Furthermore, its designation as "the vitronectin receptor" is a misnomer. Although its activity toward vitronectin was the first ligand-binding activity identified for this integrin, it is now known to have binding activity toward many ECM components including fibronectin, osteopontin, and thrombospondin. Over the last three years, additional β chains have been identified (β_4-β_7), which can form functional heterodimers in association with selected α chains within the originally described three subfamilies. Some α chains are promiscuous in that they can form functional heterodimers with β chains outside of their original subfamily designation. For example, heterodimers formed by the αv chain in association with either the β_1, β_3, or β_5 subunits have been found on selected cell types. Each particular combination of an α with a β chain appears to confer unique functional properties on the resulting heterodimer.

Integrin Adhesive Function

Integrins were initially characterized as the principal mediators of specific cellular attachment to ECM components.[27] Many integrins bind to ECM proteins and, thereby, mediate cell-ECM interactions; among the ECM ligands for integrins are fibronectin, laminin, collagen, fibrinogen, and vitronectin. Other integrins bind to cellular membrane proteins ("counter receptors") to mediate cell-cell adhesion. The major platelet integrin GPIIb/IIIa may be unique because it promotes the binding of platelets to one another through soluble multivalent mediator molecules, primarily fibrinogen. Each unique integrin heterodimer has the ability to bind to a specific set of ligands. Some integrins are capable of binding exclusively to one ligand; for example, VLA-5 ($\alpha_5\beta_1$) binds only to fibronectin; whereas others can bind to multiple alternative ligands; for example, VLA-3 ($\alpha_3\beta_1$) binds to collagen, fibronectin, or laminin. Furthermore, multiple integrins can have binding specificity for the same ligand, as exemplified by the collagen binding activities of $\alpha_1\beta_1$, $\alpha_1\beta_2$, and $\alpha_2\beta_1$. Finally, a given cell type often expresses multiple integrins. Platelets simultaneously express $\alpha_2\beta_1$ (collagen receptor), $\alpha_5\beta_1$ (fibronectin receptor), $\alpha_6\beta_1$ (laminin receptor), and both GPIIb/IIIa and the $\alpha v\beta_3$ integrin. Some of the integrin ECM ligands are known to bind to their integrin receptor through a particular tripeptide region of their amino acid sequence; that is, arginine-glycine-aspartic acid (RGD). Monoclonal antibodies, or RGD-like peptides that can block the interaction of an integrin with its ligand can abrogate the ability of a cell bearing that integrin to bind to ECM containing the ligand.[21] Through these binding interactions, integrins play critical roles in cell adhesion and migration. Certain fibronectin-binding integrins ($\alpha_3\beta_1$, $\alpha v\beta_1$) appear to mediate migration of fibroblasts across fibronectin-coated surfaces, whereas interaction through the VLA-5 fibronectin receptor inhibits migration.

Integrins as Cell-Surface Receptors for Growth and Differentiation Signals

Integrins do not merely serve as cellular anchors to the ECM but also allow the ECM to modulate the behavior of cells, and conversely, allow cells to modify the ECM. When cell-surface integrins encounter and bind to their specific ECM ligands, as yet poorly characterized signals are subsequently communicated to the intracellular regulatory machinery of the cell, informing it of the cell's precise ECM context. In many instances, the cell thereby receives a signal to undergo further differentiative events, which may include elaboration of increased amounts of proteases and ECM proteins. When synovial fibroblasts are allowed to adhere to immobilized fibronectin peptides containing the adhesive sequence RGD, there occurs a marked induction of the expression of genes encoding the secreted ECM-degrading metalloproteinases collagenase and stromelysin. Furthermore, an identical induction of expression of these genes is seen when $\alpha_5\beta_1$ specific monoclonal antibodies are used to crosslink $\alpha_5\beta_1$ (a fibronectin receptor) on fibroblasts.[28] Integrins may also play important roles in the control of cell growth. Chinese hamster ovary cells (CHO), which have been induced to over-express the $\alpha_5\beta_1$ integrin by transfection with α_5 and β_1 cDNAs (complementary deoxyribonucleic acids), undergo a profound change in cell phenotype with increased deposition of fibronectin into their ECM, decreased cell migration, and reduced ability to grow in soft agar. Furthermore, these cells are found to be nontumorigenic when injected subcutaneously into nude mice, in contrast to the behavior of control CHO cells.[29]

Role of Integrins in Construction of ECM

Integrins also appear to play central roles in the construction of the ECM. Fibroblasts in tissue culture first produce and secrete the soluble building blocks of ECM (including procollagens and fibronectin dimers), and in a second, distinct step, assemble these soluble precursors into an insoluble three-dimensional structure. Monoclonal antibodies specific for the fibroblast fibronectin receptor ($\alpha_5\beta_1$) have been found to block this second assembly step.[30-33] Apparently, fibroblast-secreted fibronectin dimers bind to the fibroblast cell-surface integrin receptors specific for fibronectin and are then incorporated into the insoluble ECM. Based on these observations, it has been proposed that certain integrins may in effect serve as "molecular knitting needles" on the surface of the fibroblast, weaving the monomeric units of their ligands into newly formed ECM.

Regulation of Integrin Function

The types, amounts, and functional status of the various integrins displayed by a particular cell type are a dynamic function of the cell's differentiated state and activation history. Integrin functionality appears to be modulated and regulated at several levels: induced alterations of the amounts of particular heterodimers on the cell surface; induced changes in the ability of an integrin to bind to a ligand (integrin activation); and alternative splicing of integrin subunits. Many of the monokines and growth factors known to modulate cellular production of ECM proteins and proteases alter integrin display. In osteoblasts, IL-1 at picomolar concentrations specifically elevates approximately 10 fold the expression of the β_1 subunit and its associated α subunits, except for the vitronectin receptor.[34] When human lung fibroblasts are stimulated with interferon (IFN), there is a twofold increase in expression of the collagen-specific integrin $\alpha_2\beta_1$, accompanied by a comparable increase in the incorporation of newly synthesized collagen into insoluble ECM.[35] These observations suggest that the $\alpha_2\beta_1$ integrin may play a role in the incorporation of soluble collagen into ECM, perhaps one similar to that of $\alpha_5\beta_1$ in fibronectin incorporation into ECM.[30-33] TGF-β, a potent inducer of ECM protein synthesis in many cell types, is a potent inducer of increased integrin display in these same cell types.[36-38]

Many, if not all, integrins must undergo specific conformational changes in the structure of their extracytoplasmic domains to acquire ligand-binding capabilities. In many instances, cellular activation induces these structural changes, perhaps secondary to perturbations of the intracytoplasmic domains of the integrin heterodimers. In other instances, the binding of a ligand to an activated integrin induces additional conformational changes in the integrin's structure, which allow the binding of other ligands to the same integrin molecule. The nature of these conformational changes and the mechanisms that induce them are under intensive study.

Several alternatively spliced variants of integrin subunits have been identified. Most of the spliced variants appear to be restricted to modifications of the structure of the short cytoplasmic tails, suggesting that these modifications, in part, regulate the association of integrins with cytoplasmic components. The functional consequences of the production of specific spliced isoforms have yet to be delineated.

Integrins in Articular Cartilage

Several lines of evidence suggest that chondrocyte integrins may play important roles in articular cartilage homeostasis. Over the past two years, we have employed a variety of analytic modalities to characterize integrin display on normal human articular chondrocytes (Table 1). In situ detection of integrins was performed in frozen sections of normal cartilage with integrin subunit-specific and heterodimer-specific monoclonal antibodies (Mabs) and polyclonal antibodies specific for the cytoplasmic domains of integrin subunits. These studies showed that chondrocytes prominently displayed the $\alpha_5\beta_1$ integrin, a specific receptor for fibronectin; the $\alpha_1\beta_1$ integrin, a collagen and laminin receptor; and the $\alpha v\beta_5$ and $\alpha v\beta_3$ vitronectin receptor integrins. Mabs specific for the αv integrin subunit reacted more intensely with chondrocytes near the articular surface than with those present in deeper layers of cartilage, whereas antibodies to the α_1 subunit and the α_5 subunit reacted equally well with cells in all layers of cartilage. The $\alpha v\beta_3$ heterodimer was strongly and exclusively expressed in the most superficial two to four layers of chondrocytes while the $\alpha v\beta_5$ heterodimer was equivalently expressed on both superficial and deep chondrocytes. Expression of the α_2 through α_4, and α_6 subunits was not detected in in situ studies.

Chondrocytes enzymatically isolated from articular cartilage were stained with the same set of antibodies and analyzed by flow cytometry. Similar to the results with the in situ studies, strong immunoreactivity was detected with antibodies specific for the β_1, α_1, α_5, and αv subunits, as well as the $\alpha v\beta_5$ heterodimer. Additionally, weak reactivity with Mab specific for the α_3 subunit was seen in these studies. Immunoprecipitation of detergent extracts of cell-surface radioiodinated and metabolically labeled primary chondrocytes with integrin-specific Mabs revealed the same types and similar quantities of integrin subunits as had been detected in the flow cytometric studies. The molecules precipitated from chondrocyte extracts demonstrated electrophoretic mobilities characteristic of the corresponding lymphocyte and platelet integrin subunits on reduced and unreduced SDS-PAGE (sodium dodecyl sulfate-polyacrylamide gel electrophoresis). These immunoprecipitation stud-

Table 1 Human articular chondrocyte integrins

Integrin form	Relative abundance on chondrocytes: In situ	Flow cytometry	Immuno-precipitation	In situ distribution	Comments*
$\alpha_5\beta_1$	3+	3+	2+	Superficial = intermediate	—
$\alpha_1\beta_1$	3+	3+	3+	Superficial = intermediate	Not detectable in situ with Mab TS2/7 only with anti-α_1-cytoplasmic domain antibody
αv subunit	3+	3+	2+	Superficial > intermediate	—
$\alpha v\beta_5$	1+	3+	2+	Superficial = intermediate	—
$\alpha v\beta_3$	1+	0	1+	Intensely stains only most superficial 1-3 layers of chondrocytes	—
$\alpha_3\beta_1$	0	1+	0	Not detectable	—
$\alpha_2\beta_1$	0	0	1+	Not detectable	—

*Mab is monoclonal antibody

ies revealed the presence of α_3 and α_5 subunits, but not α_1 subunits in immunoprecipitates prepared with anti-αv-specific Mabs.

These studies indicate that normal articular chondrocytes prominently display the $\alpha_1\beta_1$, $\alpha_5\beta_1$, and $\alpha v\beta_5$ integrin heterodimers as well as lesser amounts of the $\alpha v\beta_3$ and $\alpha_3\beta_1$ heterodimers. The αv subunit-containing integrins appear to be more readily detected on articular chondrocytes near the articular surface. Certain of these integrin forms are difficult to detect in tissue sections ($\alpha_1\beta_1$ and $\alpha_3\beta_1$), probably on the basis of restricted Mab accessibility to the relevant target epitopes. Recent studies by other investigators have reported the presence of the $\alpha_5\beta_1$ integrin in human articular chondrocytes and have demonstrated that this integrin can serve as a specific chondrocyte adhesion receptor for fibronectin.[39,40] Because these studies involved the exclusive use of in situ immunohistochemistry and did not involve use of αv-specific monoclonals, they failed to detect the presence of the $\alpha_1\beta_1$, $\alpha_3\beta_1$, and αv-containing integrins that our studies show to be present.

Potential Integrin-Ligand Interactions in Cartilage

The $\alpha_1\beta_1$ and $\alpha_3\beta_1$ integrins serve variously as receptors for collagen, laminin, and/or fibronectin in other tissues, whereas the $\alpha_5\beta_1$ integrin serves as a receptor exclusively for fibronectin. Integrin forms bearing the αv subunit can serve as receptors for vitronectin, fibronectin, thrombospondin, osteopontin, fibrinogen, and von Willebrand factor. Recent studies further indicate that αv-containing integrins can serve as receptors for collagen when suitably activated.[41]

Many of the known ligands for integrins are present in articular cartilage. Type II collagen is a specific high-affinity ligand for the $\alpha_2\beta_1$ integrin, as demonstrated in systems using platelet and fibroblast $\alpha_2\beta_1$.[42,43] Type VI collagen constitutes about 1% of the collagen of normal canine articular cartilage. After experimental transection of the anterior cruciate ligament, the quantity of type VI collagen in the resulting arthritic cartilage is markedly increased.[44] In immunolocalization experiments, type VI collagen is found to be concentrated around chondrocytes and present as an interlacing network between type II fibrils and the interterritorial matrix. In experimental osteoarthritic cartilage, in striking contrast to normal tissue, the pericellular distribution of type VI collagen appears broadened and blurred compared with controls.[45,46] Amino acid sequence studies have revealed that type VI collagen contains multiple RGD sequences, making it a candidate ligand for several of the integrins.[47]

Thrombospondin is a trimeric adhesive glycoprotein, first described in platelets and subsequently identified in articular cartilage.[48] It is synthesized by chondrocytes in monolayer culture, and preliminary evidence suggests it may be an attachment factor for chondrocytes.[46] Thrombospondin is a ligand for several integrins, including the vitronectin receptor ($\alpha v\beta_3$), and contains an RGD sequence.[49]

The RGD-containing integrin ligand fibronectin is synthesized at low levels in situ by normal human articular chondrocytes,[50] and synthesis rates are increased in osteoarthritis and in cultured chondrocytes.[50-54] The $\alpha_1\beta_1$, $\alpha_5\beta_1$, and $\alpha v\beta_3$ integrin forms are high-affinity receptors for fibronectin in a variety of tissues, and $\alpha_3\beta_1$ is a low-affinity fibronectin receptor.[55] In many tissues, alternative splicing of the fibronectin mRNA (messenger ribonucleic acid) primary transcript results in multiple mRNAs, which differ at three known regions (ED-A, ED-B, and V) and which contribute to the generation of potentially more than 200 different fibronectin isoforms.[56] In fibroblasts, TGF-β has been shown to regulate fibronectin-splicing patterns.[57] Articular chondrocytes have been shown to produce a unique alternatively spliced form of fibronectin (containing the ED-B domain), which is found in very few other adult tissues.[58] Although the fibronectin-binding integrin $\alpha_5\beta_1$ binds to the RGD motif that is present in all spliced fibronectin isoforms, it has been proposed that alternative splicing may modulate the affinity of fibronectin for this and perhaps other integrins. It therefore is possible that chondrocyte integrins have unique interactions with particular fibronectin spliced forms (ED-B) not expressed in other adult tissues. It is not known whether there is a perturbation of the production of fibronectin isoforms in aged cartilage or osteoarthritis. The presence of vitronectin in articular cartilage has not been reported.

Integrin Receptors and Aging

There are no published studies concerning age-related alterations in integrin display and function apart from those that deal with the prominent role of integrins in embryogenesis and early development. However, several observations indicate that the structure of fibronectin and its interaction with integrin receptors may alter with aging. Fibronectin isolated from early passage and late passage (in vitro aged) human fibroblasts differs in its ability to support cell adhesion and to bind to native collagens types I and II.[59-61] Furthermore, fibronectin isolated from senescent fibroblasts displays antigenic epitopes that are not present on fibronectin isolated from early passage cells.[62] Other studies strongly suggest a senescence-related alteration in the function of $\alpha_5\beta_1$ integrin. Cultures of late-passage human skin fibroblasts supplemented with radiolabeled normal plasma fibronectin take up and incorporate fibronectin into ECM at more than twice the rate observed with early-passage fibroblasts.[63] Fibroblast uptake and incorporation of fibronectin into ECM has been shown to be mediated by the fibronectin-specific integrin $\alpha_5\beta_1$ in similar systems.[30-33] We have found $\alpha_5\beta_1$ to be prominent on normal articular chondrocytes of middle-aged individuals. These observations suggest that chondrocyte integrins may play critical roles in the regulation and expression of chondrocyte function, and that age- and/or osteoarthritis-related alterations in integrin amount and function on chondrocytes may mediate some of the functional chondrocyte abnormalities observed in these conditions.

Inherited Integrin Structural Polymorphisms Might Be Predisposing Factors to the Development of Osteoarthritis

Recent studies have identified the genetic basis of a rare, inherited form of osteoarthritis. Restriction fragment length polymorphism (RFLP) study of a kindred with a precocious form of primary osteoarthritis associated with mild chondrodysplasia yields strong genetic linkage between a polymorphism in a type II procollagen gene (COL2A1) and the inherited disorder.[64] Further studies have revealed that the polymorphism results in the substitution of cystine for arginine$_{519}$ in type II collagen. These observations suggest that, in more prevalent forms of nonfamilial osteoarthritis, as yet unrecognized polymorphisms in the structure of cartilage ECM proteins may predispose to the development of osteoarthritis. Therefore, considerable effort is now directed to the search for such predisposing polymorphisms of ECM proteins. Idiopathic osteoarthritis is a common disorder and has its onset in mature individuals who lack easily identifiable pre-existing functional pathologies of connective tissue, such as developmental abnormalities, chondrodysplasia, or joint/ligamentous laxity. It is likely that any predisposing inherited structural polymorphisms would both have to be prevalent in the general population and produce only subtle alterations in the function of tissues until the age-dependent development of osteoarthritis.

Recent studies suggest that inherited abnormalities in fibronectin receptor function may be responsible for rare cases of hereditary connective tissue diseases. In one study, markedly decreased numbers of receptors for fibronectin on polymorphonuclear leukocytes were observed in four patients with Ehlers-Danlos syndrome type II (Family E), and type VI, in one patient with osteogenesis imperfecta type III, and in one patient with Marfan syndrome. Furthermore, some healthy members of the families of the patients with Ehlers-Danlos syndrome had moderately decreased fibronectin receptor numbers, suggesting that they were carriers of an abnormal gene causing this disorder. The fibronectin-binding assays used in this study were performed under conditions in which it is likely that fibronectin was binding to integrin receptors on the polymorphonuclear leukocytes,[65] and that decreased number/function of integrins were detected in these disorders. It can be proposed that certain inherited polymorphisms in integrin genes might result in the production of abnormal cartilage ECM, which is predisposed to the aging-dependent development of osteoarthritis in affected individuals.

Certain rarely occurring structural polymorphisms in human integrins result in profound defects in integrin function in homozygous individuals, including certain forms of Glanzmann's thrombasthenia (abnormalities in platelet GPIIb/IIIa), and syndromes associated with recurrent bacterial infections (inherited deficiencies/abnormality in β_2 integrins).[64] Several prevalent inherited polymorphisms of human integrin structure result in integrin proteins that are structurally and antigenically abnormal, but of grossly normal function. In the course of our studies, we identified a prevalent inherited structural polymorphism in the α_2 chain of human $\alpha_2\beta_1$ integrin, a collagen receptor displayed on many human tissues, including platelets, lymphocytes, and fibroblasts, and, in small quantities, on cultured chondrocytes. The existence of this polymorphism was subsequently confirmed by other investigators, and it is now established that this $\alpha_2\beta_1$ polymorphism (Bra/Brb or Hca/Hcb) is a diallelic system with phenotypic frequencies in both American and German populations of 21% for the Bra form and 99% for the Brb form.[66,67] Although there appear to be no gross functional abnormalities of either of these forms of $\alpha_2\beta_1$ on platelets, they can serve as potent immunogens in allotransfused patients, resulting in immune-mediated platelet destruction.

Similarly, a frequently occurring polymorphism of the integrin α_3 chain (the PlA1/PlA2 system) has been well characterized. This polymorphism has been shown to be caused by a leucine/proline amino acid polymorphism at the 33rd amino acid position of the α_3 chain.[68] The phenotypic frequencies for these two alleles have been calculated to be 98% for the PlA1 form and 28% for the PlA2 form.[69] These polymorphisms result in structurally and antigenically abnormal $\alpha v\beta_3$ vitronectin receptor, which is also a receptor for fibronectin and other ECM components, and is present on chondrocytes. Again, these structural abnormalities produce no apparent perturbation in function of GPIIb/IIIa or vitronectin receptor on platelets or keratinocytes. Finally, we have identified a structural polymorphism of the leukocyte function associated-1 (LFA-1) integrin.[70]

This polymorphism is poorly characterized as to its prevalence and appears to produce no functional abnormality of LFA-1 as manifest by susceptibility to infections. As reviewed above, evidence suggests that certain integrin receptors on chondrocytes may play important roles in the elaboration and maintenance of articular cartilage ECM. Furthermore, integrins serve important functions in normal bone homeostasis. For example, the vitronectin receptor has been implicated in the normal process of bone resorption because this integrin is expressed in large amounts on osteoclasts, is highly localized to the clear zone of osteoclast binding to bone mineral, and mediates the binding of osteoclasts to the RGD-containing bone protein osteopontin.[71] It can be proposed that subtle functional integrin abnormalities may result from the above-described prevalent $\alpha_2\beta_1$ and vitronectin receptor inherited structural polymorphisms (present in approximately 21% to 28% of the population, respectively) and that these functional abnormalities can result in production of subtly abnormal cartilage ECM, which manifests disease by an increased propensity to the development of "primary" osteoarthritis or osteoporosis in the course of aging. Studies are in progress to test this hypothesis.

References

1. Roughley PJ, White RJ, Poole AR, et al: The inability to prepare high-buoyant-density proteoglycan aggregates from extracts of normal adult human articular cartilage. *Biochem J* 1984;221:637-644.
2. Inerot S, Heinegård D, Audell L, et al: Articular-cartilage proteoglycans in aging and osteoarthritis. *Biochem J* 1978;169:143-156.
3. Mort JS, Poole AR, Roughley PJ: Age-related changes in the structure of proteoglycan link proteins present in normal human articular cartilage. *Biochem J* 1983;214:269-272.
4. Plaas AH, Sandy JD: Age-related decrease in the link-stability of proteoglycan aggregates formed by articular chondrocytes. *Biochem J* 1984;220:337-340.
5. Pelletier JP, Martel-Pelletier J, Howell DS, et al: Collagenase and collagenolytic activity in human osteoarthritic cartilage. *Arthritis Rheum* 1983;26:63-68.
6. Cartwright EC, Campbell IK, Britz ML, et al: Characterization of latent and active forms of cartilage proteinases produced by normal immature rabbit articular cartilage in tissue culture. *Arthritis Rheum* 1983;26:984-993.
7. Martel-Pelletier J, Pelletier JP, Cloutier JM, et al: Neutral proteases capable of proteoglycan digesting activity in osteoarthritic and normal human articular cartilage. *Arthritis Rheum* 1984;27:305-312.
8. Dingle JT: Heberden oration 1978. Recent studies on the control of joint damage: The contribution of the Strangeways Research Laboratory. *Ann Rheum Dis* 1979;38:201-214.
9. Roughley PJ, Mort JS: Ageing and the aggregating proteoglycans of human articular cartilage. *Clin Sci* 1986;71:337-344.
10. Saklatvala J: Tumour necrosis factor α stimulates resorption and inhibits synthesis of proteoglycan in cartilage. *Nature* 1986;322:547-549.
11. Jasin HE, Dingle JT: Human mononuclear cell factors mediate cartilage matrix degradation through chondrocyte activation. *J Clin Invest* 1981;68:571-581.

12. Tyler JA: Chondrocyte-mediated depletion of articular cartilage proteoglycans in vitro. *Biochem J* 1985;225:493-507.
13. Bocquet J, Daireaux M, Langris M, et al: Effect of an interleukin-1 like factor (mononuclear cell factor) on proteoglycan synthesis in cultured human articular chondrocytes. *Biochem Biophys Res Commun* 1986;134:539-549.
14. Huber-Bruning O, Wilbrink B, Vernooij JE, et al: Contrasting reactivity of young and old human cartilage to mononuclear cell factors. *J Rheumatol* 1986;13:1191-1192.
15. McGuire-Goldring MK, Murphy G, Gowen M, et al: Effects of retinol and dexamethasone on cytokine-mediated control of metalloproteinases and their inhibitors by human articular chondrocytes and synovial cells in culture. *Biochim Biophys Acta* 1983;763:129-139.
16. Richardson HJ, Elford PR, Sharrard RM, et al: Modulation of connective tissue metabolism by partially purified human interleukin 1. *Cell Immunol* 1985;90:41-51.
17. Goldring MB, Krane SM: Modulation by recombinant interleukin 1 of synthesis of types I and III collagens and associated procollagen mRNA levels in cultured human cells. *J Biol Chem* 1987;262:16724-16729.
18. Bayliss MT: Proteoglycan structure in normal and osteoarthrotic human cartilage. Kuettner KE, Schleyerbach R, Hascall VC (eds): *Articular Cartilage Biochemistry*. New York, Raven Press, 1986, pp 295-309.
19. Mankin HJ, Thrasher AZ: Water content and binding in normal and osteoarthritic human cartilage. *J Bone Joint Surg* 1975;57A:76-80.
20. Mankin HJ, Treadwell BV: Osteoarthritis: a 1987 update. *Bull Rheum Dis* 1986;36:1-10.
21. Ruoslahti E: Fibronectin and its receptors. *Annu Rev Biochem* 1988;57:375-413.
22. Yamada KM: Fibronectin domains and receptors. Mosher DF (ed): *Fibronectin*. San Diego, CA, Academic Press, 1989, pp 47-121.
23. Hynes RO: Integrins: a family of cell surface receptors. *Cell* 1987;48:549-554.
24. Buck CA, Horwitz AF: Integrin, a transmembrane glycoprotein complex mediating cell-substratum adhesion. *J Cell Sci* 1987;8(suppl):231-250.
25. Holzmann B, McIntyre BW, Weissman IL: Identification of a murine Peyer's patch: Specific lymphocyte homing receptor as an integrin molecule with an alpha chain homologous to human VLA-4 alpha. *Cell* 1989;56:37-46.
26. Takada Y, Huang C, Hemler ME: Fibronectin receptor structures in the VLA family of heterodimers. *Nature* 1987;326:607-609.
27. Ruoslahti E, Pierschbacher MD: New perspectives in cell adhesion: RGD and integrins. *Science* 1987;238:491-497.
28. Werb Z, Tremble PM, Behrendtsen O, et al: Signal transduction through the fibronectin receptor induces collagenase and stromelysin gene expression. *J Cell Biol* 1989;109:877-889.
29. Giancotti FG, Ruoslahti E: Elevated levels of the alpha 5 beta 1 fibronectin receptor suppress the transformed phenotype of Chinese hamster ovary cells. *Cell* 1990;60:849-859.
30. Singer II, Scott S, Kawka DW, et al: Cell surface distribution of fibronectin and vitronectin receptors depends on substrate composition and extracellular matrix accumulation. *J Cell Biol* 1988;106:2171-2182.
31. Akiyama SK, Yamada SS, Chen WT, et al: Analysis of fibronectin receptor function with monoclonal antibodies: Roles in cell adhesion, migration, matrix assembly, and cytoskeletal organization. *J Cell Biol* 1989;109:863-875.

32. Darribere T, Guida K, Larjava H, et al: In vivo analyses of integrin beta 1 subunit function in fibronectin matrix assembly. *J Cell Biol* 1990;110:1813-1823.
33. Fogerty FJ, Akiyama SK, Yamada KM, et al: Inhibition of binding of fibronectin to matrix assembly sites by anti-integrin (alpha 5 beta 1) antibodies. *J Cell Biol* 1990;111:699-708.
34. Dedhar S: Regulation of expression of the cell adhesion receptors, integrins, by recombinant human interleukin-1 beta in human osteosarcoma cells: Inhibition of cell proliferation and stimulation of alkaline phosphatase activity. *J Cell Physiol* 1989;138:291-299.
35. Clark JG, Dedon TF, Wayner EA, et al: Effects of interferon-gamma on expression of cell surface receptors for collagen and deposition of newly synthesized collagen by cultured human lung fibroblasts. *J Clin Invest* 1989;83:1505-1511.
36. Heino J, Ignotz RA, Hemler ME, et al: Regulation of adhesion receptors by transforming growth factor-beta. Concomitant regulation of integrins that share a common beta 1 subunit. *J Biol Chem* 1989;264:380-388.
37. Ignotz RA, Massague J: Transforming growth factor-beta stimulates the expression of fibronectin and collagen and their incorporation into the extracellular matrix. *J Biol Chem* 1986;261:4337-4345.
38. Ignotz RA, Heino J, Massague J: Regulation of cell adhesion receptors by transforming growth factor-beta: Regulation of vitronectin receptor and LFA-1. *J Biol Chem* 1989;264:389-392.
39. Ramachandrula A, Tiku K, Tiku ML: Tripeptide RGD-dependent adhesion of articular chondrocytes to synovial fibroblasts. *J Cell Sci* 1992;101 (Pt 4):859-871.
40. Salter DM, Hughes DE, Simpson R, et al: Integrin expression by human articular chondrocytes. *Br J Rheumatol* 1992;31:231-234.
41. Agrez MV, Bates RC, Boyd AW, et al: Arg-Gly-Asp-containing peptides expose novel collagen receptors on fibroblasts: Implications for wound healing. *Cell Regulation* 1991;2:1035-1044.
42. Zutter MM, Santoro SA: Widespread histologic distribution of the alpha 2 beta 1 integrin cell-surface collagen receptor. *Am J Pathol* 1990;137:113-120.
43. Santoro SA, Rajpara SM, Staatz WD, et al: Isolation and characterization of a platelet surface collagen binding complex related to VLA-2. *Biochem Biophys Res Commun* 1988;153:217-223.
44. McDevitt CA, Pahl JA, Ayad S, et al: Experimental osteoarthritic articular cartilage is enriched in guanidine soluble type VI collagen. *Biochem Biophys Res Commun* 1988;157:250-255.
45. Keene DR, Engvall E, Glanville RW: Ultrastructure of type VI collagen in human skin and cartilage suggests an anchoring function for this filamentous network. *J Cell Biol* 1988;107:1995-2006.
46. McDevitt CA, Miller RR: Biochemistry, cell biology, and immunology of osteoarthritis. *Curr Opin Rheumatol* 1989;1:303-314.
47. Bonaldo P, Russo V, Bucciotti F, et al: Alpha 1 chain of chick type VI collagen: The complete cDNA sequence reveals a hybrid molecule made of one short collagen and three von Willebrand factor type A-like domains. *J Biol Chem* 1989;264:5575-5580.
48. Miller RR, McDevitt CA: Thrombospondin is present in articular cartilage and is synthesized by articular chondrocytes. *Biochem Biophys Res Commun* 1988;153:708-714.
49. Albelda SM, Buck CA: Integrins and other cell adhesion molecules. *FASEB J* 1990;4:2868-2880.
50. Brown RA, Jones KL: The synthesis and accumulation of fibronectin by human articular cartilage. *J Rheumatol* 1990;17:65-72.

51. Rees JA, Ali SY, Brown RA: Ultrastructural localisation of fibronectin in human osteoarthritic articular cartilage. *Ann Rheum Dis* 1987; 46:816-822.
52. Jones KL, Brown M, Ali SY, et al: An immunohistochemical study of fibronectin in human osteoarthritic and disease free articular cartilage. *Ann Rheum Dis* 1987;46:809-815.
53. Miller DR, Mankin HJ, Shoji H, et al: Identification of fibronectin in preparations of osteoarthritic human cartilage. *Connect Tissue Res* 1984;12:267-275.
54. Evans HB, Ayad S, Abedin MZ, et al: Localisation of collagen types and fibronectin in cartilage by immunofluorescence. *Ann Rheum Dis* 1983;42:575-581.
55. Gehlsen KR, Dickerson K, Argraves WS, et al: Subunit structure of a laminin-binding integrin and localization of its binding site on laminin. *J Biol Chem* 1989;264:19034-19038.
56. Hynes R: Molecular biology of fibronectin. *Annu Rev Cell Biol* 1985;1:67-90.
57. Borsi L, Castellani P, Risso AM, et al: Transforming growth factor-beta regulates the splicing pattern of fibronectin messenger RNA precursor. *FEBS Lett* 1990;261:175-178.
58. Burton-Wurster N, Lust G, Wert R: Expression of the ED B fibronectin isoform in adult human articular cartilage. *Biochem Biophys Res Commun* 1989;165:782-787.
59. Chandrasekhar S, Sorrentino JA, Millis AJ: Interaction of fibronectin with collagen: Age-specific defect in the biological activity of human fibroblast fibronectin. *Proc Natl Acad Sci USA* 1983;80:4747-4751.
60. Chandrasekhar S, Millis AJ: Fibronectin from aged fibroblasts is defective in promoting cellular adhesion. *J Cell Physiol* 1980;103:47-54.
61. Chandrasekhar S, Norton E, Millis AJ, et al: Functional changes in cellular fibronectin from late passage fibroblasts in vitro. *Cell Biol Int Rep* 1983;7:11-21.
62. Porter MB, Pereira-Smith OM, Smith JR: Novel monoclonal antibodies identify antigenic determinants unique to cellular senescence. *J Cell Physiol* 1990;142:425-433.
63. Mann DM, McKeown-Longo PJ, Millis AJ: Binding of soluble fibronectin and its subsequent incorporation into the extracellular matrix by early and late passage human skin fibroblasts. *J Biol Chem* 1988;263:2756-2760.
64. Knowlton RG, Katzenstein PL, Moskowitz RW, et al: Genetic linkage of a polymorphism in the type II procollagen gene (COL2A1) to primary osteoarthritis associated with mild chondrodysplasia. *N Engl J Med* 1990;322:526-530.
65. Miura S, Shirakami A, Ohara A, et al: Fibronectin receptor on polymorphonuclear leukocytes in families of Ehlers-Danlos syndrome and other hereditary connective tissue diseases. *J Lab Clin Med* 1990;116:363-368.
66. Kiefel V, Santoso S, Katzmann B, et al: A new platelet-specific alloantigen Bra: Report of four cases with neonatal alloimmune thrombocytopenia. *Vox Sang* 1988;54:101-106.
67. Woods VL Jr, Kurata Y, Montgomery RR, et al: Autoantibodies against platelet glycoprotein Ib in patients with chronic immune thrombocytopenic purpura. *Blood* 1984;64:156-160.
68. Newman PJ, Derbes RS, Aster RH: The human platelet alloantigens, PI^{A1} and PI^{A2}, are associated with a $leucine^{33}/Proline^{33}$ amino acid polymorphism in membrane glycoprotein IIIa, and are distinguishable by DNA typing. *J Clin Invest* 1989;83:1778-1781.

69. Shulman NR, Aster RH, Leitner A, et al: Immunoreactions involving platelets: V. Post-transfusion purpura due to a complement-fixing antibody against a genetically controlled platelet antigen. A proposed mechanism for thrombocytopenia and its relevance in "autoimmunity." *J Clin Invest* 1961;40:1597-1620.
70. Pischel KD, Marlin SD, Springer TA, et al: Polymorphism of lymphocyte function-associated antigen-1 demonstrated by a lupus patient's alloantiserum. *J Clin Invest* 1987;79:1607-1614.
71. Heinegård D, Oldberg Å: Structure and biology of cartilage and bone matrix noncollagenous macromolecules. *FASEB J* 1989;3:2042-2051.

Future Directions

Clinical Aspects

Define the role of aging processes as risk factors for loss of joint function.

With increasing age there is an increasing prevalence of impaired joint function (pain with use, decreased range of motion, or, more rarely, instability); yet the relationship between increasing age and altered joint function remains unclear. Little information exists about the natural history of function of normal joints or of joints with disease and injury. For example, although in some cases joint function progressively deteriorates following injury or the onset of joint disease, there is also evidence that many diseased or injured joints do not progress to end-stage disease.

Studies are thus needed to define the relationship between aging processes and changing joint function in normal joints and in diseased and injured joints. Groups of patients need to be identified who have normal joints and who have injured and diseased joints to be followed longitudinally with periodic assessment of joint function and symptoms. The studies should include investigation of the conditions that may affect joint function, such as bone disease, metabolic factors, neurologic conditions, nutrition, genetic history, history of injury, and activity level.

Identify diagnostic criteria for joint disease.

Currently it is not possible to define sharply the diagnostic criteria for osteoarthritis or other forms of degenerative joint disease. For example, radiographic or serum diagnostic observations may not reliably predict the presence of clinical criteria. The criteria to be defined might be clinical variables (eg, changes in range of joint motion, pain patterns), measurement of joint function (eg, joint torques, dynamic gait analysis), changes in imaging (eg, plain radiographs, MRI, MR spectroscopy), and molecules present in body fluids (eg, fragments of cartilage matrix molecules or enzymes).

Clinical and experimental studies are needed to define the relationships between these variables. A potential result of this work is that different diagnostic criteria would be validated and the actual onset of disease in different situations would be defined. It is also possible that diagnostic criteria based on different variables would not agree. For example, radiographic diagnosis might not consistently agree with clinical diagnoses.

Identify markers for the pathologic processes leading to joint disease.

Diagnostic criteria indicate the presence or absence of disease. The pathologic processes that lead to disease might be detected before the disease becomes manifest. Markers for the pathologic processes that lead to joint disease would make it possible to identify high-risk individuals, to follow the course of the pathologic process, and to follow the response to intervention. Clearly, some patients with impaired joint function develop a progressive condition, yet reliable markers do not exist that would allow orthopaedic surgeons to identify them or to monitor the processes that lead to joint disease in these individuals. Neither is it possible to distinguish normal age-related changes in joint function from pathologic processes that cause progressive loss of joint function with time.

Physicians need to identify the presence or absence of the underlying pathologic processes leading to joint disease. The markers for the pathologic processes could be clinical variables, imaging variables, and body fluid markers of joint cartilage turnover.

Morphology

Investigate age-related changes in joint morphology and their relationship to changes on imaging studies and to morphologic evidence of cartilage repair.

Joint morphology, encompassing gross, microscopic, and molecular anatomy, is a marker of function. Changes in morphology with age are markers of changes in function with age. To use morphology as an indicator of function, the two should be studied in conjunction. These studies could best be performed on human autopsy specimens for different age groups and should include the following investigations: (1) Quantitative (gross and microscopic) morphologic studies using modern technology should be done for each tissue of each joint and for groups from each decade. Only against this background can the morbid anatomy of disease be evaluated. (2) Universal, age-related regional morphologic differences observed in particular joint structures should be correlated with local functional demands. Such information is necessary as a background for the elucidation of joint pathology. (3) Occasional morphologic changes outside of normal aging need to be clearly identified and characterized by site, age of onset, and etiology. (4) Noninvasive, longitudinal studies on large populations require data relating gross and microscopic morphology to various imaging modalities. (5) The age-related changes in the processes of injury and repair need to be clearly differentiated in morphologic studies.

Biomechanics

Study changes in the physical properties of human articular cartilage with age, at specific sites, and in different joints.

The biomechanical, electromechanical, and physicochemical properties of articular cartilage have been shown to change with age in a limited number of studies. More information on site-specific and joint-specific changes with age is needed to provide a baseline for age-dependent matrix structure-function studies. In addition, studies of failure mechanisms are needed.

Study the microscopic material properties of the extracellular matrix.

The quality, structure, and function of matrix macromolecules change with increasing age. However, these changes are not always reflected in macroscopic material properties of cartilage (eg, tensile and compressive moduli). Micromechanical property changes may thus better reflect and determine the consequences of altered matrix molecules on tissue function. It is important, therefore, to develop better methods for measuring matrix material properties at micro-structural levels to examine local heterogeneities (eg, surface, middle, and deep zone properties) and, ultimately, at the scale of the individual cell.

Develop experimental systems to investigate the age-related functional significance of specific matrix collagens and proteoglycans.

A limited number of studies have focused on the relationship between the physical properties of cartilage and the biochemical composition of the matrix. Age-dependent relationships have received less attention and should be studied in greater detail. In addition, techniques are becoming available that may allow controlled alteration or deletion of specific matrix molecules (eg, the use of continuous cell lines and transgenic animals). These techniques may allow the contribution to function of specific molecules to be understood better.

Investigate the effect of physical forces on the metabolism of human chondrocytes as a function of age (eg, explant and cell culture).

Progressive remodeling of the extracellular matrix with age may result in a significantly altered structure of the pericellular, territorial, and interterritorial matrix domains. In turn, an altered local mechanical and physicochemical environment of the cells could modify their ability to respond to mechanical signals and, hence, to injury and aging.

Develop structural theoretical models of cartilage to interpret the physiologic levels of stress, strain, and lubrication characteristics within the intact joint.

Structural models are needed to aid in the understanding of loading conditions in vivo, and to interpret the effects of physical stimuli on chondrocyte metabolic responses specific to certain joints and including age-related changes. Measurements of loading, surface geometry, and contact area in vivo are needed to validate the model.

Articular Chondrocytes

Characterize distinctive phenotypes of articular chondrocytes, the manner in which they are regulated, and how they vary with age.

Chondrocytes exhibit a gradient of morphologic and biosynthetic phenotypes progressing from the superficial to the calcified zones. They are associated with differences in the composition of the extracellular matrix (ECM) with depth. Furthermore, the pattern of these cellular and matrix gradients differs from joint to joint. Further studies are needed to understand the mechanisms that regulate chondrocytes in distinct tissue regions.

Define sensory mechanisms by which chondrocytes interact with their environment.

The function of integrins and other cellular receptors for ECM components, cytokines, and growth/differentiation factors, and the influence of metabolic and nutritional factors on chondrocytes require more study.

Define how chondrocytes sense the mechanical forces that act through the surrounding ECM.

Potential mechanochemical transducing mechanisms might be revealed through carefully controlled experiments to understand the ability of chondrocytes to respond to mechanical forces.

Determine how chondrocytes respond to mechanical forces when the ECM is perturbed.

Chondrocyte expression of ECM components, proteases, cytokines, integrins, and intracellular mediators need to be studied in various experimental systems. Are there age-related variations in such properties?

While it may be easier to study such properties in chondrocytes freed from ECM, it would be physiologically more relevant to understand chondrocytes in intact tissue. Experimental systems for doing this, for example culturing chondrocytes in restricting gels, have potential.

Determine the role of chondrocyte proliferation in cartilage homeostasis.

Chondrocytes freed from tissue proliferate in tissue culture, but their capacity to do so is decreased when they are obtained from older individuals. It is unclear, however, whether chondrocytes normally proliferate in adult cartilage in vivo. Extracellular matrix homeostasis and wound repair in many tissues is accompanied by cellular proliferation. If chondrocytes are required to divide, either to maintain normal cartilage or to mount a limited remodeling response, then a decreased proliferative capacity with age could impair such homeostatic properties. Experimental approaches that can address these questions are needed.

Extracellular Matrix

Identify matrix macromolecules, in addition to type II collagen and

aggregan, that change in their abundance with age.

As yet uncharacterized, or still poorly understood, molecules may play critical roles in maintaining matrix integrity and tissue function. Explorative methods might involve two-dimensional mapping techniques of matrix proteins to profile variations in abundance. Subtractive hybridization of cDNA libraries could reveal variations in synthesis.

Define microanatomic changes in the extracellular matrix of articular cartilage with age.

Information is needed on changes with depth from the articular surface, distance from the chondrocyte, and variations between individual joints. Investigations might involve immunohistochemical studies, for example using antibodies that recognize all forms of a given macromolecule and antibodies that recognize only specifically modified forms. Other methods that allow microdissection of the cartilage matrix could also be applied.

Identify mechanisms responsible for the age-related proteolytic cleavage of matrix constituents.

Such mechanisms could involve proteinases or nonenzymic agents. Proteolytic cleavage is of particular importance as it results in an alteration of matrix integrity and is likely to be detrimental. Identification of such processes would be a first step in any attempt at their prevention. Proteolytic cleavage sites that occur in vivo could be identified and compared with cleavages mediated by specific agents in vitro. Immunochemical reagents that recognize specific regions of matrix macromolecules would be useful in such studies.

Identify age-related changes in macromolecular interactions.

Potential changes include those resulting from altered macromolecular synthesis and those induced through chemical changes in the existing extracellular matrix, for example through cross-linking reactions. Of particular interest are events that result in altered material or biologic properties. Examples include altered glycosaminoglycan structures and covalent changes brought about by Maillard reactions.

If possible, distinguish normal age-related changes in the adult from pathologic consequences of osteoarthritis (OA).

If such changes can be distinguished, they may provide clinically useful markers. Distinctions between normal aging and osteoarthritis might be found in proteolytic mechanisms of degradation giving rise to unique products, or in the synthesis of matrix constituents not found in the normal adult.

Section Three

Skeletal Muscle

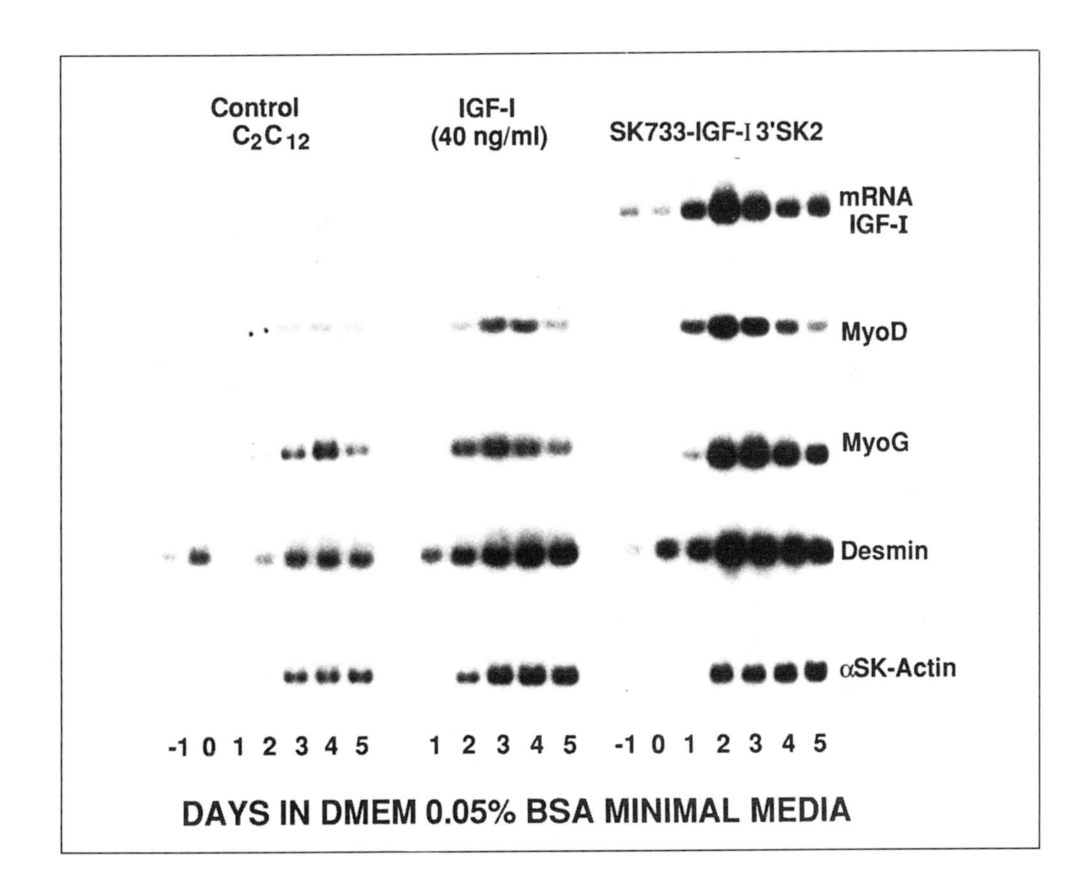

Section Leader

Frank W. Booth PhD

Section Contributors

Frank W. Booth, PhD
Susan V. Brooks, PhD
Marybeth Brown, PT, PhD
David R. Clemmons, MD
Francesco DeMayo, PhD
William J. Evans, PhD
John A. Faulkner, PhD
Raymond Hintz, MD
Z.V. Kendrick, PhD
Heung Man Lee
David T. Lowenthal MD, PhD
David Powell, MD
Scott Powers, PhD, EdD
Robert J. Schwartz, PhD
L.E. Underwood, MD
Steven H. Weeden, BA
Kuo Chang Yin, MS

Overview

Skeletal muscle is the largest organ/tissue in the nonobese person in the third decade of life. However, from the age of 25 years, skeletal muscle mass is lost with aging; 4% of existing muscle is lost per decade from ages 25 to 50 years. Thereafter, the rate of loss increases. Approximately 10% of existing skeletal muscle is lost per decade from ages 50 to 80 years. In the elderly, with very low plasma growth hormone and insulin-like growth factor-1 (IGF-1) levels, skeletal muscle may be lost at the rate of 35% per decade. Fewer contractions against high loads, a loss of growth hormone factors, and other unknown changes may contribute to the age-related loss of skeletal muscle.

The loss of skeletal muscle with age is obligatory; however, strength/exercise, growth hormone/muscle myogenic factors, and nutrition, alone or in combinations, can slow muscle loss with aging. In a study of strength training, 69-year-old men who were strength-trained had cross-sectional areas of muscle equivalent to those of sedentary humans who were 28 years old. However, the 69-year-old strength-trained people had muscle masses that would have been smaller than those of 28-year-olds, if they had strength trained. Although strength training cannot maintain the maximal possible mass of skeletal muscle during aging, it shifts the relationship between age and strength upward.

Administration of growth hormone with IGF-1 reversed a negative nitrogen balance and improved pulmonary function in patients. Improved nitrogen balance improved strength gains in very old subjects undergoing strength training. Thus, various interventions may delay muscle atrophy occurring with aging from adulthood to senescence, but loss of skeletal muscle mass with aging seems immutable.

The observation that strength training could be beneficial in its delay of muscle loss justifies examination of the effects of strength exercises on skeletal muscle of older subjects. However, strength training can cause muscle injury. Information suggests that skeletal muscles from older mice are more susceptible to, and recover less well from, injury produced by lengthening contractions than do muscles from younger rats. Thus, even though skeletal muscle from older subjects can increase strength during training, its ability to return mass to that of 20-year-olds is diminished with aging. The assessment of these data led to the suggestions that additional interventions to counter aging-associated atrophy of skeletal muscle should be considered. In this regard, the use of transgenic animals should be considered to delineate the mechanisms responsible for the age-related atrophy. Moreover, gene transfer techniques could be employed in efforts to introduce into skeletal muscle regulatory factors that would counter the muscle atrophy that occurs with aging. Such methodologic approaches might provide clues to the processes initiating and continuing the loss of skeletal muscle mass from the age of 25 years to senescence. Although no chapters are included on age-related changes in muscle structure, matrix, and protein quality, further work is needed in these areas.

The loss of skeletal muscle is probably related to decreased mobility, which could further enhance the aging initiation of muscle atrophy. The eventual effect of major losses in mobility will be a loss of the quality of life.

Chapter 13

Age-Related Immobility: The Roles of Weakness, Fatigue, Injury, and Repair

John A. Faulkner, PhD
Susan V. Brooks, PhD

Introduction

Impairments in any of a number of different organ systems contribute significantly to the age-related increase in immobility. The organ systems include the nervous system, musculoskeletal system and joints, circulatory system, and respiratory system. Any loss of mobility immediately reduces participation in the physical activities of daily living, which decreases the quality of life.

All movement occurs as a result of contractions of muscles. Different activities require varying proportions of three types of muscle contractions. If the load and the force are equal, or if the load is immovable, the muscle remains at the same length (isometric contraction). Under circumstances of a load less than the force developed by the muscle, the muscle shortens (shortening contraction). When the load is greater than the force developed, the muscle is lengthened (lengthening contraction).[1]

The properties of muscle function that contribute to mobility are force, velocity, and power. Force (newtons), which may be measured during each of the three types of contractions (Fig. 1), is least during shortening contractions and greatest during lengthening contractions. Force is normalized per unit of cross-sectional area (kN/m^2). Normal muscles in adult animals develop a maximum isometric tetanic force (F_o) of ~280 kN/m^2 regardless of fiber type,[2] while twofold higher forces are observed during lengthening contractions. The velocity of shortening is a function of fiber length and relative load. Consequently, velocities are normalized by fiber length and compared at zero load (maximum velocity of unloaded shortening).

During a single contraction, the power is a product of the average force (N) and the average velocity (m/s). Power is measured in watts (W, where 1 W = 1 N·m/s) and is normalized by muscle mass (W/kg). The optimum velocity for the development of power is about one third of the maximum velocity of unloaded shortening.

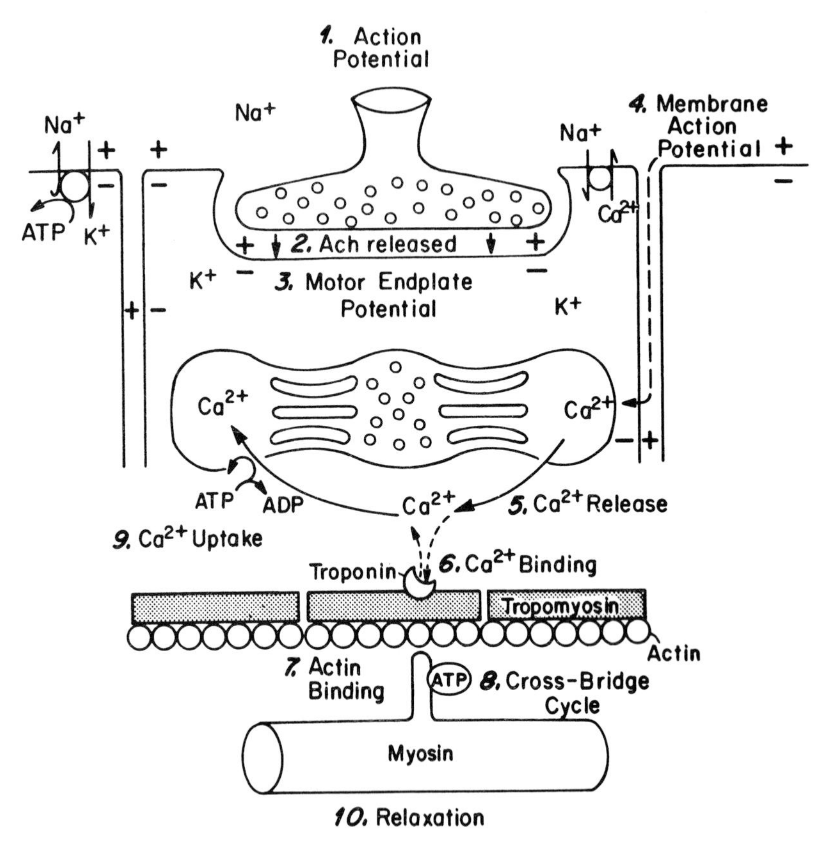

Fig. 2 *A diagramatic representation of a neuromuscular junction, transverse tubular system, sarcoplasmic reticulum, and myofilament. The numbers indicate ten possible sites of fatigue. (Adapted with permission from Vander, Sherman, Luciano,* Human Physiology, *McGraw-Hill, 1980.)*

significantly, decreases the force per cross-bridge.[12] The concentration of free calcium in the cytosol is a function of calcium release, calcium uptake, and calcium binding. Each of these is affected by metabolities, particularly hydrogen ions.[15] In addition, calcium sensitivity may be reduced.

Compared with the sustained power developed by muscles of young animals, the muscles of old animals demonstrate a 50% decrease in sustained power.[16] The decreased sustained power of muscles in old compared with young animals is primarily a function of the ability of muscles in old animals to increase the duty cycle of contractions.[16] Apparently, the enzyme systems associated with oxidative metabolism are rate limiting and are unable to maintain an energy balance at high duty cycles. Because many activities require repetitive contractions, the impaired capacity of muscles in old animals to sustain force and power constitutes a threat to the successful performance of the activities of daily living.

Contraction-Induced Injury

Skeletal muscles may be injured by their own contractions.[17,18] Fibers are more likely to be injured when muscles are lengthened during a contraction than during isometric or shortening contractions, or passive lengthening.[19] In addition, muscles in old animals are more susceptible to injury than are muscles in young or adult animals.[20] Within a whole muscle some fibers are injured and some are not.[19] The injury to fibers is highly focal with only a few sarcomeres involved in series. The injury may include only a few sarcomeres in parallel, or the injury may spread across the whole cross section of the fiber.[17]

Although morphologic evidence of injury by histologic or electron microscopic techniques is imperative, the time course of the injury and recovery is followed most successfully by measuring force deficits. A force deficit is calculated from the following equation: [1 minus (maximum force one minute after a lengthening contraction protocol divided by maximum force prior to the lengthening contraction protocol)] multiplied by 100. The force, or power, deficit during a series of repeated lengthening contractions may be caused by fatigue or by injury. Recovery from fatigue is usually complete after three hours and the initial injury may be estimated by the force deficit at this time.

The initial injury appears to be mechanical in nature with some sarcomeres being stretched beyond overlap.[21] Our hypothesis is that during a protocol of lengthening contractions, weak sarcomeres in which the cross-bridges cannot bear the high loads are stretched beyond overlap by stronger sarcomeres in series and are injured. This hypothesis is supported indirectly by a mathematical model of the behavior of weak and strong sarcomeres[22] and electron micrographs of focal damage.[17] Following the mechanical injury, a secondary injury occurs (Fig. 3). The peak in the secondary injury occurs between one and five days. The concept of a secondary injury is supported by peak values in the force deficit,[19,20] direct morphological evidence of damage to fibers,[17-20] creatine kinase and other enzyme effluxes from muscles,[18] and, in human beings, reports of muscle pain.[18] The secondary injury can be eliminated in young animals and significantly reduced in old animals by treatment with polyethylene glycol superoxide dismutase, which is consistent with the hypothesis that the secondary injury is caused primarily by oxygen free radicals.[19]

Repair and Regeneration

Following the peak in the severity of the secondary injury, a gradual recovery of normal muscle structure and function is observed. The force deficit decreases steadily, and restoration of fiber integrity is noted in both histologic[18,20] and electron microscopic sections.[17] Myotubes are observed by five days after the lengthening

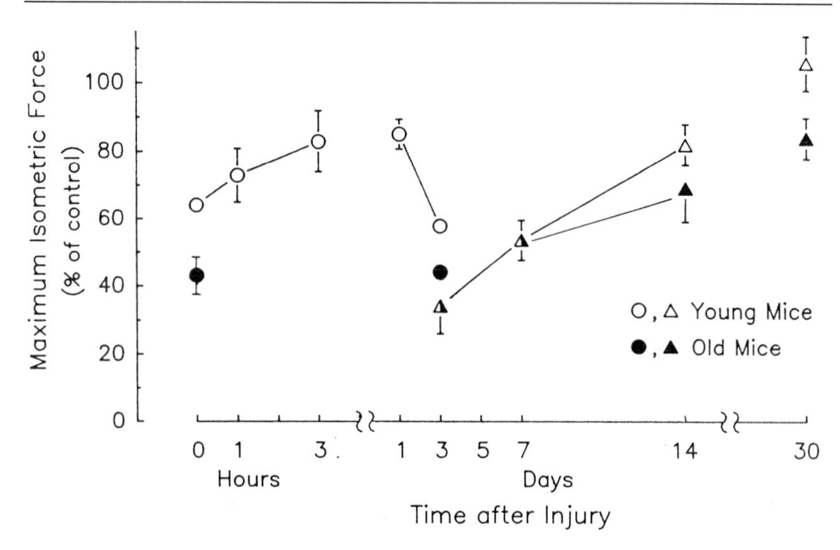

Fig. 3 *Maximum isometric tetanic force of extensor digitorum longus muscles from young and old mice at selected times after lengthening contractions. Values are expressed as a percentage of the control value. (Adapted with permission from Brooks SV, Faulkner JA: Contraction-induced injury: Recovery of skeletal muscles in young and old mice.* Am J Physiol *1990;258(3Pt1):C436-C442, and from Zerba E, Komorowski TE, Faulkner JA: Free radical injury to skeletal muscles of young, adult, and old mice.* Am J Physiol *1990;258(3Pt1):C429-C435.)*

contraction protocol,[20] and recovery is virtually complete by 14 days (Fig. 3). The recovery from contraction-induced injury appears to elicit the same cascade of cellular events as other types of injury, such as sharp or blunt trauma, ischemia, myotoxic agents (Marcaine, lidocaine), and a variety of muscle diseases (in particular the muscular dystrophies). The cascade of events includes: influx of mononuclear cells (phagocytes and macrophages), activation and mitotic division of satellite cells, and subsequent development of myoblasts, myotubes, and differentiated fibers.[23] In addition, while muscles in young animals recovered completely within 14 days of injury, those in old animals showed deficits in force and mass several months[24] following an injury of similar magnitude.

Age-Related Weakness and Fatigue

We have hypothesized that the increased susceptibility of muscles to injury in old compared with young animals, coupled with the impaired ability to regenerate, may contribute to the increased weakness and fatigability of muscles in old animals. In muscles of mixed fiber types in human beings[25] and in rats,[26] indirect evidence supports a loss of fast motor units with an increase in the number of fibers in the slow motor units. The mechanism has been attributed to an age-related, selective denervation of fast fibers.[26] We propose that contraction-induced injury to fibers produces a temporary denervation, which is more likely to become a permanent denervation in muscles of old animals.

Acknowledgments

We acknowledge the support of NIH grant AG-06157, and NIH grant AG-00114, which provided fellowship support.

References

1. Faulkner JA, Brooks SV, Zerba E: Skeletal muscle weakness and fatigue in old age: Underlying mechanisms, in Cristofalo VJ, Lawton MP (eds): *Annual Review of Gerontology and Geriatrics*. New York, Springer Publishing, 1990, vol 10, chap 9, pp 147-166.
2. Brooks SV, Faulkner JA: Contractile properties of skeletal muscles from young, adult and aged mice. *J Physiol (Lond)* 1988;404:71-82.
3. Grimby G, Salton B: The ageing muscle. *Clin Physiol* 1983;3:209-218.
4. Phillips SK, Bruce SA, Woledge RC: In mice, the muscle weakness due to age is absent during stretching. *J Physiol (Lond)* 1991;4137:63-70.
5. Carlson BM, Faulkner JA: Muscle transplantation between young and old rats: Age of host determines recovery. *Am J Physiol* 1989;256(6Pt1):C1262-C1266.
6. Faulkner JA, White TP: Adaptations of skeletal muscle to physical activity, in Bouchard C, Shepherd RJ, Stephens T, et al (eds): *Exercise, Fitness and Health: A Consensus of Current Knowledge*. Champaign, IL, Human Kinetics Publishers, 1990, pp 265-279.
7. Moore DH II: A study of age group track and field records to relate age and running speed. *Nature* 1975;253:264-265.
8. Porter R, Whelan J (eds): *Human Muscle Fatigue: Physiological Mechanisms*. London, England, Pitman Medical, 1981.
9. Faulkner JA, Brooks SV: Fatigability of mouse muscles during constant length, shortening, and lengthening contractions: Interactions between fiber types and duty cycles, in Sargeant T, Kernell D (eds): *Neuromuscular Fatigue*. Amsterdam, Netherlands, Elsevier Publishers, 1992.
10. Sargeant AJ, Kernel D (eds): *Current Problems of Neuromuscular Fatigue*. Amsterdam, The Netherlands, Royal Netherlands Academy of Science, 1992.
11. Wilkie DR: Muscular fatigue: Effects of hydrogen ions and inorganic phosphate. *Fed Proc* 1986;45:2921-2923.
12. Metzger JM, Moss RL: Effects of tension and stiffness due to reduced pH in mammalian fast- and slow-twitch skinned skeletal muscle fibres. *J Physiol (Lond)* 1990;428:737-750.
13. Metzger JM, Moss RL: Greater hydrogen ion-induced depression of tension and velocity in skinned single fibres of rat fast than slow muscles. *J Physiol (Lond)* 1987;393:727-742.
14. Goldman YE: Kinetics of the actomyosin ATPase in muscle fibers. *Annu Rev Physiol* 1987;49:637-654.
15. Lee JA, Westerblad H, Allen DG: Changes in tetanic and resting $[Ca^{2+}]_i$ during fatigue and recovery of single muscle fibres from xenopus laevis. *J Physiol (Lond)* 1991;43:307-320.
16. Brooks SV, Faulkner JA: Maximum and sustained power of extensor digitorum longus muscles from young, adult, and old mice. *J Gerontol* 1991;46:B28-B33.

17. Fridén J, Sjöström M, Ekblom B: Myofibrillar damage following intense eccentric exercise in man. *Int J Sports Med* 1983;4:170-176.
18. Jones DA, Newham DJ, Round JM, et al: Experimental human muscle damage: Morphological changes in relation to other indices of damage. *J Physiol (Lond)* 1986;375:435-448.
19. McCully KK, Faulkner JA: Injury to skeletal muscle fibers of mice following lengthening contractions. *J Appl Physiol* 1985;59:119-126.
20. Zerba E, Komorowski TE, Faulkner JA: Free radical injury to skeletal muscles of young, adult, and old mice. *Am J Physiol* 1990;258(3Pt1):C429-C435.
21. Higuchi H, Yoshioka T, Maruyama K: Positioning of actin filaments and tension generation in skinned muscle fibres released after stretch beyond overlap of the actin and myosin filaments. *J Muscle Res Cell Motil* 1988;9:491-498.
22. Morgan DL: New insights into the behavior of muscle during active lengthening. *Biophys J* 1990;57:209-221.
23. Carlson BM, Faulkner JA: The regeneration of skeletal muscle fibers following injury: A review. *Med Sci Sports Exerc* 1983;15:187-198.
24. Brooks SV, Faulkner JA: Contraction-induced injury: Recovery of skeletal muscles in young and old mice. *Am J Physiol* 1990;258(3Pt1):C436-C442.
25. Campbell MJ, McComas AJ, Petito F: Physiological changes in ageing muscles. *J Neurol Neurosurg Psychiatry* 1973;36:174-182.
26. Kanda K, Hashizume K: Changes in properties of the medial gastrocnemius motor units in aging rats. *J Neurophysiol* 1989; 61:737-746.

Chapter 14

Structural Aspects of Aging Human Skeletal Muscle

Frank W. Booth, PhD
Steven H. Weeden, BA

Introduction

A loss of skeletal muscle mass occurs during aging from 25 years to senescence. A specific loss of type II muscle fibers is responsible for much of the age-related decline in muscle mass. Data imply that a process of denervation and the failure to reinnervate type II muscle fibers may be the causal factor for the loss of muscle fibers with aging. Evidence suggests that the aging-related loss in muscle mass is inexorable but may be attenuated by an increase in high-intensity exercise by type II muscle fibers.

Loss of Muscle Mass

Decreases in muscle size/mass occur as humans age from adulthood. Muscle loss begins to occur as early as 25 years of age. The rate of loss appears to have two phases. Lexell and associates[1] reported that approximately 10% of muscle mass is lost from ages 25 to 50 years. Thereafter, they found that muscle atrophy accelerates. From ages 50 to 80 years, an additional loss of 40% was observed. Thus, by the age of 80 years, one half of skeletal muscle had been lost. In this study, Lexell and associates[1] examined cross sections of the vastus lateralis muscles of 43 men between 15 and 83 years of age. All 43 men were normally active until their sudden deaths. None had systemic disease. The authors believe that their sample represents a cross-section of the healthy male population in Sweden.[1] These findings confirm earlier observations. Tzankoff and Norris[2] found a biphasic decrease in 24-hour creatinine excretion in men from ages 22 to 80 years. The absolute value of 24-hour creatinine excretion is directly related to the muscle mass in the body. From ages 22 to 50 years, creatinine excretion decreased 12%. Creatinine excretion declined an additional 29% from ages 50 to 80 years. Another study has reported that the loss of muscle mass with aging is greater in legs than in arms.[3] Thus, atrophy with aging varies with different muscles. The decrease in muscle mass in humans beginning from 25 years of age is the result of both the decrease in size of type II muscle fibers and the loss of type II muscle fibers in men and women.[1,3]

Loss of Type II Fiber Diameter

Skeletal muscle is composed of muscle fibers of different biochemical composition and function. Type I muscle fibers contract relatively slowly; that is, they have low adenosine triphosphatase (ATPase); are fatigue resistant; and have relatively high mitochondrial density. Type II muscle fibers contract relatively rapidly, that is, have high ATPase. Type II muscle fibers can be divided into at least two subcategories: high mitochondrial density (type IIa) and low mitochondrial density (type IIb). The diameter of type I muscle fibers is not altered by aging until senescence. On the other hand, type II fibers show a decrease in fiber diameter with aging, the onset of which may be gender specific. The fiber area of types IIa and IIb fibers decreases progressively in females from ages 20 to 80 years.[3] However, in males the diameter of types IIa and IIb muscle fibers does not begin to decrease until the age of 60 years. In contrast, a later publication reported a progressive reduction in type II fiber diameter in men from 20 to 80 years of age.[1] The average reduction in fiber size was 26% from 20 to 80 years.

Loss of Muscle Fiber Number

A 39% reduction in the number of total fibers in the vastus lateralis muscle occurred from ages 20 to 80 years in apparently healthy men who died suddenly.[1] A biphasic loss of muscle fibers was noted. Absolute numbers of muscle fibers were highest at 24.2 years and decreased very slightly to age 50 years; thereafter, most muscle fiber loss occurred from 50 to 80 years of age.[1]

Offsetting of Age-Associated Loss in Muscle Mass

Strength training has been shown to offset the age-related loss of muscle mass from ages 50 to 70 years.[4] In this study, men aged 50 to 70 years strength-trained three times per week for an average of 12 to 17 years before measurements were taken at an age of 68 years. Their values were compared to four other groups: 28-year-old men, 68-year-old sedentary men, 69-year-old swimmers, and 70-year-old runners. Consistent with the strength-trained groups, swimmers and runners averaged three bouts of exercise per week for the previous 12 to 17 years. On the average, strength-trained subjects performed three sets of ten lifts with weights for legs, torso, shoulders, and arms; swim-trained subjects swam 800 to 1,000 meters each session; and run-trained subjects ran 9 to 12 kilometers. Only the elderly strength-trained groups had cross-sectional areas of the quadriceps femoris and the biceps brachii muscles equal to the values of the 28-year-old group. The elderly controls, swimmers, and runners all had 20% to 24% smaller cross sections of these two muscles as compared to the 28-year-old controls and the elderly strength-trained groups. Similar directional results among the groups were obtained for comparisons of muscle fiber areas. Only strength training maintained the area of type IIa and IIb muscle fibers at the size of the 28-year-old group in the elderly groups. Uniquely noted, the mean cross-sectional area of type IIb fibers in the vastus lateralis muscle was significantly greater in the elderly strength-trained group than in the 28-year-old sedentary group. In the elderly control, swimmer, and runner groups, mean fiber area of type IIa and IIb fibers was 17% to 26% smaller than in the 28-year-old group. These data suggest that strength training offsets the aging-related loss of muscle mass and muscle fiber area up to the age of 70 years in men. Additional data are needed to determine if strength training extended to 80 and 90 years maintains size of skeletal muscle.

Increased Denervation of Muscle Fibers

A progressive process of denervation and reinnervation of limb muscles with increasing age beyond 50 years is suggested from indirect evidence.[5] The size of motor units increases with age; that is, each motor unit innervates an increased number of muscle fibers with age. Two potential interpretations of this observation are made by Howard and associates.[6] They suggest that the increased size of motor units with age could be caused by the selective loss of small-sized motor units or by an ongoing, gradual denervation with compensatory reinnervation. Based on the fact that the loss of muscle mass and muscle fiber size is offset with strength training in 50 to 70 year olds, we speculate the following, that high threshold motor nerves innervating type II muscle fibers are infrequently recruited past the age of 50 years, and that the decreased motor nerve firing leads to denervation of the muscle fiber.

Decreased Mobility is Related to Increased Mortality in the Elderly

Orthopaedic events (most commonly hip, vertebral, or pelvic fracture) in the elderly are one of the causes of continuous bed rest. The following three occurrences are associated with continuous bed rest in the elderly: (1) increased mortality; (2) decreased quality of life; and (3) being underweight. According to Clark and associates,[7] almost half of the elderly who became bedridden died within six months. Nearly all those who survived regained ambulation.[7] Continuous bed rest also is connected with a worsening of the ability to perform activities of daily living (ADL), that is, transfer ability, bladder and bowel continence, toileting, and feeding. In one nursing home study, 68% of nonambulatory residents were underweight by at least 20% of their normal body weight, although 83% of the ambulatory residents were overweight by 15%.[8] However, it remains uncertain which came first, being underweight or the immobility. It is known that losses of 9% to 10% in calf and thigh muscle volumes occur after 30 days of continuous bed rest in adult men.[9] Based on this information, it is likely that nonambulatory and underweight nursing home patients had increased rates of muscle atrophy after the decrease in their mobility. Nevertheless, future studies should quantify both the loss of skeletal muscle in nonambulatory patients and the amount of muscle mass regained in those patients who regain ambulation. Because a minimal amount of muscle mass is required for mobility and decreases in mobility reduce the quality of life, skeletal muscle atrophy and the ability to rehabilitate are important components of aging that require a greater research emphasis.

Connective Tissue Changes in Muscle With Aging

Several changes in the collagen content of skeletal muscle have been noted in laboratory animals as they age. The total collagen content increases from birth to 10 months of age and then levels off as rats age to senescence.[10] The distribution of muscle connective tissue also changes with aging. The muscle area occupied by endomysin connective tissue increases from 10 months to senescence in the rat.[11] Type IV collagen increases from birth to 4 months and then remains relatively constant in the aging rat.[10] Muscle stiffness increases with age, as shown by passive length-strain plots for muscle.[12] Whether these increases in muscle collagen are associated with changes in muscle stiffness is not known. Kovanen and Suominen[13] found that stiffness increased with age in both slow-twitch (soleus) and fast-twitch (rectus femoris) muscles, with the majority of the increase in ultimate tensile strength, tangent modulus, and elastic efficiency occurring in the rat's growth phase from 1 to 4 months of age. Most of the decreased 0.45 M NaCl-soluble collagen also happened from 1 to 4 months of age, which suggests a correlative role with the aforementioned changes. Little additional change in ultimate tensile strength, tangent modulus, elastic efficiency, and 0.45 M NaCl-soluble collagen occurred from 4 to 24 months of age in these rat muscles. Strain and load-relaxation were unchanged when aging from 1 to 24 months in the rat. Life-long daily treadmill running affected slow muscle more than fast muscle, with the soleus muscle of trained rats showing increased ultimate tensile strength, tangent modulus, and elastic efficiency, while the same trained soleus muscles exhibited decreased strain and load-relaxation.[13] The concentration of the basement membrane component, laminin, decreases with age in rats.[10] All of these changes affect the mechanical as well as the regenerative ability of animals in an age-related fashion.

Soluble and insoluble collagen concentrations change in skeletal muscle with aging. Soluble collagen concentration of muscle increases from birth to the fifth week and then decreases from 5 weeks of age to 4 months of age.[10] Thereafter, soluble collagen concentration remains constant to senescence. Insoluble collagen concentration increases to the fifth month of life in rat skeletal muscle and remains unchanged thereafter.

Following ischemic injury to skeletal muscle, it is likely that muscle stiffness could increase in old, but not young rats. The young animal replaces muscle with muscle after ischemic injury, but the old rat replaces some muscle with collagen during repair.[14] From these findings, it is possible to speculate that after orthopaedic maneuvers that sever skeletal muscle in the octogenarian, regenerated muscle would include some connective tissue. The process of regeneration would result in increased stiffness because of the increases in the series elastic element.

As previously stated, the total collagen concentration increases in skeletal muscle with aging.[10] These papers also showed that collagen synthesis as well as degradation decreases as the animal ages. However, from the finding that total collagen concentration increases in skeletal muscle with age, the percentage decrease in collagen degradation must be greater than the percentage decrease in the collagen synthesis.

Conclusions

The loss of muscle mass begins at the age of 25 years in sedentary human beings, and accelerates after the age of 50 years in humans. By the age of 80 years, one half of skeletal muscle mass has been lost. A key component in the loss of muscle mass is the failure to reinnervate type II muscle fibers after their denervation. This may result in a loss of muscle fiber number with aging. Interestingly, strength training prevents type II fiber atrophy but probably has no effect on fiber loss in men between the ages of 50 and 70 years. This suggests that decreased high-intensity exercise by sedentary men aged 50 to 70 years plays a major role in the loss of fiber diameter.

Acknowledgment

This work was supported by US Public Health Service Grant AR 19393.

References

1. Lexell J, Taylor CC, Sjostrom M: What is the cause of the ageing atrophy? Total number, size, and proportion of different fiber types studied in the whole vastus lateralis muscle from 15- to 83-year-old men. *J Neurol Sci* 1988;84:275-294.
2. Tzankoff SP, Norris AH: Effect of muscle mass decrease on age-related BMR changes. *J Appl Physiol* 1977;43:1001-1006.

3. Grimby G, Danneskiold-Samsoe B, Hvid K, et al: Morphology and enzymatic capacity in arm and leg muscles in 78–81 year old men and women. *Acta Physiol Scand* 1982;115:125-134.
4. Klitgaard H, Mantoni M, Schiaffino S, et al: Function, morphology and protein expression of ageing skeletal muscle: A cross-sectional study of elderly men with different training backgrounds. *Acta Physiol Scand* 1990;140:41-54.
5. Jennekens FG, Tomlinson BE, Walton JN: Histochemical aspects of five limb muscles in old age: An autopsy study. *J Neurol Sci* 1971;14:259-276.
6. Howard JE, McGill KC, Dorfman LJ: Age effects on properties of motor unit action potentials: ADEMG analysis. *Ann Neurol* 1988;24:207-213.
7. Clark LP, Dion DM, Barker WH: Taking to bed: Rapid functional decline in an independently mobile older population living in an intermediate-care facility. *J Am Geriatr Soc* 1990;38:967-972.
8. Selikson S, Damus K, Hamerman D: Risk factors associated with immobility. *J Am Geriatr Soc* 1988;36:707-712.
9. Duvoisin MR, Convertino VA, Buchanan P, et al: Characteristics and preliminary observations of the influence of electromyostimulation on the size and function of human skeletal muscle during 30 days of simulated microgravity. *Aviat Space Environ Med* 1989;60:671-678.
10. Kovanen V: Effects of ageing and physical training on rat skeletal muscle. An experimental study on the properties of collagen, laminin, and fibre types in muscles serving different functions. *Acta Physiol Scand* 1989;577(suppl):1-56.
11. Boreham CA, Watt PW, Williams PE, et al: Effects of ageing and chronic dietary restriction on the morphology of fast and slow muscles of the rat. *J Anat* 1988;157:111-125.
12. Goldspink G: Cellular and molecular aspects of adaptation in skeletal muscle, in Komi PV (ed): *Strength and Power in Sport*. Oxford, Blackwell Scientific Publications, 1991, pp 211-229.
13. Kovanen V, Suominen H: Effects of age and life-long endurance training on the passive mechanical properties of rat skeletal muscle. *Compr Gerontol A* 1988;2:18-23.
14. Ullman M, Ullman A, Sommerland H, et al: Effects of growth hormone on muscle regeneration and IGF-I concentration in old rats. *Acta Physiol Scand* 1990;140:521-525.

Chapter 15

The Effects of Exercise Training on Skeletal Muscle in Humans and Aging Fischer 344 Rats

David T. Lowenthal, MD, PhD
Scott Powers, PhD, EdD
Z. V. Kendrick, PhD

Introduction

A growing body of evidence suggests that many of the metabolic and contractile changes that occur with senescence may not stem from aging exclusively, but may also be the result of decreased physical activity.[1,2] In this regard, numerous studies have demonstrated that regular endurance and resistance exercise results in significant remodeling of both bone and skeletal muscle in young adult and aging humans and animals. This chapter will provide a brief overview of the effects of exercise on the musculoskeletal system in aging humans and animals.

Exercise and the Aging Musculoskeletal System

Aging and Muscular Strength

With increasing age there is a loss of muscle mass and muscle strength.[3,4] Age-related reductions in skeletal muscle mass correlate highly with diminished muscle strength. Indeed, muscle mass has been shown to be the major determinant of both age and gender differences in muscle strength.[5] The relationship between muscle strength and muscle mass in older adults is independent of whether the muscle group under study was from the upper or lower body.[5] Age-related declines in muscle strength have been shown to have a significant inverse correlation with walking speed in frail men and women.[6]

The mechanisms responsible for the age-related decrease in muscle strength continue to be debated; however, likely contributory factors are decreased physical activity, hormonal influences, and neuromuscular alterations. For example, inactivity has been shown to cause muscle atrophy in animals and humans of all ages.[7] Therefore, a reduction in physical activity with aging could contribute to the observed reduction in skeletal muscle mass observed in senescence.[7] Furthermore, alterations in blood levels of testosterone and thyroxine in senescence may impact skeletal muscle metabolism by reducing muscle mass and altering myosin isoforms.[8] Independently or collectively, these hormonal alterations can reduce the amount of muscle force produced. Finally, aging causes structural and functional changes in both spinal motor neurons and the neuromuscular junction.[3,4] Because of these changes, the number of motor units recruited decreases, resulting in a significant reduction in the force produced.[3,4,9-11]

Exercise-Induced Improvements in Muscular Strength

Progressive resistance exercise training (that is, weight training) increases muscle mass and improves muscle strength in aging humans.[6,12,13] The increase in muscle strength experienced by elderly humans who are involved in a regular exercise program are related to at least two factors: (1) an increased recruitment of motor units, approaching maximal levels in the working muscle group, and (2) structural and metabolic changes in the muscle tissue causing increases in contractile proteins.[13-16]

Animal studies also indicate that strength training results in modest increases in muscle mass in aging animals. For example, Goldspink and Howells[17] studied muscle hypertrophy in male hamsters of different ages. The animals were subjected to a weight-lifting program, after which hypertrophy of the forelimb muscles was measured. Significant muscle hypertrophy was revealed in animals of all age groups. These data further support the theory that senescent muscle is plastic and capable of adaptation.

Exercise, Aging, and Bone Mineral Density

Both bone and joint disease occur with age. Osteoarthritis and osteoporosis, especially in postmenopausal women, may severely limit a person's ability to complete physical work tasks. Inactivity and illness contribute to osteoporosis in a vicious cycle, causing further bone demineralization or bone matrix loss.[18-22]

It is equally obvious that physical activity and the resultant stress imposed on the skeletal system improves bone mineralization.[21] Smith[21] studied demineralization of the midshaft of the radius in 30 elderly women. Approximately 50% of these women participated in a chair exercise program for three years. Over time, bone mineralization decreased in the control group but increased in the exercising subjects. The mechanism for enhanced bone growth most likely includes improved circulation to bone and increased gravitational and muscle stress, which influence the bone's cellular activity. Although regular exercise may be important in lessening the degree of osteoporosis, it is currently unclear whether exercise will prevent osteoporosis.

Bone mineral density is correlated with muscle strength in older men and women[23-26] Several studies have demonstrated that bone mineral density is improved by exercise programs such as swimming and tennis that increase muscle strength and/or endurance[27] and place additional repetitive stresses on bone.[28-30] In general, regular physical activity has been shown to improve bone mineral density.[27,31,32]

Aging and Exercise-Induced Alterations in Muscular Endurance

Senescent animals and humans have lower skeletal muscle endurance and oxidative capacity than young adults.[33-35] This age-related reduction in muscle mitochondrial volume does not appear to be a function of aging per se, but may be linked to a decrease in locomotion during senescence. In rats, for example, spontaneous activity declines markedly between 9 and 28 months of age.[34,35] Cartee and Farrar[33] presented convincing arguments that an increase in physical activity can reverse the age-associated decline in the oxidative capacity of hindlimb muscle. Indeed, these authors demonstrated that regular treadmill exercise increases the oxidative capacity of hindlimb muscle of old animals to a level that equals or exceeds that achieved by young animals exercising at the same absolute work rate. Similar results have been reported by others.[36,37] Collectively, these data support the theory that skeletal muscle bioenergetic plasticity is not abolished in aging.

Oxygen free-radical generation is ubiquitous in all tissues, including skeletal muscle. During oxidative stress, reactive oxygen species (for example, superoxide anions and hydroxyl radicals) can elicit widespread damage to cell constituents, such as membrane lipids, mitochondrial enzymes, and DNA.[38,39] Recent studies have suggested that increased rates of oxidative phosphorylation in muscle (as results from exercise, for example) increase the rate of radical formation and may be potentially harmful to muscle cells.[40,41] Indeed, a growing body of evidence suggests that free radicals play an important role as mediators of skeletal muscle injury during oxidative stress.[39,42] As a safeguard against oxidative stress, skeletal muscle contains the antioxidant enzymes superoxide dismutase, catalase, and glutathione peroxidase, which protect muscle against free-radical mediated injury to the cells. Although with aging there is a general decline in antioxidant enzyme capacity in some vital organs, such as the liver and heart, the antioxidant capacity of skeletal muscle is enhanced despite a general decline in muscle oxidative capacity.[42] Furthermore, there is evidence that regular endurance-type exercise in aging animals significantly increases the oxidative capacity of skeletal muscle.[36,42,43]

The fact that endurance training increases both the oxidative and antioxidant capacity in skeletal muscle of aging animals is physiologically important. For example, an increase in muscle mitochondrial volume would likely improve muscular endurance by increasing the muscle's ability to metabolize fat and therefore reduce the rate of glycogen degradation.[44,45] Furthermore, exercise-induced increases in muscle antioxidant capacity would improve the muscle's ability to defend against free radicals and hydroperoxides.[42] This ability is important because recent evidence has linked free radicals to muscle fatigue.[46]

Aging Skeletal Muscle and Endurance Exercise-Induced Fiber Type Alterations

The total number of muscle fibers decreases with age;[11,47] however, controversy exists concerning the influence of aging on skeletal muscle fiber type. For example, Florini and Ewton[48] reported no age-related change in fiber type in rat locomotor muscles. In contrast, other investigations have reported both increases and decreases in type I (slow twitch) fibers in senescent rats.[49-51] The reasons for these discrepancies are not clear but may be due to differences in animal care between studies (that is, type of animal housing and nutrition).[52,53]

The effects of exercise training on skeletal muscle fiber type in senescence are not well documented. In one of the few studies on this topic, Lowenthal[51] examined the effects of endurance exercise (treadmill exercise) and aging on the muscle fiber distribution in the plantaris and soleus muscles of the barrier-raised Fischer 344 rat with an age range from adult through senescence. Aging caused a progressive decline in the type I (slow twitch oxidative) muscle fibers in the plantaris muscle of sedentary animals. Exercise training caused a small increase in the percentage of both type IIa (fast twitch oxidative) and type I muscle fibers. These data further support the theory that endurance exercise training increases the oxidative capacity of skeletal muscle while maintaining type I slow oxidative muscle fibers during aging.

Summary

Both human and animal aging is associated with a loss of bone and muscle mass, a decrease in muscular strength, and a reduction in muscle endurance. Recent evidence suggests that these changes which occur with senescence may not stem from aging per se, but may be the result of decreased physical activity. In this regard, it is clear that regular strength-type exercise training promotes increases in muscle mass, bone mass, and muscle strength. Similarly, regular endurance exercise results in an increase in muscle oxidative capacity and antioxidant capacity. Collectively, these findings support the hypothesis that aging does not impair muscle metabolic plasticity.

Acknowledgments

The authors acknowledge the contributions of plantaris and soleus muscles by Dr. Joseph Starnes, University of Texas, Austin,

Texas. This work was supported by National Institute of Aging Grant AG-03213, National Institute of General Medical Sciences Grant GM 21524, and a Research Grant from the American Federation for Aging Research.

References

1. Grimby G: Physical activity and effects of muscle training in the elderly. *Ann Clin Res* 1988;20:62-66.
2. Goldberg A, Hagberg JM: Physical exercise in the elderly, in Finch C, Hayflick L (eds): *Handbook of the Biology of Aging*. New York, Van Nostrand Reinhold, 1977, pp 445-469.
3. Campbell MJ, McComas AJ, Petito F: Physiological changes in ageing muscles. *J Neurol Neurosurg Psychiatry* 1979;36:174-182.
4. McCarter RJ: Age-related changes in skeletal muscle function. *Aging* 1990;2:27-38.
5. Frontera WR, Hughes VA, Lutz KJ, et al: A cross-sectional study of muscle strength and mass in 45 78-year-old men and women. *J Appl Physiol* 1991;71:644-650.
6. Bassey EJ, Bendall MJ, Pearson M: Muscle strength in the triceps surae and objectively measured customary walking activity in men and women over 65 years of age. *Clin Sci* 1988;74:85-89.
7. Roy RR, Baldwin KM, Edgerton VR: The plasticity of skeletal muscle: Effects of neuromuscular activity. *Exerc Sport Sci Rev* 1991;19:269-312.
8. Florini JR: Hormonal control of muscle growth. *Muscle Nerve* 1987;10:577-598.
9. Tomlinson BE, Walton JN, Rebeiz JJ: The effects of ageing and of cachexia upon skeletal muscle: A histopathological study. *J Neurol Sci* 1969;9:321-346.
10. McCarter R: Effects of age on contraction of mammalian skeletal muscle, in Kaldor G, DiBattista WJ (eds): *Aging in Muscle*. New York, Raven Press, 1978, vol 6, pp 1-21.
11. Lexell J, Taylor CC, Sjostrom M: What is the cause of the ageing atrophy? Total number, size, and proportion of different fiber types studied in whole vastus lateralis muscle from 15 83-year-old men. *J Neurol Sci* 1988;84:275-294.
12. deVries HA: Physiological effects of an exercise training regimen upon men aged 52 to 88. *J Gerontol* 1970;25:325-336.
13. Moritani T: Training adaptations in the muscles of older men, in Smith EL, Serfass RC (eds): *Exercise and Aging: The Scientific Basis*. Hillside, NJ, Enslow Publishers, 1981, pp 149-166.
14. Tomonaga M: Histochemical and ultrastructural changes in senile human skeletal muscle. *J Am Geriatr Soc* 1977;25:125-131.
15. Larsson L: Morphological and functional characteristics of the ageing skeletal muscle in man: A cross-sectional study. *Acta Physiol Scand Suppl* 1978;457:1-36.
16. Vandervoort AA: Effects of ageing on human neuromuscular function: Implications for exercise. *Can J Sport Sci* 1992;17:178-184.
17. Goldspink G, Howells KF: Work-induced hypertrophy in exercised normal muscles of different ages and the reversibility of hypertrophy after cessation of exercise. *J Physiol Lond* 1974;239:179-193.
18. Bassett CAL, Becker RO: Generation of electric potentials by bone in response to mechanical stress. *Science* 1962;137:1063-1064.
19. Donaldson CL, Hulley SB, Vogel JM, et al: Effect of prolonged bed rest on bone mineral. *Metabolism* 1970;19:1071-1084.

20. Smith EL Jr, Reddan W, Smith PE: Physical activity and calcium modalities for bone mineral increase in aged women. *Med Sci Sports Exerc* 1981;13:60-64.
21. Smith EL: Bone change in the exercising older adult, in Smith EL, Serfass RC (eds): *Exercise and Aging. The Scientific Basis*. Hillside, NJ, Enslow Publishers, 1981, pp 179-186.
22. Smith EL: Exercise for prevention of osteoporosis: A review. *Phys Sports Med* 1982;10:72-83.
23. Evans WJ: Exercise, nutrition, and aging. *J Nutr* 1992;122(3 suppl): 796-801.
24. Sinaki M, Offord KP: Physical activity in postmenopausal women: Effect on back muscle strength and bone mineral density of the spine. *Arch Phys Med Rehabil* 1988;69:277-280.
25. Pocock N, Eisman J, Gwinn T, et al: Muscle strength, physical fitness, and weight but not age predict femoral neck bone mass. *J Bone Miner Res* 1989;4:441-448.
26. Snow-Harter C, Bouxsein M, Lewis B, et al: Muscle strength as a predictor of bone mineral density in young women. *J Bone Miner Res* 1990;5:589-595.
27. Davee AM, Rosen CJ, Adler RA: Exercise patterns and trabecular bone density in college women. *J Bone Miner Res* 1990;5:245-250.
28. Orwoll ES, Ferar J, Oviatt SK, et al: The relationship of swimming exercise to bone mass in men and women. *Arch Intern Med* 1989;149:2197-2200.
29. Huddleston AL, Rockwell D, Kulund DN, et al: Bone mass in lifetime tennis athletes. *JAMA* 1980;244:1107-1109.
30. Pirnay F, Bodeux M, Crielaard JM, et al: Bone mineral content and physical activity. *Int J Sports Med* 1987;8:331-335.
31. Brewer V, Meyer BM, Keele MS, et al: Role of exercise in prevention of involutional bone loss. *Med Sci Sports Exerc* 1983;15:445-449.
32. Jacobson PC, Beaver W, Grubb SA, et al: Bone density in women: College athletes and older athletic women. *J Orthop Res* 1984; 2:328-332.
33. Cartee GD, Farrar RP: Muscle respiratory capacity and VO_2 max in identically trained young and old rats. *J Appl Physiol* 1987;63:257-261.
34. Fitts RH, Troup JP, Witzmann FA, et al: The effect of aging and exercise on skeletal muscle function. *Mech Ageing Dev* 1984; 27:161-172.
35. Goodrick CL, Ingram DK, Reynolds MA, et al: Effects of intermittent feeding upon growth, activity, and lifespan in rats allowed voluntary exercise. *Exp Aging Res* 1983;9:203-209.
36. Hammeren J, Powers S, Lawler J, et al: Exercise training-induced alterations in skeletal muscle oxidative and antioxidant enzyme activity in senescent rats. *Int J Sports Med* 1992;13:412-416.
37. Powers SK, Criswell D, Lieu FK, et al: Exercise-induced cellular alterations in the diaphragm. *Am J Physiol* 1992;263:R1093-R1098.
38. Jenkins RR: Free radical chemistry: Relationship to exercise. *Sports Med* 1988;5:156-170.
39. Sjodin B, Hellsten-Westing Y, Apple FS: Biochemical mechanisms for oxygen free radical formation during exercise. *Sports Med* 1990; 10:236-254.
40. Alessio HM: Exercise-induced oxidative stress. *Med Sci Sports Exerc* 1993;25:218-224.
41. Davies KJA, Quintanilha AT, Brooks GA, et al: Free radicals and tissue damage produced by exercise. *Biochem Biophys Res Commun* 1982;107:1198-1205.

42. Ji LL: Antioxidant enzyme response to exercise and aging. *Med Sci Sports Exerc* 1993;25:225-231.
43. Powers SK, Lawler J, Criswell D, et al: Alterations in diaphragmatic oxidative and antioxidant enzymes in the senescent Fischer 344 rat. *J Appl Physiol* 1992;72:2317-2321.
44. Gollnick PD, Riedy M, Quintinskie JJ, et al: Difference in metabolic potential of skeletal muscle fibres and their significance for metabolic control. *J Exp Biol* 1985;115:191-199.
45. Holloszy JO, Coyle EF: Adaptations of skeletal muscle to endurance exercise and their metabolic consequences. *J Appl Physiol* 1984; 56:831-838.
46. Reid MB, Haack KE, Franchek KM, et al: Reactive oxygen in skeletal muscle: I. Intracellular oxidant kinetics and fatigue in vitro. *J Appl Physiol* 1992;73:1797-1804.
47. Taylor AW, Noble EG, Cunningham DA, et al: Ageing, skeletal muscle contractile properties and enzyme activities with exercise, in Sato Y, Poortmans J, Hashimoto I, et al (eds): *Integration of Medical and Sports Sciences*. Basel, Karger Publishing, 1992, pp 109-125.
48. Florini JR, Ewton DZ: Skeletal muscle fiber types and myosin ATPase activity do not change with age or growth hormone administration. *J Gerontol* 1989;44:B110-B117.
49. Caccia MR, Harris JB, Johnson MA: Morphology and physiology of skeletal muscle in aging rodents. *Muscle Nerve* 1979;2:202-212.
50. Ansved T, Larsson L: Effects of ageing on enzyme-histochemical, morphometrical and contractile properties of the soleus muscle in the rat. *J Neurol Sci* 1989;93:105-124.
51. Lowenthal DT: The effect of aging, exercise, and calorie restriction on skeletal muscle histochemistry in Fischer 344 rats. Unpublished doctoral dissertation. 1986, Temple University, Philadelphia, Pennsylvania.
52. Rumsey WL, Kendrick ZV, Starnes JW: Bioenergetics in the aging Fischer 344 rat: Effects of exercise and food restriction. *Exp Gerontol* 1987;22:271-287.
53. McCarter RJ, Masoro EJ, Yu BP: Rat muscle structure and metabolism in relation to age and food intake. *Am J Physiol* 1982;242:R89-R93.

Chapter 16

Physical and Orthopaedic Limitations to Exercise in the Elderly

Marybeth Brown, PT, PhD

The benefits of exercise for the older adult include increases in strength, balance, and flexibility[1-3] and significant improvements in glucose tolerance, hypertension, and physiologic work capacity, or Vo_2max.[4-6] Less clear, however, is whether improvements in physical and physiologic function in older adults come at the expense of skeletal tissues.

In their study of 70- to 79-year-old men who jogged three times per week for 24 weeks, Pollock and associates[7] found the incidence of injury to be 57%. Pollock and associates[8-10] also examined injury rates among joggers and walkers 20 to 79 years of age, and their observations suggest that the older you are, the greater the likelihood of injury, particularly among women. The increasing incidence of osteoarthritis with age[11] may be related to the findings of Pollock and associates.

Whether alternate forms of exercise to enhance aerobic capacity are less likely to result in injury has not been addressed significantly. A few investigators have examined walking programs and report the incidence of injury to be considerably lower than in jogging, between 12% and 14%.[9,12] It is not known whether alternate forms of aerobic exercise that have little or no impact force can promote fitness without injury in the older adult. Additionally, the incidence of injury with low-intensity activity to promote strength and flexibility has not been examined. Consequently, the primary purpose of this investigation was to examine the occurrence of injury or painful episodes that limit participation in two forms of exercise activity—low-intensity mobility exercise and moderate-intensity aerobic training. Physical and orthopaedic factors believed to be related to injury also were examined.

Methods

To be admitted into the study, subjects had to be sedentary nonsmokers who were free of heart disease and willing to exercise five days per week. The average age, height, and weight of the subjects for each exercise program are presented in Table 1. Two hundred and three men and women between the ages of 60 and 72 years participated in three months of flexibility and strengthening exercise. Three participants had to withdraw from the exercise program because pain due to pre-existing conditions (two with neck pain and one with back pain) was aggravated by the exercise and required treatment outside the scope that we could provide. Of the 200 subjects who successfully completed the low-intensity activity program, 126 subsequently went on to complete 12 months of moderate-intensity training, which consisted primarily of fast walking and jogging.

The three-month low-intensity exercise program was performed with the subject sitting, standing, prone, side-lying, supine, on all fours, and sitting Indian style. Exercises were performed slowly, using the weight of the body part and gravity as resistance. As the subject's ability to accomplish the exercises improved, the number of repetitions was increased, the exercises were modified to make them more difficult (such as lifting both legs simultaneously instead of one leg at a time), and subjects were asked to move the body part even more slowly. A more complete description of the exercise program has been presented.[13]

Before participation in the moderate-intensity training program, all subjects were screened for obvious postural deformities and range of motion and strength deficits of the lower extremities. Information regarding low bone mass, previous lower extremity joint surgeries and injuries, obesity, and osteoarthritis were obtained. Based on this information, subjects were placed into jog and nonjog groups.

Moderate-intensity training began initially with walking and fast walking on an indoor track. If appropriate, participants later incorporated jogging into the routine. For those who did not jog, stationary bicycling, Nordic Track, or uphill treadmill walking were added.

A typical exercise program for a jogger might be five minutes of warm-up walking, ten minutes of fast walking, alternating jogging one track revolution and walking five track revolutions during a 20-minute period (the circumference of the track is one-twentieth of a mile), ten minutes of fast walk, and a cool-down walk. A typical exercise program for a nonjogger might be five minutes of warm-up walking, ten minutes of moderate or fast walking, 25 minutes of stationary bicycling (five minutes each at 75, 100, 125, 100, and 75 watts), ten minutes of walking and a cool down.

Training began at approximately 60% of $\dot{V}O_2max$ and was increased to approximately 85% of $\dot{V}O_2max$ as subjects were able. All of the individuals who entered the moderate-intensity training program had completed the three-month stretching/strengthening phase of activity. Because the incidence of injury per se was low in

Table 1 Subject characteristics

Subjects	Number	Age	Height (cm)	Weight (kg)
		Low-intensity activity program*		
Females	116	64.7 ± 3.2	160.5 ± 5.8	65.6 ± 11.9
Males	84	64.9 ± 3.3	174.6 ± 7.8	83.5 ± 13.8
		Moderate-intensity training program+		
Females	77	65.4 ± 2.3	163.1 ± 1.64	66.0 ± 11.1
Males	49	65.4 ± 3.3	177.6 ± 5.0	81.5 ± 12.7

*200 subjects participated.
+126 subjects participated.

both programs, the term painful episode was chosen to characterize those events that caused the subject to stop or alter the exercise program for more than two consecutive days.

Results

Low-Intensity Exercise

Of the 200 participants in the low-intensity exercise program, 146 (73%) progressed through the flexibility/strengthening program without experiencing anything beyond transient muscle soreness. The remaining 54 (27%) subjects had complaints of joint discomfort, primarily at the knee (Table 2). Twenty-five painful episodes were reported by 24 men and 34 episodes were reported by 30 women, for an average of 0.3 episodes of pain per person (Table 3). These painful episodes resolved quickly once conservative measures were instituted, such as modification of the exercises, rest, ice, and wrapping with an elastic bandage.

Moderate Intensity Training Exercise

Of the 49 men and 77 women who participated in one year of moderate-intensity aerobic exercise training, 16 (33%) and 31 (40%), respectively, did not experience any painful episodes. Twenty-four (49%) of the male joggers and 28 (36%) of the female joggers had one or more painful episodes during the training period. Participants with pre-existing orthopaedic conditions who had been placed in the nonjogging group had a higher incidence of painful episodes than did those who jogged; 27 of 29 participants in the fast walk, treadmill walk, or bicycle group had one or more painful episodes. An average of 0.64 painful episodes per person occurred among those who jogged, whereas an average of 1.40 painful episodes per person occurred among those who did not jog (Table 3). The knee became painful far more often than any other site, and the foot and ankle complex was the second most often affected. An outline of painful episodes with aerobic training is presented in Table 4.

Table 2 Painful episodes with mobility exercise

	Females (N=116)	Males (N=84)
No pain	86	60
Pain		
Knee	20	13
Feet	1	1
Back	4	8
Neck	4	2
Hip	5	1
	34	25

Table 3 Painful episodes per person

	Males	Females	Total
Low-intensity exercise	0.30	0.29	0.30
Moderate-intensity exercise	1.04	1.13	1.09
With jogging	0.76	0.58	0.64
Without jogging	1.44	1.32	1.40

Table 4 Incidence of painful episodes by area with moderate-intensity exercise training

	Joggers (N=98)	Nonjoggers (N=27)
Knee pain*	18	10
Foot pain	13	10
Plantar fasciitis	4	5
Anterior compartment	1	1
Posterior compartment	3	0
Hip pain†	13	8
Back pain	7	2
Muscle	4	1
	63	37

*Includes probable osteoarthritis, capsulitis, iliotibial band tendinitis, and pes anserina tendinitis
†Includes subtrochanteric bursitis, probable osteoarthritis, sacroiliac joint pain

Of the 126 participants, ten sustained an actual injury: six muscle strains (hamstrings or adductors), three stress fractures, and one fall. Two men and three woman withdrew from the program with painful conditions that had been exacerbated by the exercise—two participants with knee pain, two with back pain, and one with painful feet.

For the remaining 121 participants, all of the painful episodes were treated successfully using conservative methods. Treatment consisted of exercise modification, stretching, proper shoes, orthotic management, taping, rest, ice, elastic wrapping, and nonsteroidal anti-inflammatory medication. Those with injuries also successfully completed the exercise program following proper periods of rest, exercise modification, wrapping, gentle stretching, self-administered massage, and ice.

A number of problems prevalent in an older adult population arose that had to be addressed. Approximately 5% of the women had complaints of stress incontinence and were unable to fast walk or jog without leakage of urine. Three of the men had hernias, which precluded jogging as a mode of exercise. Three participants experienced dizziness, three had claudication-like symptoms, and three had insensitivity of the feet.

Orthopaedic and Physical Factors Related to Painful Episodes

Obesity, history of previous joint injury and osteoarthritis, foot deformity, limited range of dorsiflexion, and hip flexor tightness emerged as factors strongly related to the development of joint pain. Obesity in this study was defined as being heavier than 125% of ideal body weight.[14] Men and women who were clearly obese usually had a history of osteoarthritis and thus were placed in the nonjogging group. Nearly all of these individuals had painful episodes. Foot deformities included bunions, hallux valgus, calcaneal valgus and varus, subluxed metatarsals, and flattening of the arch. Painful episodes generally presented before appropriate footwear and/or orthotics were obtained. Proper shoes, taping of the foot, and footwear modification eliminated almost all of the painful episodes of the ankle and foot complex. Two of four individuals with plantar fasciitis and all three subjects with posterior calf pain had less than neutral dorsiflexion range of motion. Pain disappeared after dorsiflexion was restored to neutral or above with stretching exercises. Of those with back, hip, and knee pain, 76% had hip flexor tightness of 10° or more (average 14°). In all but one case, reducing hip flexor tightness by stretching to less than 10° eliminated the pain.

Factors believed to be associated with the development of painful episodes are weakness of the calf, quadriceps, hip abductors, and hip extensors; slow contraction times (for example, time to peak torque); and limited hip extension and ankle dorsiflexion range of motion.

Discussion

The data derived from this study support the hypothesis that older adults are more likely to experience a painful episode with moderate-intensity training exercise than with low-intensity mobility exercise.[7-10,12] One quarter of those involved in the flexibility/strengthening program experienced a painful episode, while nearly three quarters of those in the training program had pain. Although impact forces are a likely contributor to the increase in painful episodes with training, as suggested by Pollock and associates,[7] other factors also are related, because those whose training did not include jogging experienced more difficulty than did the joggers.

In this type of older adult population, injury, defined as an acute exercise-related event resulting in immediate cessation of activity, does not appear to be likely. Rather, mild to moderate orthopaedic discomfort may occur, most of which is readily treatable. Given the overwhelmingly positive exercise outcomes from this study,[4-6,13] we strongly endorse flexibility/strengthening and training exercise for older men and women. The positive impact of exercise on risk factors for disease far outweighs the negative aspect of exercise-related painful episodes.

Although our finding of a 64% incidence of painful episodes with jogging is similar to the 57% reported by Pollock and associates[7] for men and women 70 to 79 years of age, several differences are apparent. Subjects in the study by Pollock and associates jogged three times per week.[7] Participants in our study were encouraged to train five times per week and averaged 4.1 days per week attendance. Thus, subjects in our study trained for more hours per week than those in the study by Pollock and associates.[7] Women joggers in our study did not have more painful episodes than men, a finding at variance with those in the latter study. One reason for the possible discrepancy is that in our study more women than men were placed in the nonjogging group, based on information obtained at screening. Another explanation might be that the women in our study gained a significant amount of flexibility and strength during the three-month mobility program.[13] The women in this study might have been better prepared physically for the rigors of jogging than the women in the study by Pollock and associates.[7] Age also might be a factor, because women in this study were an average of six years younger than those studied by Pollock and associates.[7] Percentage of body fat was similar for women in the two studies (approximately 38%) and, thus, it would appear that body mass does not help explain the difference in findings between the two groups.

The older adults in this study were healthy and probably more physically capable than many men and women in the same age group. A high proportion of the participants experienced a painful episode, however. Clearly, painful episodes should be expected during exercise in older adults, and a resource person, such as a physical therapist, should be available to treat participants as soon as the episode occurs. An older adult is likely to adhere more strongly to an exercise program if someone is available to help in the event of a painful episode.

A number of investigators have reported that the most common site of injury with jogging is the knee.[15-19] Data from this study are consistent with this finding, because the knee was affected more often than any other joint. Painful episodes most often involved the knee in all participants of the three exercise groups; those who jogged, those who trained using the treadmill and stationary bicycle, and those who completed the low-intensity flexibility/strengthening program. The incidence of osteoarthritis at the knee with advancing age appears higher than at other lower extremity articulations,[20] which may be related to our findings.

Physical screening prior to exercise revealed a number of subjects who we believed were at risk for a painful episode during jogging because of pre-existing orthopaedic problems, obesity, reduced strength, and range-of-motion deficits. Placing these people in a nonjogging group did not seem to reduce the number of painful episodes, but did appear to provide a means for exercising at-risk subjects without apparent harm. Painful episodes in this group resolved quickly with conservative treatment. Exercise days lost due to pain appeared to be fewer in this cohort than in the jogging group, but hard data were not collected to substantiate this observation.

Physical factors, such as obesity, pre-existing joint injury and osteoarthritis, limited range of hip and ankle motion, and deformity, appeared to be strongly related to the development of pain, but no causal relationship can be assigned. Future study should focus on the question of who can exercise safely and under what conditions. The possible relationship of strength and speed of torque development to injury and painful episodes needs to be examined.

Conclusions

Painful episodes with exercise are likely in an older adult population, more so with aerobic training than with lower-intensity flexibility/strengthening activity. Although a large proportion of exercising older adults probably will experience painful episodes, conservative measures, such as exercise modification, rest, taping, nonsteroidal anti-inflammatory drugs, and stretching, are likely to be very successful in resolving them. Physical factors that appear to be related to the development of painful episodes include obesity, pre-existing joint injury and osteoarthritis, joint deformity, limited hip and ankle range of motion, and, possibly, weakness. No causal relationship was defined. If painful episodes are treated appropriately, 60- to 71-year-old adults can exercise successfully.

Acknowledgment

Supported by Program Project Grant AG05562 from the National Institutes of Health, John O. Holloszy, MD, principal investigator.

References

1. Frontera WR, Meredith CN, O'Reilly KP, et al: Strength conditioning in older men: Skeletal muscle hypertrophy and improved function. *J Appl Physiol* 1988;64:1038-1044.
2. Agre JC, Pierce LE, Raab DM, et al: Light resistance and stretching exercise in elderly women: Effect upon strength. *Arch Phys Med Rehabil* 1988;69:273-276.
3. Larsson L: Physical training effects on muscle morphology in sedentary males at different ages. *Med Sci Sports Exerc* 1982;14:203-206.

4. Holloszy JO, Spina RJ, Kohrt WM: Health benefits of exercise in the elderly, in Sato Y, Poortmans J, Hashimoto I, et al (eds): *Integration of Medical and Sports Sciences*. Basel, Switzerland, S Karger, 1992, vol 37, pp 91-108.
5. Kohrt WM, Malley MT, Coggan AR, et al: Effects of gender, age, and fitness level on response of VO_2max to training in 60 71-yr olds. *J Appl Physiol* 1991;71:2004-2011.
6. Martin WH III, Ogawa T, Kohrt WM, et al: Effects of aging, gender, and physical training, on peripheral vascular function. *Circulation* 1991;84:654-664.
7. Pollock ML, Carroll JF, Graves JE, et al: Injuries and adherence to walk/jog and resistance training programs in the elderly. *Med Sci Sports Exerc* 1991;23:1194-1200.
8. Pollock ML, Gettman LR, Milesis CA, et al: Effects of frequency and duration of training on attrition and incidence of injury. *Med Sci Sports Exerc* 1977;9:31-36.
9. Pollock ML, Miller HS Jr, Janeway R, et al: Effects of walking on body composition and cardiovascular function of middle-aged men. *J Appl Physiol* 1971;30:126-130.
10. Pollock ML, Dawson GA, Miller HS Jr, et al: Physiologic responses of men 49 to 65 years of age to endurance training. *J Am Geriatr Soc* 1976;24:97-104.
11. Matheson GO, Macintyre JG, Taunton JE, et al: Musculoskeletal injuries associated with physical activity in older adults. *Med Sci Sports Exerc* 1989;21:379-385.
12. Carroll JF, Pollock ML, Graves JE, et al: Incidence of injury during moderate and high-intensity walking training in the elderly. *J Gerontol* 1992;47:M61-M66.
13. Brown M, Holloszy JO: Effects of a low intensity exercise program on selected physical performance characteristics of 60 to 71 year olds. *Aging* 1991;3:129-139.
14. Metropolitan Life Insurance Standards for Ideal Body Weight.
15. Lane NE, Bloch DA, Wood PD, et al: Aging, long-distance running, and the development of musculoskeletal disability: A controlled study. *Am J Med* 1987;82:772-780.
16. Macera CA, Pate RR, Powell KE, et al: Predicting lower-extremity injuries among habitual runners. *Arch Intern Med* 1989;149:2565-2568.
17. Kilbom A, Hartley LH, Saltin B, et al: Physical training in sedentary middle-aged and older men: I. Medical evaluation. *Scand J Clin Lab Invest* 1969;24:315-322.
18. Kannus P, Niittymaki S, Jarvinen M, et al: Sports injuries in elderly athletes: A three-year prospective, controlled study. *Age and Ageing* 1989;18:263-270.
19. Koplan JP, Powell KE, Sikes RK, et al: An epidemiologic study of the benefits and risks of running. *JAMA* 1982;248:3118-3121.
20. Panush RS, Brown DG: Exercise and athritis. *Sports Med* 1987;4:54-64.

Chapter 17

Sarcopenia: The Age-Related Loss in Skeletal Muscle Mass

William J. Evans, PhD

Introduction

Advancing adult age is associated with profound changes in body composition. Using total body potassium as an index of fat-free mass, Novak[1] assessed the body composition of more than 500 men and women between the ages of 18 and 85 years. He determined that body fat increased from 18% to 36% in men and from 33% to 44% in women. Cohn and associates,[2] using total body neutron activation procedures, determined that the principal component of the decline in fat-free mass was a decrease in muscle mass, with minimal change in nonmuscle mass. Age-related loss in skeletal muscle has been referred to as sarcopenia.[3] Cohn and associates[2] also observed that total body nitrogen declined in very close association with total body calcium, which suggests a link between sarcopenia and osteopenia.

Skeletal muscle is the largest reservoir of protein in the body. Age-related reduction in muscle is a direct cause of the age-related decrease in muscle strength.[4-8] Our laboratory[5] recently examined muscle strength and mass in 1,200 healthy 45- to 78-year-old men and women, and concluded that muscle mass (not function) is the major determinant of age- and sex-related differences in strength. This relationship is independent of muscle location (upper versus lower extremities) and function (extension versus flexion). Reduced muscle strength is a major cause of the increased prevalence of disability in the elderly.[9] Muscle strength is also closely related to functional capacity. Bassey and associates[10] measured muscle strength and amount and speed of customary walking in a large sample of men and women older than 65 years. They found an age-related decline in muscle strength and a significant negative correlation between strength and chosen walking speed for both sexes ($r = -0.41$, $p < 0.001$ for men; $r = -0.36$, $p < 0.01$ for women). Fiatarone and associates[11] observed a closer relationship between quadriceps strength and walking time ($r = 0.745$, $p < 0.01$) and number of steps taken during a 6-meter walk ($r = 0.717$, $p < 0.01$) in a group of frail, institutionalized men and women (ages 86 to 96 years). In these subjects, fat-free mass ($r = 0.732$) and regional muscle mass estimated by computed tomography ($r = 0.752$) were correlated with muscle strength. In the same population, it was recently demonstrated that leg power is closely associated with functional performance.[12] In older, frail women, leg power was highly correlated with walking speed ($r = 0.93$, $p < 0.001$). These data suggest that with advancing age and with the very low activity levels seen in the elderly, muscle strength is a critical component of walking ability. That the high prevalence of falls among the institutionalized elderly may be a consequence of their lower muscle strength was suggested by Whipple and associates[13] who examined nursing home residents and found a significant reduction in muscle strength in elderly fallers compared with nonfallers.

Daily energy expenditure declines progressively throughout adulthood.[14] In sedentary individuals, the main determinant of energy expenditure is fat-free mass,[15] which declines by about 15% between the third and eighth decades of life, contributing to a lower basal metabolic rate in the elderly.[2] Tzankoff and Norris[16] saw that 24-hour creatinine excretion (an index of muscle mass) was closely related to basal metabolic rate at all ages. These data indicate that preservation of muscle mass and prevention of sarcopenia can help prevent the decrease in metabolic rate.

In addition to its role in energy metabolism, skeletal muscle and its age-related decline may contribute to such age-associated changes as reduction in bone density,[17-19] insulin sensitivity,[20] and aerobic capacity.[21] For these reasons, strategies for preservation of muscle mass with advancing age and for increasing muscle mass and strength in the previously sedentary elderly may be an important way to increase functional independence and decrease the prevalence of many age-associated chronic diseases.

Disuse

The remarkable similarities between the effects of disuse and those of aging on skeletal muscle lead to the hypothesis that many age-related changes in muscle are due more to disuse than to age. Reduced activity levels will cause loss of protein from muscle cells. This reduced protein content results from increased rates of protein turnover, with the rate of breakdown exceeding the rate of synthesis. The primary characteristic of muscle disuse atrophy is a reduction in fiber size. The degree of atrophy of muscle during immobilization is highly dependent on the fixed length of the muscle. Muscles that are fixed at greater than resting length have a far slower rate of atrophy than muscle in the shortened position. The susceptibility of specific fiber types to immobilization atrophy remains in question. Some investigators have demonstrated that the greater number of corticosteroid receptor proteins will cause higher mobilization of sarcoplasmic proteins from type II fibers during stress, such as immobilization or starvation.[22] Muscle strength is directly associated with cross-sectional area; however, disuse atrophy appears to cause not only a loss in absolute strength, but also a loss in strength per cross-sectional area. The loss in strength per muscle is related to the disproportionate loss of contractile protein compared to other cell components.

Disuse is also associated with decreased capacity to utilize substrates, such as fatty acids, glucose, and pyruvate. In addition, immobilization rapidly decreases glucose tolerance. Glucose tolerance is significantly decreased after one to three weeks of bed rest in normal subjects.[23] In mice, there is a decrease in the insulin responsiveness of the soleus (type I) muscle after only one day of immobilization. Bed rest may lead to a false-positive glucose tolerance test and diagnosis of diabetes.[24] Age-related changes in skeletal muscle, such as decreased insulin sensitivity, mass, strength, endurance, and glycogen content, may be a product of the reduced activity often seen with advancing age.[25]

Strength Training in the Elderly

Strength conditioning generally is defined as training in which the resistance against which a muscle generates force is increased progressively over time. Muscle strength has been shown to increase in response to training between 60% and 100% of the one repetition maximum (1 RM), which is the maximum amount of weight that can be lifted with one contraction. Strength conditioning will result in an increase in muscle size, which largely is the result of an increase in contractile proteins. The mechanisms by which the mechanical events stimulate an increase in ribonucleic acid (RNA) synthesis and subsequent protein synthesis are not well understood. Lifting a weight requires that a muscle shortens as it produces force. This is called a concentric contraction. Lowering the weight, on the other hand, forces the muscle to lengthen as it produces force. This is an eccentric muscle contraction. These lengthening muscle contractions can produce ultrastructural damage, which may stimulate increased muscle protein turnover.

The capacity of older men and women to respond to strength training is preserved into very old age. Our laboratory[26] measured the effects of a progressive-resistance, strength-conditioning program of the thigh muscles in healthy men aged 60 to 72 years. Unlike other studies that examined the effects of strength training in older persons,[27,28] we used a very high intensity stimulus (80% of the 1RM maximum) during the training. After 12 weeks, knee extensor and flexor strength increased by 107% ($p < 0.0001$) and 227% ($p < 0.0001$), respectively, and total muscle area, measured by computed tomography, increased by 11.4%. Type I and type II muscle fiber area (biopsy from the vastus lateralis) increased significantly (33.5% and 27.6%, respectively). Daily excretion of urinary

3-methylhistidine (indicating actin-myosin breakdown rates) increased by 41%. The adaptations to strength training observed in this study are very similar to those reported for younger subjects after training of similar intensity and duration.

Half of the men who participated in this study were given a daily protein/calorie supplement (S), providing an extra 560 ± 16 kcal/day (16.3% as protein, 43.6% as carbohydrate, and 40.1% as fat) in addition to their normal ad lib diet. The rest of the subjects received no supplement (NS) and consumed an ad lib diet. By the 12th week of the study, dietary calorie (2,960 ± 230 in S versus 1,620 ± 80 kcal in NS) and protein (118 ± 10 in S versus 72 ± 11 g/day in NS) intake was significantly different between the S and NS groups. Composition of the midthigh, estimated by computed tomography, showed that the S group had greater gains in muscle than did the NS men. In addition, urinary creatinine excretion was greater at the end of the training in the S group than in the NS group,[29] indicating a greater muscle mass in the S group. There was no difference in strength gains between the two groups.

Frontera and associates[30] also measured upper body and lower body VO_2 max before and after 12 weeks of strength training. Leg VO_2 max showed a slight but significant increase, but arm VO_2 max was unchanged in response to leg-muscle strength training. The strength training caused an increase in citrate synthase activity of the vastus lateralis, but it had no effect on cardiovascular function as measured by hemoglobin content, red blood cell volume, plasma volume, or total blood volume. These data indicate that muscle hypertrophy may cause an increase in maximal aerobic power.

Fiatarone and associates[11] examined the effects of a high-intensity progressive resistance training program on a group of institutionalized men and women aged 86 to 96 years. They saw that after only eight weeks, knee extensor strength increased by an average of 174% ± 31%. Midthigh muscle area increased by 9.0% ± 4.5%. The average tandem gait speed improved by 48% after training. These studies indicate that progressive resistance training will stimulate large increases in muscle size and strength at any age.

Studies on Skeletal Muscle Damage

Our laboratory has been examining the metabolic responses to eccentric exercise, particularly in the differences between younger and older subjects.[31-34] Eccentric exercise is a natural component of virtually all physical activity and is a major component of a progressive resistance training program. Following one bout of high-intensity eccentric exercise,[35] previously sedentary men (average age = 64 years) showed a prolonged increase in the rate of muscle protein breakdown, evidenced by an increase in urinary 3-methylhistidine/creatinine, which peaked ten days later. In addition, we measured an increase in circulating interleukin (IL-1) levels in these subjects three hours after the exercise. Endurance-trained men in the same age group, performing the same exercise, did not display increased circulating IL-1 levels. However, their pre-exercise plasma IL-1 levels were significantly higher than those seen in the untrained subjects. The temporal sequence of changes seen in this study suggests that the myofibrillar damage resulting from the performance of eccentric exercise induced a response similar to the acute phase response to infection and that this response modulates changes in muscle protein turnover involved in repair.

Damage to tissue, like infection, stimulates a wide range of defense reactions, known as the acute phase response.[36] The acute phase response is critical for its antiviral and antibacterial actions, as well as for promoting the clearance of damaged tissue and subsequent repair. Within hours of injury or exercise,[32] the number of circulating neutrophils can increase manyfold. Neutrophils migrate to the site of injury, where they phagocytize tissue debris and release factors known to increase protein breakdown, such as lysozyme and oxygen radicals.[37] Greater neutrophil increases have been observed after eccentric exercise than after concentric exercise.[38] While neutrophils have a relatively short half-life (one or two days within tissue),[39] the life span of monocytes may be one to two months after migration to damaged tissue.[40] Substantial monocyte accumulation in skeletal muscle was found after completion of a marathon. Following eccentric exercise, monocyte accumulation in muscle was not seen until four to seven days later.[41,42] In addition to their ability to phagocytize damaged tissue, monocytes secrete cytokines, such as IL-1 and tumor necrosis factor (TNF). These and other cytokines mediate a wide range of metabolic events, which have an effect on virtually every organ system in the body.

Elevated cytokine levels during infection or injury have different and selective effects. IL-1 mediates an elevated core temperature during infection.[43] In laboratory animals, IL-1 and TNF increase muscle proteolysis and liberation of amino acids,[44] possibly providing substrate for increased hepatic protein synthesis. While circulating IL-1 has been shown to increase acutely as a result of eccentric exercise,[35] it returned to resting levels by 24 hours after the exercise. Biopsies of the vastus lateralis taken before, immediately after, and five days after downhill running showed an immediate and prolonged increase in IL-1β.[45] This study implicates muscle IL-1β in the postexercise change in protein metabolism. Cytokines influence some of their target tissues via induction of prostaglandin E_2 (PGE_2), which also serves as a negative feedback inhibitor of cytokine production.[46]

IL-1 has been shown to stimulate muscle proteolysis in vitro.[47] In vivo, IL-1 can influence muscle protein metabolism by stimulating the production of PGE_2, which, in turn, is known to increase muscle lysosome function.[48] Our laboratory recently observed a significant longitudinal correlation between changes in IL-1 secretion and changes in 3-methylhistidine excretion over a 12-day period following a single bout of eccentric exercise in previously sedentary men.[31] On day 12 following the exercise, a significant correlation was found between 3-methylhistidine excretion and both IL-β and PGE_2 secretion. There is evidence that IL-1 activates branched chain α-keto acid dehydrogenase[44] (a rate-limiting enzyme for amino acid oxidation in skeletal muscle) and increases muscle protein breakdown, as determined by leucine tracer studies.[49] The results of the investigation by Cannon and associates[31] are consistent with the concept that mononuclear cell products, including PGE_2 and IL-1β, are regulators of muscle proteolysis.

Protein Metabolism

Eccentric exercise induces increases in muscle hydrolase activity,[50,51] the intracellular concentration of Ca^{2+},[52] and IL-1β. Urinary 3-methylhistidine levels indicate that muscle protein turnover also is increased. Fielding and associates[33] used a primed, constant infusion of 1-^{13}C-leucine as an indicator of whole body protein metabolism, before, immediately after, and ten days after a single bout of high-intensity eccentric exercise in previously sedentary old and young men. They observed a 9% increase in leucine flux immediately following exercise and persisting for up to ten days, suggesting an increase in protein breakdown. In addition, the rate of leucine oxidation increased in a similar manner. Expressed per kilogram of body weight, there were no apparent differences in the rate of leucine metabolism between the younger and older men. Leucine oxidation and flux were significantly elevated (compared to pre-exercise samples) at both postexercise time points, indicating a prolonged increase in protein turnover. However, the ratio of 3-methylhistidine to creatinine rose sooner and remained elevated longer in the older men. Consistent with these changes in protein degradation, ultrastructural examination of muscle biopsies obtained from the young and old men revealed nearly six times more myofibrillar damage in muscle of the older men.[34] The reason for the greater amounts of damage seen in the older subjects may be due to their lower fitness levels. It is interesting to note that despite the much greater level of muscle damage seen in the older men, the increase in circulating creatinine kinase (CK) activity (normalized for muscle mass) over the days after exercise was no different between young and old. These results suggest that older subjects have an exaggerated proteolytic response following a single bout of eccentric exercise. These data also indicate that exercise that results in muscle damage has a long-lasting effect on protein metabolism, which can cause an increased need for dietary protein. Previously sedentary young men consuming 1 g protein/kg/day show an increased urinary nitrogen excretion and a prolonged period of negative nitrogen balance when beginning a vigorous exercise program.[53] It has also been demonstrated that endurance trained men (ages 20 to 58 years) who spent between four and 18 hours per week exercising had a relatively high protein requirement.[54] Using nitrogen balance to assess protein needs, this study found that the current recommended daily allowance of 0.8 g protein/kg/day was inadequate to maintain balance, even when the diet provided sufficient calories. These studies indicate that the need for dietary protein may be high at the initiation of training and remain somewhat elevated as long as training continues.

Eccentric Exercise and Vitamin E

Our laboratory has recently examined the effects of vitamin E on exercise-induced muscle damage in old and young men.[31,32] It was our hypothesis that eccentric exercise results in an increase in the production of neutrophil-generated oxygen radicals and causes the exercise-induced increase[55,56] in lipid peroxidation. We expected that increased rates of lipid peroxidation would then contribute to increased membrane permeability and increased leakage of muscle proteins. Because of the well-known properties of vitamin E as an antioxidant and oxygen radical scavenger, it might reduce the eccentric exercise-caused increase in circulating CK activity.

We examined older (55- to 74-year-old) and younger (22- to 29-year-old) subjects after some of them had received 800 IU of vitamin E for 48 days in a double-blind placebo-controlled study. The subjects ran downhill on an inclined treadmill (−16% grade) for 45 minutes. The results were contrary to our expectations. There was a clear difference in CK release and neutrophilia between the younger and older placebo groups, with the younger subjects demonstrating a significantly greater circulating CK activity and neutrophil count following the eccentric exercise. However, vitamin E affected the older subjects by increasing their responses, which eliminated the differences between the two age groups. At the time of peak concentrations in the plasma, circulating CK activity correlated ($r = 0.751$, $p < 0.001$) with superoxide release from neutrophils.

The striking difference between the younger and older placebo groups points to the fact that certain metabolic responses to eccentric exercise may be attenuated with advancing age. The association of circulating skeletal muscle enzyme activity with neutrophil mobilization and function supports the concept that neutrophils are involved in the delayed increase in muscle membrane permeability after damaging exercise. These data indicate that vitamin E supplementation may affect the rate of repair of skeletal muscle following muscle damage, and that these effects may be more pronounced in older subjects.

Summary

Sarcopenia, defined as the age-related loss in skeletal muscle mass, results in decreased strength, aerobic capacity, and functional capacity. Sarcopenia is also closely linked to age-related losses in nonmineral, basal metabolic rate and increased body fat content. Through physical exercise and training, especially a combination of aerobic exercise and resistance training, it may be possible to prevent sarcopenia and the remarkable array of associated abnormalities, such as type II diabetes, coronary artery disease, hypertension, osteoporosis, and obesity. Using an exercise program of sufficient frequency, intensity, and duration, it is possible to increase muscle strength and endurance, even in the oldest elderly. No pharmacologic intervention holds greater promise of improving health and promoting independence in the elderly than exercise.

References

1. Novak LP: Aging, total body potassium, fat-free mass, and cell mass in males and females between ages 18 and 85 years. *J Gerontol* 1972;27:438-443.
2. Cohn SH, Vartsky D, Yasumura S, et al: Compartmental body composition based on total-body nitrogen, potassium, and calcium. *Am J Physiol* 1980;239:E524-E530.
3. Evans WJ, Rosenberg IH, Thompson J (eds): *Biomarkers: The Ten Determinants of Aging You Can Control*. New York, NY, Simon & Schuster, 1991.
4. Bruce SA, Newton D, Woledge RC: Effect of age on voluntary force and cross-sectional area of human adductor pollicis muscle. *Q J Exp Physiology* 1989;74:359-362.
5. Frontera WR, Hughes VA, Lutz KJ, et al: A cross-sectional study of muscle strength and mass in 45- to 78-year-old men and women. *J Appl Physiol* 1991;71:644-650.
6. Larsson L, Grimby G, Karlsson J: Muscle strength and speed of movement in relation to age and muscle morphology. *J Appl Physiol* 1979;46:451-456.
7. Rice CL, Cunningham DA, Paterson DH, et al: Strength in an elderly population. *Arch Phys Med Rehabil* 1989;70:391-397.
8. Young A, Stokes M, Crowe M: Size and strength of the quadriceps muscles of old and young women. *Eur J Clin Invest* 1984;14:282-287.
9. Jette AM, Branch LG: The Framingham Disability Study: II. Physical disability among the aging. *Am J Public Health* 1981;71:1211-1216.
10. Bassey EJ, Bendall MJ, Pearson M: Muscle strength in the triceps surae and objectively measured customary walking activity in men and women over 65 years of age. *Clin Sci* 1988;74:85-89.
11. Fiatarone MA, Marks EC, Ryan ND, et al: High-intensity strength training in nonagenarians: Effects on skeletal muscle. *JAMA* 1990;263:3029-3034.
12. Bassey EJ, Fiatarone MA, O'Neill EF, et al: Leg extensor power and functional performance in very old men and women. *Clin Sci* 1992;82:321-327.
13. Whipple RH, Wolfson LI, Amerman PM: The relationship of knee and ankle weakness to falls in nursing home residents: An isokinetic study. *J Am Geriatr Soc* 1987;35:13-20.
14. McGandy RB, Barrows CH Jr, Spanias A, et al: Nutrient intakes and energy expenditure in men of different ages. *J Gerontol* 1966; 21:581-587.
15. Ravussin E, Lillioja S, Anderson TE, et al: Determinants of 24-hour energy expenditure in man: Methods and results using a respiratory chamber. *J Clin Invest* 1986;78:1568-1578.

16. Tzankoff SP, Norris AH: Longitudinal changes in basal metabolism in man. *J Appl Physiol* 1978;45:536-539.
17. Bevier WC, Wiswell RA, Pyka G, et al: Relationship of body composition, muscle strength, and aerobic capacity to bone mineral density in older men and women. *J Bone Miner Res* 1989;4:421-432.
18. Sinaki M, McPhee MC, Hodgson SF, et al: Relationship between bone mineral density of spine and strength of back extensors in healthy postmenopausal women. *Mayo Clin Proc* 1986;61:116-122.
19. Snow-Harter C, Bouxsein M, Lewis B, et al: Muscle strength as a predictor of bone mineral density in young women. *J Bone Miner Res* 1990;5:589-595.
20. Kolterman OG, Insel J, Saekow M, et al: Mechanisms of insulin resistance in human obesity: Evidence for receptor and postreceptor defects. *J Clin Invest* 1980;65:1272-1284.
21. Fleg JL, Lakatta EG: Role of muscle loss in the age-associated reduction in VO2max. *J Appl Physiol* 1988;65:1147-1151.
22. Evans WJ: Exercise and muscle metabolism in the elderly, in Hutchinson ML, Munro HN (eds): *Nutrition and Aging*. Orlando, Academic Press Inc, 1986, chap 13, pp 179-191.
23. Lutwak L, Whedon GD: The effect of physical conditioning on glucose tolerance. *Clin Res* 1959;7:143-144.
24. Steinke J, Soeldner JS: Diabetes mellitus, in Thorn GW, Adams RD, Braunwald E, et al (eds): *Harrison's Principles of Internal Medicine*. New York, McGraw-Hill, 1977, chap 95, pp 563-583.
25. Bortz WM II: Disuse and aging. *JAMA* 1982;248:1203-1208.
26. Frontera WR, Meredith CN, O'Reilly KP, et al: Strength conditioning in older men: Skeletal muscle hypertrophy and improved function. *J Appl Physiol* 1988;64:1038-1044.
27. Aniansson A, Gustafsson E: Physical training in elderly men with special reference to quadriceps muscle strength and morphology. *Clin Physiol* 1981;1:87-98.
28. Moritani T, deVries HA: Potential for gross muscle hypertrophy in older men. *J Gerontol* 1980;35:672-682.
29. Meredith CN, Frontera WR, O'Reilly KP, et al: Body composition in elderly men: Effect of dietary modification during strength training. *J Am Geriatr Soc* 1992;40:155-162.
30. Frontera WR, Meredith CN, O'Reilly KP, et al: Strength training and determinants of VO2 max in older men. *J Appl Physiol* 1990;68:329-333.
31. Cannon JG, Meydani SN, Fielding RA, et al: Acute phase response in exercise. II. Associations between vitamin E, cytokines, and muscle proteolysis. *Am J Physiol* 1991;260:R1235-R1240.
32. Cannon JG, Orencole SF, Fielding RA, et al: Acute phase response in exercise: Interaction of age and vitamin E on neutrophils and muscle enzyme release. *Am J Physiol* 1990;259:R1214-R1219.
33. Fielding RA, Meredith CN, O'Reilly KP, et al: Enhanced protein breakdown after eccentric exercise in young and older men. *J Appl Physiol* 1991;71:674-679.
34. Manfredi TG, Fielding RA, O'Reilly KP, et al: Plasma creatine kinase activity and exercise-induced muscle damage in older men. *Med Sci Sports Exerc* 1991;23:1028-1034.
35. Evans WJ, Meredith CN, Cannon JG, et al: Metabolic changes following eccentric exercise in trained and untrained men. *J Appl Physiol* 1986;61:1864-1868.

36. Kampschmidt RF: Leukocytic endogenous mediator/endogenous pyrogen, in Powanda MC, Canonico PG (eds): *Infection: The Physiologic and Metabolic Responses of the Host*. Elsevier/North-Holland, Amsterdam, Biomedical Press, 1981, chap 3, pp 55-74.
37. Babior BM, Kipnes RS, Curnutte JT: Biological defense mechanisms: The production by leukocytes of superoxide, a potential bactericidal agent. *J Clin Invest* 1973;52:741-744.
38. Smith LL, McCammon M, Smith S, et al: White blood cell response to uphill walking and downhill jogging at similar metabolic loads. *Eur J Appl Physiol* 1989;58:833-837.
39. Bainton DF: Phagocytic cells: Developmental biology of neutrophils and eosinophils, in Gallin JI, Goldstein IM, Snyderman R (eds): *Inflammation: Basic Principles and Clinical Correlates*. New York, Raven Press Ltd, 1988, chap 16, pp 265-280.
40. Johnston RB Jr: Current concepts: Immunology: Monocytes and macrophages. *N Engl J Med* 1988;318:747-752.
41. Jones DA, Newham DJ, Round JM, et al: Experimental human muscle damage: Morphological changes in relation to other indices of damage. *J Physiol (London)* 1986;375:435-448.
42. Round JM, Jones DA, Cambridge G: Cellular infiltrates in human skeletal muscle: Exercise induced damage as a model for inflammatory muscle disease? *J Neurol Sci* 1987;82:1-11.
43. Cannon JG, Kluger MJ: Endogenous pyrogen activity in human plasma after exercise. *Science* 1983;220:617-619.
44. Nawabi MD, Block KP, Chakrabarti MC, et al: Administration of endotoxin, tumor necrosis factor, or interleukin 1 to rats activates skeletal muscle branched-chain α-keto acid dehydrogenase. *J Clin Invest* 1990;85:256-263.
45. Cannon JG, Fielding RA, Fiatarone MA, et al: Increased interleukin-1 in human skeletal muscle after exercise. *Am J Physiol* 1989; 257:R451-R455.
46. Kunkel SL, Scales WE, Spengler R, et al: Dynamics and regulation of macrophage tumor necrosis factor-a (TNF), interleukin-1-a (IL-1a) and interleukin-1-B (IL-1B) gene expression by arachidonate metabolites, in Powanda MC, Oppenheim JJ, Kluger MJ, et al (eds): *Monokines and Other Non-Lymphocytic Cytokines*. New York, Alan R Liss, 1988, pp 61-66.
47. Baracos V, Rodemann HP, Dinarello CA, et al: Stimulation of muscle protein degradation and prostaglandin E2 release by leukocytic pyrogen (interleukin-1): A mechanism for the increased degradation of muscle proteins during fever. *N Engl J Med* 1983;308:553-558.
48. Rodemann HP, Goldberg AL: Arachidonic acid, prostaglandin E2 and F2 alpha influence rates of protein turnover in skeletal and cardiac muscle. *J Biol Chem* 1982;257:1632-1638.
49. Flores EA, Bistrian BR, Pomposelli JJ, et al: Infusion of tumor necrosis factor/cachectin promotes muscle catabolism in the rat: A synergistic effect with interleukin 1. *J Clin Invest* 1989;83:1614-1622.
50. Vihko V, Salminen A, Rantamäki J: Acid hydrolase activity in red and white skeletal muscle of mice during a two-week period following exhausting exercise. *Pfluegers Archiv* 1978;378:99-106.
51. Vihko V, Salminen A, Rantamäki J: Exhaustive exercise, endurance training, and acid hydrolase activity in skeletal muscle. *J Appl Physiol* 1979;47:43-50.
52. Duan C, Delp MD, Hayes DA, et al: Rat skeletal muscle mitochondrial [Ca2+] and injury from downhill walking. *J Appl Physiol* 1990; 68:1241-1251.

53. Gontzea I, Sutzescu P, Dumitrache S: The influence of muscle activity on nitrogen balance and the need of man for proteins. *Nutr Rep Int* 1974;10:35-43.
54. Meredith CN, Zachin MJ, Frontera WR, et al: Dietary protein requirements and body protein metabolism in endurance-trained men. *J Appl Physiol* 1989;66:2850-2856.
55. Davies KJ, Quintanilha AT, Brooks GA, et al: Free radicals and tissue damage produced by exercise. *Biochem Biophys Res Commun* 1982;107:1198-1205.
56. Dillard CJ, Litov RE, Savin WM, et al: Effects of exercise, vitamin E, and ozone on pulmonary function and lipid peroxidation. *J Appl Physiol* 1978;45:927-932.

Chapter 18

Role of Growth Hormone and IGF-I in Mediating the Anabolic Response

David R. Clemmons, MD
L.E. Underwood, MD

Introduction

The precise role of the insulin-like growth factors (IGFs) in maintaining muscle mass and strength in elderly subjects has not been determined. Previous studies have shown that in a significant number of normal elderly persons, growth hormone (GH) and IGF-I levels in serum are reduced.[1] Because the administration of either growth hormone or IGF-I to GH-deficient animals under experimental conditions results in increased muscle and connective tissue mass, it has been proposed that these hormones may be important in maintaining muscle mass. The recent availability of biosynthetic GH and IGF-I in sufficient quantities to be administered to humans has made testing of the therapeutic efficacy of these hormones in this age group a feasible goal.

The IGFs are small peptides that bear a strong structural homology to insulin.[2] Unlike insulin, however, the IGFs circulate in blood in substantially higher concentrations than are necessary to stimulate a maximum mitogenic effect. IGF-I secretion is under the control of GH. Patients deficient in GH have low serum IGF-I concentrations, which rise to normal following GH administration.[3] Patients with GH-producing pituitary tumors have elevated levels of both GH and IGF-I.[4] The soft tissue and skeletal enlargement that occurs in these patients is believed to be mediated by IGF-I.

The studies of IGF-I action were performed using in vitro test systems because insufficient material was available for in vivo testing. These initial studies showed that IGF-I would stimulate both DNA and protein synthesis in muscle and skeletal tissues.[5] More recently, IGF-I infusion studies have shown that IGF-I stimulates long bone growth and muscle protein synthesis in hypophysectomized animals.[6,7] Likewise, transgenic mice in which GH secretion has been selectively destroyed and in which the IGF-I transgene has been expressed have been shown to grow normally and to have body size and tissue mass comparable to that of normal mice.[8]

Therefore, IGF-I appears to mimic most, if not all, of the growth-promoting actions of GH. The somatomedin hypothesis of GH action has been established recently in humans, in that patients with total resistance to the growth-promoting actions of GH, caused by mutations of the GH receptor, have been shown to grow normally over periods ranging between three and 12 months in response to daily IGF-I administration.[9] These findings suggest that almost all of the growth-promoting actions of GH could be mediated by IGF-I.

In addition to the growth-promoting actions of GH, IGF-I can also mimic many of its metabolic effects. Administration of IGF-I enhances protein synthesis in animals in whom protein synthesis has been attenuated by starvation, administration of glucocorticoids, or hypophysectomy. The animals respond to IGF-I infusion with an inhibition of protein breakdown and some stimulation of protein synthesis.[10,11] However, studies in rodents suggest that GH is a better stimulant of protein synthesis than IGF-I, although both substances appear to have equivalent effects on inhibiting protein breakdown.[12] Therefore, IGF-I can at least hypothetically be considered as a potential substitute for GH in elderly subjects who have diminished muscle mass and strength and who may respond to GH.

Anabolic Effects of GH

Skeletal muscle mass declines with age. This absolute decline in the amount of skeletal mass is associated with increasing frailty, which can result in a propensity to falling. Likewise, in several diseases of aging, such as arthritis, that result in reduced mobility, decreased muscle mass in combination with restricted movement can further accentuate the frailty and inability to maintain normal functional activity. This loss of muscle mass is associated with an increase in adiposity such that body fat content is doubled at age 60 years compared with body fat content at 30 years, even if weight is maintained at a constant level. Several observations suggest that exogenous administration of GH or other anabolic substances could potentially reverse this process in the elderly. Specifically, in men older than 40 years of age and in women older than 60 years of age, there is a marked decrease in GH secretion.[13] Although this is not uniform, it occurs in a substantial number of subjects, such that at least a third of men older than 60 years have such low levels of GH secretion that they can be classified as GH deficient. Their serum IGF-I concentrations are reduced into the GH deficient range.[13]

In addition to its muscle trophic properties, GH is known to be lipolytic. Acute administration of GH results in an increase in free fatty acids. This lipolytic response can be maintained over several weeks of injections.[14] Because of these properties, it has been suggested that the physiologic decrease that occurs in GH secretion with aging is associated with the relative increase in fat mass and the relative loss of muscle mass that occur with aging. Similarly, GH stimulates bone turnover, so that the decrease in the relative rates of bone formation and bone reabsorption that occur during aging has been proposed to be caused by the decrease in GH secretion.

GH has been administered to a variety of subjects under various test conditions that would be associated with an increased catabolic response. The catabolic processes that have been studied most extensively in humans are the obligatory nitrogen loss that occurs in response to dietary restriction,[14] induction of general anesthesia and surgical procedures,[15] glucocorticoid treatment,[16] and chronic obstructive lung disease in the elderly,[17] as well as the normal aging process.[18] Administration of GH to patients with intact pituitary function who are undergoing surgery has been shown to improve the negative nitrogen balance. In one study, short-term administration increased protein synthesis and decreased protein breakdown.[19] All of these effects occurred in spite of hypocaloric feeding, which should have further enhanced catabolism. The most extensively studied catabolic model has been that of dietary restriction. Specifically, restriction of nutrients to 33%, 50%, or 66% of normal caloric intake induces negative nitrogen balance. In spite of these restrictions, however, administration of GH to such subjects has increased nitrogen retention, varying between 110 and 193 mmol of nitrogen per day. This effect can be maintained easily for two weeks, and sometimes for periods as long as four weeks, of dietary restriction.[14,19,20] Following four weeks of caloric restriction, however, the response attenuates and there is evidence that subjects so restricted become refractory to the nitrogen-sparing effects of GH and/or IGF-I. Studies using the glucocorticoid-induced model of protein catabolism have shown that GH is a potent stimulator of protein synthesis and inhibitor of protein breakdown.[16] A response that is equivalent to that seen in calorie-restricted normal volunteers can be demonstrated. Several studies have shown an effect of GH in postoperative patients who are receiving parenteral feedings. Ward and associates[19] treated postoperative patients with GH (0.1 mg/kg/day) and observed a 70 mmol/day increase in nitrogen retention. Likewise, Shernan and associates[15] observed increases in the range of 150 to 180 mmol/day in postoperative patients receiving full nutritional support.

GH Effects on Muscle Mass and Function in Humans

Although these models clearly suggest that GH improves nitrogen metabolism, none of these studies directly measures changes in muscle mass or muscle function. Two longer term studies have addressed the effect of GH on muscle mass changes. Each study shows an improvement during the administration of GH. Specifically, Salomon and associates[21] administered GH in a randomized, placebo-controlled manner to GH-deficient adults who were not elderly or frail, most of whom had undergone resection of pituitary tumors resulting in GH deficiency. All of these subjects had an increased percentage of body fat and reduced muscle mass at the initiation of the study. Administration of GH to these subjects resulted in a 22% increase in muscle mass when administered over a three-month period as compared with control subjects who received placebo injections.[21] Likewise, fat mass was decreased, as determined by underwater weighing. These changes in body composition were lost when GH administration was discontinued, suggesting that maintenance of the altered metabolic state would require continued administration. Importantly, a 21% increase in exercise capacity was observed in these subjects. A parallel study was conducted in normal, elderly men who were GH deficient. Specifically, eight elderly men who had reduced GH secretion, in response to provocative testing and low IGF-I values, received six months of GH therapy followed by six months of placebo therapy. These men all had normal androgen secretion. GH caused an increase in muscle mass and a decrease in fat mass, both of which were statistically significant and were maintained throughout the treatment interval.[18] Although this study showed an increase in muscle mass, it did not measure a change in muscle function.

Our laboratory recently completed a study of seven elderly subjects who had chronic obstructive lung disease and markedly reduced muscle mass. These subjects were between 69% and 88% of ideal body weight because of severe emphysema and all had lost at least 3 kg in the year prior to the initiation of the study. Subjects were placed on a metabolic ward and were given a normal control diet for one week. This diet did not result in significant weight gain, although mean nitrogen balance was positive; eg, 1.6 g/day. The administration of GH for three weeks was associated with an additional 1.8 g/day improvement in nitrogen balance, and a substantial increase in serum IGF-I concentrations.[17] Specifically, nitrogen balance increased from +1.6 g/day to +3.4 g/day, and this increase was maintained throughout the three weeks of treatment. Likewise, weight increased by approximately 2.2 kg. A two-week washout period showed that at least 2 kg of this weight was due to an increase in solid tissue mass. Most importantly, there was a 27% increase in inspiratory force and a 13% increase in expiratory force. Both of these changes were statistically significant. The increase in inspiratory force occurred in six of seven subjects and the only subject who did not respond was the only subject who had a normal inspiratory force at the initiation of the study. We have concluded that administration of GH to elderly patients with chronic obstructive lung disease appears to result in an increase in muscle mass and an improvement in muscle strength, as evidenced by increased inspiratory force and increased weight gain. The results suggest that GH may be a useful adjunct to training regimens of subjects with chronic obstructive pulmonary disease in order to improve the muscle strength of the chest wall and their breathing mechanics.

In spite of proven effectiveness of GH, multiple problems have been associated with administering it to catabolic subjects. The most severe of these is refractoriness, or failure to respond to GH because of extreme catabolism. States such as acute burns,[22] bacterial sepsis,[23] or severe traumatic injury may result in such massive cytokine release that it is not possible for GH to exert its anabolic actions. Less severe complications include the induction of hyperglycemia, particularly in elderly subjects with a predisposition to glucose intolerance and in elderly and nonelderly subjects who are taking corticosteroids. Because GH can partially antagonize the catabolic effect of corticosteroids, this induction of hyperglycemia is a severe drawback to its use in such patients. Finally, almost all subjects develop some mild degree of fluid retention, although this problem may be minimal.

Trophic Effects of IGF-I in Humans

A possible treatment alternative for patients with severe hyperglycemia caused by glucocorticoid therapy or a genetic predisposition to diabetes is the use of IGF-I. As noted in the studies outlined previously, IGF-I is a potent anabolic factor that can mimic most of the growth-promoting actions of GH in vivo. However, unlike GH, IGF-I causes an increase in glucose transport and a decrease in blood sugar.[24] Therefore, IGF-I administration might obviate the hyperglycemic response to GH that occurs in some patients.

For these reasons, we have undertaken pilot studies to determine whether administration of IGF-I can reverse the catabolic response to caloric restriction that occurs in normal human subjects. Seven volunteers who were of normal weight underwent seven days of dietary restriction during which caloric intake was restricted to 20 kcal/kg of ideal body weight. This degree of caloric restriction was associated with an induction of negative nitrogen balance, and the subjects had a mean daily nitrogen balance of -2.8 g/day. Following infusion of IGF-I (12 μg/kg/hr for 16 hrs daily), nitrogen balance increased from -236 ± 45 mmol to a mean of -65 ± 40 mmol/day ($p > 0.001$), during the last four days of the infusion.[25] A similar effect was produced by GH. Markedly increased blood levels of IGF-I were required to achieve this response, however. Serum IGF-I concentrations rose to peak values of 1,270 ng/ml on day two of the infusion and then fell to 925 ng/ml on day six. In contrast, maximum IGF-I levels after GH were 450 ng/ml. Therefore, total IGF-I levels in the IGF-I treatment group markedly exceeded those in the GH treatment group. This major increase in total IGF-I also was associated with a disproportionate increase in free IGF-I. Free IGF-I values increased 3.5-fold in the subjects given IGF-I, but only by 40% in the subjects treated with GH.

These changes appeared to be caused by changes in insulin-like growth factor-binding protein 3 (IGFBP-3). Specifically, IGFBP-3 was inhibited by approximately 30% in the IGF-I treated subjects, whereas it increased by approximately 10% in the GH-treated subjects. This resulted in a major change in the number of IGF binding sites and suggested that the number of binding sites present in the

IGF-I treated patients was inadequate to bind all the serum IGF-I, thus allowing for some of the material to be present in the free form.[26] This free IGF-I had striking effects on glucose metabolism. Specifically, all subjects on all days of the infusion were hypoglycemic, with morning blood sugars that averaged 52 mg/dl. Likewise, there were 16 episodes of symptomatic hypoglycemia that required discontinuation of the IGF-I infusion. To further substantiate that IGF-I was substituting for insulin in such patients, insulin and c-peptide levels were measured. In the patients treated with IGF-I, serum insulin was suppressed from 75 to 16 pM during the IGF-I infusion at the same time that they were hypoglycemic. In the subjects treated with GH, there was an increase in serum insulin (from 75 to 109 pM) and an increase in blood glucose, suggesting that the subjects were developing a resistance to insulin. Measurement of c-peptide confirmed that these changes were caused by changes in insulin secretion. These findings strongly suggested that IGF-I could be used in catabolic patients to increase nitrogen retention; however, because of its potent hypoglycemic promoting actions, its use may be limited to glucocorticoid-treated patients or patients in whom severe hyperglycemia develops as a result of GH administration.

Because of the induction of significant changes in glucose homeostasis in catabolic patients following administration of either GH or IGF-I, we wondered whether administering GH plus IGF-I in combination might obviate some of the problems that were observed when either agent was administered alone. The rationale is that GH and IGF-I have opposite effects on glucose homeostasis, therefore their combined effect might be to maintain relatively normal glucose metabolism. Furthermore, animal studies have suggested that GH exerts some of its anabolic effects by mechanisms that are independent of IGF-I.[27] Thus, administration of the combination might result in a true synergism of anabolic-promoting activity and the degree of anabolic response might be markedly increased in patients who appear to be partially resistant to the anabolic effects of GH and/or IGF-I.

Because of these differences, it appeared that GH might be an ideal substance to promote the nitrogen-retaining effects of IGF-I and yet antagonize its hypoglycemic actions. To determine whether this was the case, seven normal volunteers who were within 10% of ideal body weight were selected and calorically restricted as in the previous study. The subjects were restricted for seven days, which resulted in the induction of a negative nitrogen balance to a level similar to that in the previous study, that is −205 mmol/day. Administration of IGF-I alone resulted in nitrogen retention of 108 ± 20 mmol/day, and this was maintained throughout the study interval. In contrast, the combination of GH and IGF-I markedly improved nitrogen retention to 262 ± 43 mmol/day. Substantial urinary potassium conservation also was noted. Urinary potassium excretion was reduced by 34 ± 8 mmol/day in the combination treated group, whereas there was no significant reduction in the group that received IGF-I alone. This suggested that most of the nitrogen retained during combination therapy entered the muscle and/or the skeletal tissue. Most importantly, GH markedly attenuated the hypoglycemic response induced by IGF-I. Mean blood glucose determined at six intervals per day was 4.4 mmol/L when IGF-I plus GH was administered versus 3.8 mmol/L when IGF-I alone was administered ($p < 0.001$). IGF-I caused a marked decline in c-peptide from 1,165 pM to 475 pM, whereas the combination maintained levels at 2,280 pM. This suggests that the combination resulted in better maintenance of normal carbohydrate metabolism. Combination therapy produced far higher serum IGF-I concentrations, however, resulting in an increase to 1,854 ng/ml versus 1,092 ng/ml on IGF-I alone.

The ability to tolerate such an increase in total serum IGF-I levels without developing hypoglycemia may be due to differences in changes in IGF-binding proteins. Specifically, in the study discussed above, GH plus IGF-I in combination increased the serum concentration of IGFBP-3 from 4.5 μg/ml to 6.4 μg/ml, whereas IGF-I alone suppressed it from 4.5 to 2.85 μg/ml. Therefore, there was a substantial difference in the IGF-binding capacity of the serum in the two treatment groups at the end of the study, which, if applied to the clinical setting, would have resulted in a markedly disproportionate increase in free IGF-I concentrations in subjects who received IGF-I alone. IGFBP-2 was increased to a greater extent in the group that received IGF-I alone, but the degree of increase was insufficient to offset the magnitude of the changes in IGFBP-3.

The mechanism accounting for this difference in IGFBP-3 appears to be related to increased clearance. IGFBP-3 forms a ternary (stable) complex with IGF-I or II and a third protein called acid labile subunit (ALS). This complex has a much greater half-life than free IGFBP-3 and, therefore, its formation is associated with prolongation of the IGF-I half-life. IGF-I infusion was associated with a decrease in ALS from 33 μg/ml to 19 μg/ml, whereas the GH plus IGF-I combination maintained ALS levels. This would have allowed IGFBP-3 to be maintained in a stable high-molecular-weight complex and for total IGF-binding capacity to be maintained in the combined treatment group. Therefore, administration of GH with IGF-I would have slowed IGFBP-3 and IGF-I clearance, thus allowing maintenance of more stable plasma concentrations.

The side effects that were noted were minimal, although six of seven subjects did complain of tenderness in the area of the parotid glands. The findings indicate definitively that the combination of IGF-I plus GH induces more of an anabolic response and that this anabolic response occurs in muscle and connective tissues and is substantially greater than that induced by IGF-I alone. More importantly, from a safety aspect, combination therapy is much better tolerated than administration of IGF-I alone, which suggests that this combination might be able to be administered safely on an outpatient basis. Further studies will be needed to determine the efficacy of this therapy in elderly subjects with chronic muscle wasting.

Both IGF-I and GH are potent trophic factors when administered to humans. Because elderly subjects have decreased muscle and skeletal tissue mass and are predisposed to injuries and fractures, administration of these hormones could help to improve mobility and reduce the incidence of fractures. Clinical studies to determine whether the use of these hormones results in functional improvement are warranted.

Acknowledgments

The authors thank Jennifer O'Lear and Leigh Elliott who prepared the manuscript. This work was supported by a grant (HD 28801) from the National Institutes of Health.

References

1. Zapf J, Froesch ER, Humbel RE: The insulin-like growth factors (IGF) of human serum: Chemical and biological characterization and aspects of their possible physiological role. *Curr Top Cell Regul* 1981; 19:257-309.
2. Rinderknecht E, Humbel RE: Primary structure of human insulin-like growth factor II. *FEBS Lett* 1978;89:283-286.
3. Copeland KC, Underwood LE, Van Wyk JJ: Induction of immunoreactive somatomedin-C human serum by growth hormone: Dose response relationships and effect on chromatographic profiles. *J Clin Endocrinol Metab* 1980;50:690-697.
4. Clemmons DR, Van Wjk JJ, Ridgway EC, et al: Evaluation of acromegaly by radioimmunoassay of somatomedin-C. *N Engl J Med* 1979;301:1138-1142.
5. Salmon WD Jr, DuVall MR: In vitro stimulation of leucine incorporation into muscle and cartilage protein by a serum fraction with sulfation factor activity: Differentiation of effects from those of growth hormone and insulin. *Endocrinology* 1970;87:1168-1180.
6. Schoenle E, Zapf J, Hauri C, et al: Comparison of in vivo effects of insulin-like growth factors I and II and of growth hormone in hypophysectomized rats. *Acta Endocrinol (Copenh)* 1985;108:167-174.
7. Schoenle E, Zapf J, Humbel RE, et al: Insulin-like growth factor I stimulates growth in hypophysectomized rats. *Nature* 1982;296:252-253.
8. Behringer RR, Lewin TM, Quaife CJ, et al: Expression of insulin-like growth factor I stimulates normal somatic growth in growth hormone-deficient transgenic mice. *Endocrinology* 1990;127:1033-1040.
9. Walker JL, Ginalska-Malinowska M, Romer TE, et al: Effects of infusion of insulin-like growth factor-I in a child with GH insensitivity syndrome (Laron dwarfism). *N Engl J Med* 1991;324:1483-1488.
10. Jacob R, Barrett E, Plewe G, et al: Acute effects of insulin-like growth factor-I on glucose and amino acid metabolism in the awake fasted rat. *J Clin Invest* 1989;83:1717-1723.
11. Douglas RG, Gluckman PD, Ball K, et al: The effects of infusion of insulin-like growth factor (IGF-I) I, IGF-II and insulin on glucose and protein metabolism in fasted lambs. *J Clin Invest* 1991;88:614-622.

12. Jacob RJ, Sherwin RS, Bowen L, et al: Metabolic effects of IGF-I and insulin in spontaneously diabetic BB/w rats. *Am J Physiol* 1991;260(2Pt1):E262-268.
13. Rudman D, Kutner MH, Rogers CM, et al: Impaired growth hormone secretion in the adult population: Relation to age and adiposity. *J Clin Invest* 1981;67:1361-1369.
14. Snyder DK, Clemmons DR, Underwood LE: Treatment of obese, diet-restricted subjects with growth hormone for 11 weeks: Effects on anabolism, lipolysis and body composition. *J Clin Endocrinol Metab* 1988;67:54-61.
15. Shernan SK, Demling RH, Lalonde C, et al: Growth hormone enhances reepithelialization of human split-thickness skin graft donor sites. *Surg Forum* 1989;40:37-39.
16. Horber FF, Haymond MW: Human growth hormone prevents the protein catabolic side effects of prednisone in humans. *J Clin Invest* 1990;86:265-272.
17. Pape GS, Friedman M, Underwood LE, et al: The effect of growth hormone on weight gain and pulmonary function in patients with chronic obstructive lung disease. *Chest* 1991;99:1495-1500.
18. Rudman D, Feller AG, Nagraj HS, et al: Effects of human growth hormone in men over 60 years old. *N Engl J Med* 1990;323:1-6.
19. Ward HC, Halliday D, Sim AJW: Protein and energy metabolism with biosynthetic human growth hormone after gastrointestinal surgery. *Ann Surg* 1987;206:56-61.
20. Zeigler TR, Young LS, Manson JMcK, et al: Metabolic effects of recombinant human growth hormone in patients receiving parenteral nutrition. *Ann Surg* 1988;208:6-16.
21. Salomon F, Cuneo RC, Hesp R, et al: The effects of treatment with recombinant human growth hormone on body composition and metabolism in adults with growth hormone deficiency. *N Engl J Med* 1989;321:1797-1803.
22. Belcher HJ, Mercer D, Judkins KC, et al: Biosynthetic human growth hormone in burned patients: A pilot study. *Burns* 1989;15:99-107.
23. Dahn MS, Lange MP, Jacobs LA: Insulinlike growth factor 1 production is inhibited in human sepsis. *Arch Surg* 1988;123:1409-1414.
24. Guler H-P, Zapf J, Froesch ER: Short-term metabolic effects of recombinant human insulin-like growth factor-I in healthy adults. *N Engl J Med* 1987;317:137-140.
25. Clemmons DR, Smith-Banks A, Underwood LE: Reversal of diet-induced catabolism by infusion of recombinant insulin-like growth factor-I (IGF-I) in humans. *J Clin Endocrinol Metab* 1992;75:234-238.
26. Clemmons DR, Dehoff ML, Busby WH, et al: Competition for binding to insulin-like growth factors (IGF) binding protein 2, 3, 4, and 5 by the IGFs and IGF analogs. *Endocrinology* 1992;131:890-895.
27. Nilsson A, Isgaard J, Lindahl A, et al: Regulation by growth hormone of number of chondrocytes containing IGF-I in rat growth plate. *Science* 1986;233:571-574.

Chapter 19

Myogenic Vector Expression of Human Insulin-Like Growth Factor I in C_2C_{12} Myoblasts Potentiates Myogenesis

Robert J. Schwartz, PhD
Kuo Chang Yin, MS
Heung Man Lee
David Powell, MD
Raymond Hintz, MD
Francesco DeMayo, PhD

Growth factors play a pivotal role in myogenic programming of mesodermal cells and the progression of committed myoblasts into terminally differentiated muscle. The early stages of embryonic muscle cell commitment appear to be induced by activin,[1] a member of the transforming growth factor-beta (TGF-β) family, and basic fibroblast growth factor (FGF).[2,3] These growth factors direct the expression of myogenic determination genes subsequently leading to the up regulation of myogenic specific genes. At later stages in the myogenic differentiation pathway, the addition of mitogen-rich media, FGFs, and TGFs to cultured myoblasts acts to inhibit myogenesis and myoblast fusion.[4,5] In striking contrast to these growth factors, which inhibit late stage muscle differentiation, insulin-like growth factors (IGF)-I/II have been shown to stimulate myogenic differentiation.[6] Biosynthesis and secretion of IGF-I/II and IGF-binding proteins naturally occurs with the onset of myoblast fusion[7]; coincident with the appearance of muscle-specific gene products. IGF activity on muscle appears to be mediated by interactions with the IGF-I receptor, a ligand-activated tyrosine-specific protein kinase. Recent studies[8] demonstrated that antisense oligo deoxynucleotides complementary to the N-terminal sequences of IGF-I and IGF-II suppressed spontaneous differentiation of myogenic cell lines. Thus, expression and secretion of IGFs by myoblasts appear to have a positive role in promoting muscle terminal differentiation.

Recent studies also indicate that IGF-I plays a role in late stage muscle growth. When IGF-I was included in the maintenance media of cultured primary myofibers, the myofibers were larger in diameter, and displayed a near doubling in myosin content and a substantial increase in protein stability and synthesis in comparison with untreated myogenic cultures.[9] In animal studies, administration of growth hormone to hypophysectomized rats resulted in a significant increase in IGF-I mRNA of skeletal muscle. Implantation of growth hormone-secreting cells in non-growing rats caused a seven-fold increase in IGF-I mRNA and a 50% increase in the mass of the gastrocnemius muscles.[10] Increased expression of IGF-I genes through passive mechanical stretch and acute exercise shows a correspondence between muscle hypertrophy and IGFs.[11] The growth-promoting properties of IGF-I have also been shown to play an important role in muscle regeneration and repair, in which IGF-I acts as a powerful stimulant of muscle precursor cell proliferation and differentiation.[6,12]

Previous studies have suggested that specific IGF-binding proteins play an important role in localizing the IGF synthesized within a tissue, potentiating its effects on target cells and possibly in targeting the IGF in blood to specific tissue beds.[13] Recent studies have demonstrated that myogenic cell lines secrete a 30 kDA IGF binding protein (IGFBP), which is up regulated very early in the differentiation process, and coincides with IGF-I/II secretion.[7] Synthesis of this IGFBP, identified as IGFBP-5, by BC3H-I myoblasts has been reported to be decreased by basic FGF and TGF-B, two factors that are known to inhibit myogenic differentiation.[14]

We hypothesize that the insulin-like growth factors and the cognate binding protein, BP5, are the major trophic growth factor complex for stimulating muscle differentiation, growth, and hypertrophy. This pathway proceeds through biosynthesis and secretion of IGFs/BPs from muscle stimulation of IGF-I cell surface receptors and a tyrosine kinase signaling cascade which directs muscle growth. One way to test this hypothesis is to construct expression vectors that appropriately overexpress IGF-I in developing muscle cells. We could then determine if the expression of IGF-I is sufficient to drive myogenesis to terminal differentiation under stringent growth conditions. We could also determine if the biosynthetic and secretory activity of IGFBP-5 was linked to the levels of IGF-I secreted by fusing myoblasts.

Construction of a Myogenic Vector System

We developed a myogenic vector system (MyVS) based on unique regulatory elements of the skeletal α-actin gene. The switching of actin gene expression studied during myogenesis,[15,16] indicated that paired vertebrate sarcomeric cardiac and skeletal α-actin genes are up regulated sequentially during early muscle development, whereas only skeletal α-actin gene activity is maintained at high levels of adult skeletal muscle.[16] In fact, by accurate dot hybridization analysis, we determined that skeletal α-actin mRNA composes 8% of the poly(A)-enriched RNA and 0.09% of the total cellular RNA of adult avian muscle[15]; the highest level of any mRNA species in skeletal muscle. We reasoned that by first identifying and then assembling the appropriate regulatory components of the skeletal α-actin gene, it might be possible to construct a myogenic expression vector with activity approaching the potency of the endogenous striated actin gene. In this regard we have identified not only promoter positive and negative cis-acting elements, which direct tissue-specific transcription of the skeletal actin gene,[17,18] but also sequences such as the 3′ untranslated regions, which improve the stability of the skeletal actin mRNA in muscle tissue. Three MyVS-IGF-I constructions were made, and, by adding various non-coding regulatory sequences of the α-actin gene, we could evaluate their capacity to increase IGF-I mRNA content in muscle cells. Schematic representation of these constructions is shown in Figure 1.

Our first construction contained the skeletal (Sk202) actin promoter[17] linked to the human IGF-I cDNA.[19] This construction was made so that the SV40 poly(A) addition site and the SV40 small t-intron were linked to the 3′ untranslated regions of the IGF-I cDNA. The SV40 sequences were added with the designed purpose of increasing the stability of nuclear IGF-I RNA transcripts, as they have been used in a number of expression vectors. However, Gorman's laboratory recently reported that the small t-intron could cause aberrant splicing within nascent transcripts[20]; thus producing deletions in the spliced mRNA and possibly defective translated products. The second vector, Sk733IGF-I, contains ~411 nucleotides of the skeletal α-actin promoter, the natural cap site, 5′ untranslated leader and the first intron. An NcoI site was engineered to create a unique insertion cloning site for the IGF-I cDNA, in which the initiation ATG was also converted to an NcoI site. The Sk733IGF-I construction utilizes the IGF-I poly(A) site. In the third construct, the 3′ end of IGF-I was replaced with a DNA fragment which incorporated the skeletal α-actin 3′ untranslated regions, poly(A) addition site, and transcriptional terminating sequences.

Expression of IGF-I in Stably Transfected Myoblasts

We studied these vectors by making a broad population of stably transfected C_2C_{12} myoblasts. IGF-I expression levels could be directly evaluated in these stable myoblast cell lines. Each IGF-I construction was co-transfected with the drug selectable vector EMSV-Hygromycin into mouse C_2C_{12} cells, and, following two weeks of selection, a population of stable myoblasts was selected. A population of C_2C_{12} myoblasts stably transfected only with EMSV-Hygromycin served as the controls. Normally, changing culture medium from growth media (10% fetal calf serum) to 2% horse serum initiates the differentiation process in which over a period of four days control C_2C_{12} myoblasts fuse.

We determined that Sk733IGF-I 3′Sk, containing the skeletal 3′ UTR, is effective in driving IGF-I expression in muscle cells, and assayed the presence of this factor by both radioimmunoassays of tissue culture media and immunoperoxidase staining of cells. We found increased levels of IGF-I during the fusion of several of our muscle cultures. The levels of IGF-I in control cultures was in the

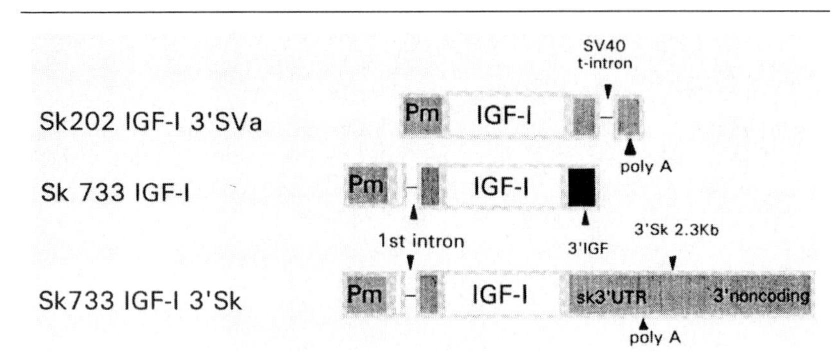

Fig. 1 *Schematic representation of three skeletal α-actin promoter-driven IGF-I constructions with different 3′ untranslated regions.*

range of 0.2 to 0.5 ng/ml. In comparison, cultures transfected with vector Sk733IGF-I 3′Sk have levels of IGF-I at least 100 times greater than control myoblasts. In comparison to Sk733IGF-I 3′Sk and the SV40 containing vector (Sk202IGF-I 3′SVa), the replacement of the 3′ UTR with skeletal 3′ UTR corresponded to an increase of 20-fold in IGF-I secreted into the culture media (Table 1). Because very low levels of IGF-I were detected in transcripts from SV40-containing vectors, we assume that either primary transcripts were not processed correctly or that the absence of α-actin leader sequences and the natural cap site, as contained in the other two constructions, might have reduced translational activity.

Timing of muscle gene expression was examined by RNA blotting with a radioactive IGF-I DNA probe. Figure 2 shows an appropriate increase in IGF-I mRNA accumulation when myoblasts are switched from growth (G) to differentiation (D) media. Approximately two orders of magnitude increase in IGF-I mRNA was detected in myogenic cells containing Sk733IGF-I 3′Sk. Vectors containing SV40 sequences were inappropriately expressed; in which most of the transcripts accumulated in replicating myoblasts, not myotubes. Thus, the 3′ SV40 sequences can adversely influence the developmental timing and transcriptional activity of the skeletal α-actin promoter.

The role of the skeletal α-actin 3′ UTR on mRNA stability was examined in stably transfected C_2C_{12} myogenic cells. A transcription blocker, actinomycin D (8 mg/ml) was added to myogenic cultures under differentiation conditions and timed samples were removed for RNA blotting analysis as shown in Figure 3. Transcripts containing the natural IGF-I 3′ UTR were found to turn over rapidly (half-life of less than 1 hour) in differentiation media in contrast to transcripts which contain contiguous skeletal α-actin 3′ UTR, which showed a high level of stability (half-life greater than 18 hours). Transcripts containing SV40 sequences were intermediate in stability between skeletal actin and short lived IGF-I 3′ UTRs.

Table 1 IGF-I Levels in Stably Transfected C_2C_{12} Myoblasts

Construction	IGF-I (ng/ml of media)/ 4 days/10^5 cells
Control C_2C_{12} (EMSV-Hygromycin)	0.5
Sk202IGF-I	4.4
Sk733IGF-I	3.4
Sk733IGF-I 3′Sk	79.0

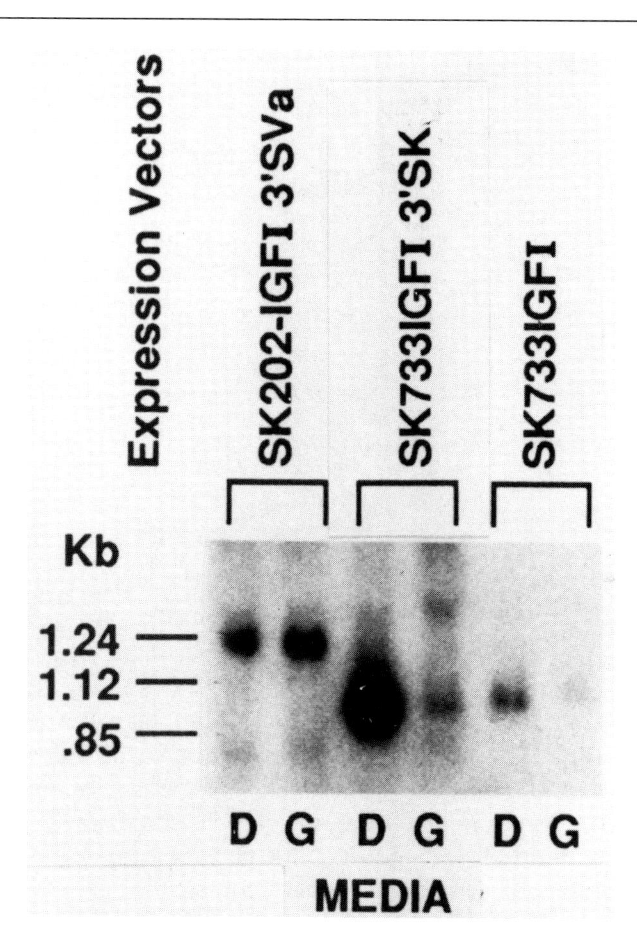

Fig. 2 *Accumulation of IGF-I mRNA in stably transfected C_2C_{12} cells depends upon IGF-I constructs containing the skeletal actin 3′ UTR. Relative levels of IGF-I mRNA were revealed by hybridization of total RNA (10 mg/lane) isolated from growing (G) and differentiated (D) C_2C_{12} myoblasts probed with 32P labeled IGF-I cDNA (109 cpm/mg) on RNA blots.*

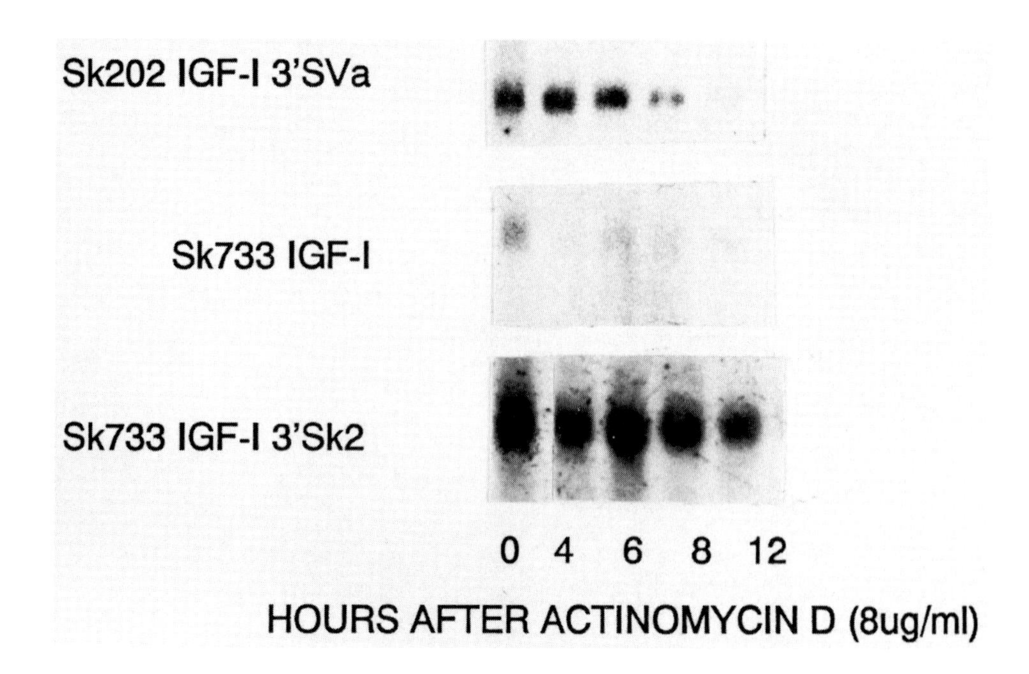

Fig. 3 *IGF-I mRNA stability is increased with contiguous skeletal α-actin 3' UTR. Stable populations of differentiated C_2C_{12} myotubes were treated with a potent transcription blocker, actino-mycin D, for short periods of time. Relative levels of IGF-I mRNA content were assessed by RNA blot analysis as shown in Figure 2.*

In myogenic cultures, the expression of IGF-I appears to potentiate myoblast fusion and drives late stage terminal differentiation. We observed that myoblasts containing Sk733IGF-I or Sk733IGF-ISk fused at least two to three days earlier than C_2C_{12} EMSV-Hygromycin control myoblasts. The phenotype of the myoblasts containing Sk733IGF-ISk, the strongest expressing vector, formed extensive and very thick myotubes in comparison to C_2C_{12} controls, which were poorly fused and thin. At the gene expression levels as shown in Figure 4, we observed a correspondence between elevated IGF-I expression and a stimulation (range of 10- to 100-fold) of myogenic determination factors, such as MyoD and myogenin, and myogenic specified genes, such as intermediate filament protein, desmin, and the α-actin gene. The observation that IGF-I stimulates myogenic differentiation by stimulating expression of the myogenic basic helix-loop-helix factors such as MyoD and myogenin is in general agreement with Florini and associates.[8] A similar degree of myogenic gene activity was also observed in well-differentiated myotubes that were challenged with IGF-I in stringent growth media, thus eliminating the concern that a stimulation of cell differenti-

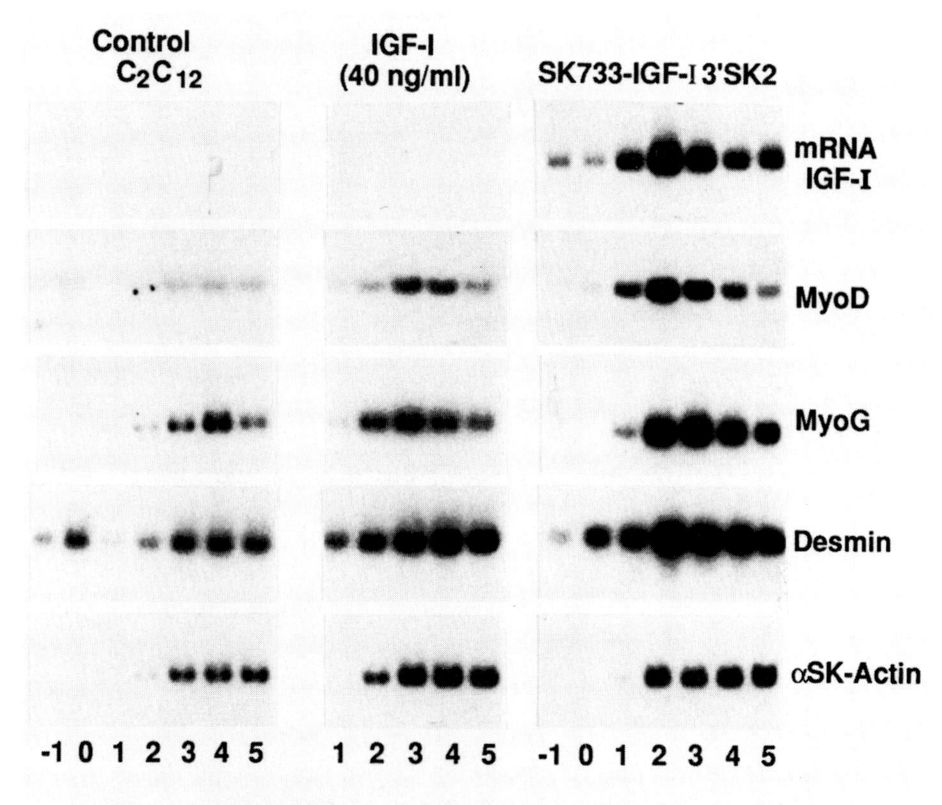

Fig. 4 *Myogenic gene expression depends upon IGF-I. Relative levels of IGF-I, MyoD, myogenin (MyoG), desmin, and skeletal α-actin mRNA were assessed by Northern RNA blot analysis (10 mg RNA/lane) from C_2C_{12} control and Sk733IGF-I Sk cultures. C_2C_{12} myoblasts were grown in 10% fetal calf serum (−1 day) and then switched to serum-free media at day 0.*

ation or gene activity was a secondary consequence of the greater cell numbers resulting from the mitogenic action of IGF-I.

We also determined a direct correspondence between elevated levels of IGF-I in the media of C_2C_{12} muscle cells and the appearance of IGF-I binding protein 5 mRNA. Concordance between IGF-I expression and BP5 mRNA levels is shown in Figure 5. Similarly, elevated secretion of BP5 was detected by radioligand 125I IGF binding assays, in which secreted BP5 was dependent upon elevated levels of IGF-I. Usually, IGF binding proteins 1 and 3 are predominant in serum and form complexes with IGF. They may serve to provide a storage pool and prolong the biologic half-life of the IGFs in circulation, by complexing with the growth factor.[13] Hypothetically, BP5 synthesis and secretion from muscle is enhanced by IGF-I overexpression and as a complex might further potentiate the ac-

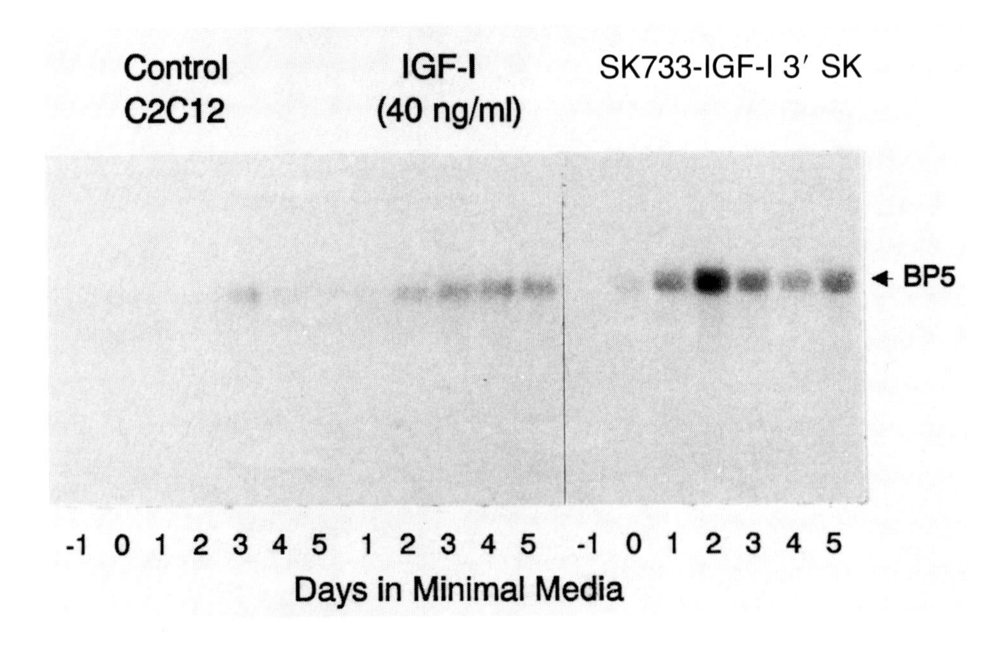

Fig. 5 *IGF-I binding protein 5 gene activity is dependent upon IGF-I. Relative levels of BP5 mRNA were determined by Northern RNA blot analysis as in Figure 4.*

tion of IGF-I with the cell surface receptors and provide a positive regulatory function on muscle hypertrophy. Possibly, binding proteins 1 and 3, which are enriched in serum, might function in the opposing manner and, by binding IGF-I, actually inhibit IGF action on muscle growth. The most direct way of testing the role of IGFBPs will be to either add different IGFBPs directly to C_2C_{12} cells along with stimulatory levels of IGF (40 ng/ml), or overexpress several different IGFBPs directly in myoblasts.

Summary

In summary, expression of IGF-I in C_2C_{12} myoblasts appeared to influence their myogenic phenotype by enhancing fusion with extensive myotube formation. Furthermore, under stringent serum-free growth conditions, only myoblasts expressing elevated levels of IGF-I proceeded to undergo terminal differentiation; and to accumulate elevated levels of MyoD and myogenin factors, intermediate filaments, sarcomeric contractile proteins, and the cognate IGF-I binding protein transcripts. This study suggests that the expression and secretion of IGFs by myoblasts is a central developmental regulatory event that allows myogenesis to proceed to its terminally differentiated state. Currently, we are exploring the role of MyVS-

IGF-I in transgenic mouse lines and by direct gene transfer into muscle as a vehicle to promote muscle growth.

Acknowledgments

These studies were supported by an Advanced Research Program grant from Texas Higher Education Coordinating Board, Austin, Texas and by funds from Gene Medicine, Inc.

References

1. Mitrani E, Ziv T, Thomsen G, et al: Activin can induce the formation of axial structures and is expressed in the hypoblast of the chick. *Cell* 1990;63:495-501.
2. Kimelman D, Kirschner M: Synergistic induction of mesoderm by FGF and TGF-beta and the identification of an mRNA coding for FGF in the early Xenopus embryo. *Cell*1987;51:869-877.
3. Melton DA: Pattern formation during animal development. *Science* 1991;252:234-241.
4. Brennan TJ, Edmondson DG, Li L, et al: Transforming growth factor-beta represses the actions of myogenin through a mechanism independent of DNA binding. *Proc Natl Acad Sci USA*1991; 88:3822-3826.
5. Olwin BB, Hauschka SD: Fibroblast growth factor receptor levels decrease during chick embryogenesis. *J Cell Biol* 1990;110:503-509.
6. Florini JR, Magri KA: Effects of growth factors on myogenic differentiation. *Am J Physiol* 1989;256(4Pt1):C701-C711.
7. Tollefsen SE, Larjara R, McCusker RH, et al: Insulin-like growth factors (IGF) in muscle development: Expression of IGF-1, the IGF-1 receptor, and an IGF binding protein during myoblast differentiation. *J Biol Chem* 1989;264:13810-13817.
8. Florini JR, Ewton DZ, Roof SL: Insulin-like growth factor-I stimulates terminal myogenic differentiation by induction of myogenin gene expression. *Mol Endocrinol* 1991;5:718-724.
9. Vandenburgh HH, Karlisch P, Shansky J, et al: Insulin and IGF-I induce pronounced hypertrophy of skeletal myofibers in tissue culture. *Am J Physiol*1991;260(3Pt1):C475-C484.
10. Turner JD, Rotwein P, Novakofski J, et al: Induction of mRNA for IGF-I and -II during growth hormone-stimulated muscle hypertrophy. *Am J Physiol*1988;255(4Pt1):E513-E517.
11. DeVol DL, Rotwein P, Sadow JL, et al: Activation of insulin-like growth factor gene expression during work-induced skeletal muscle growth. *Am J Physiol*1990;259(1Pt1):E89-E95.
12. Grounds MD: Towards understanding skeletal muscle regeneration. *Pathol Res Pract* 1991;187:1-22.
13. Clemmons DR: Insulin-like growth factor binding proteins: Roles in regulating IGF physiology. *J Dev Physiol* 1991;15:105-110.
14. McCusker RH, Camacho-Hubner C, Clemmons DR: Identification of the types of insulin like growth factor binding proteins that are secreted by muscle cells in vitro. *J Biol Chem* 1989;264:7795-7800.
15. Schwartz RJ, Rothblum KN: Gene switching in myogenesis: Differential expression of the chicken actin multigene family. *Biochemistry* 1981;20:4122-4129.
16. Hayward LJ, Schwartz RJ: Sequential expression of chicken actin genes during myogenesis. *J Cell Biol* 1986;102:1485-1493.

17. Bergsma DJ, Grichnik JM, Gossett LM, et al: Delimitation and characterization of cis-acting DNA sequences required for the regulated expression and transcriptional control of the chicken skeletal alpha-actin gene. *Mol Cell Biol* 1986;6:2462-2475.
18. Chow KL, Schwartz RJ: A combination of closely associated positive and negative cis-acting promoter elements regulates transcription of the skeletal alpha-actin gene. *Mol Cell Biol* 1990;10:528-538.
19. Jansen M, van Schaik FM, Ricker AT, et al: Sequence of cDNA encoding human insulin-like growth factor I precursor. *Nature* 1983;306:609-611.
20. Huang MT, Gorman CM: The simian virus 40 small-t intron, present in many common expression vectors, leads to aberrant splicing. *Mol Cell Biol* 1990;10:1805-1810.

Future Directions

Determine if exercise, when initiated in the later years of life, can be used to improve function without causing further deterioration in previously injured or diseased musculoskeletal tissues.

Older adults with preexisting musculoskeletal injury or disease may be more likely to develop additional exercise-related tissue deterioration. Longitudinal study would determine if selected forms of exercise (eg, weightbearing versus nonweightbearing) would provide desired exercise outcomes, such as increased endurance strength and maximal oxygen uptake, without causing further tissue damage. In addition to volutional exercise, other forms need to be tested for efficacy, validity, and safety.

Develop guidelines for exercise prescription for older adults to maximize function.

Currently, most exercise prescription for the older adult is done empirically to achieve a desired outcome, such as improved strength. There are no guidelines to indicate if one particular type of exercise program would be better than another to promote strength and endurance, particularly if certain orthopaedic limitations exist. There are no guidelines to assist clients or practitioners in choosing an activity program to enhance aerobic capacity when physical and orthopaedic limitations are present. For example, those with genu valgus probably should not be encouraged to pursue walking or jogging as their exercise mode, but how much valgus (if any) can be present before walking or jogging would be harmful is unknown. Older women with forward shoulders and kyphosis may be at risk for rotator cuff injury and, therefore, may not benefit from selected forms of weight training. Suggested exercise modes and contraindications to each form of physical activity need to be established for those with orthopaedic limitations.

Develop minimum standards of strength, flexibility, and postural stability for participation of older adults in physical activity.

Not enough information is currently available to determine if an individual can safely participate in an exercise program from a musculoskeletal perspective. Such information would allow exercise prescription on a more thoughtful basis. Although not well documented, it appears that older adults with weakness, limitations in range of motion, and instability are at greater risk for injury during exercise than those without complicating factors. For example, it is not uncommon for individuals who are attempting to walk for exercise to complain of back pain. There appears to be a relationship between hamstring length and injury with jogging and fast walking. Those with instability might be at risk for falling with a walking program.

Identify determinants of exercise adherence.

Anecdotal evidence suggests that a greater percentage of older adults are exercising. Increasing numbers of older adults are purchasing exercise equipment and participating in activity classes, mall walking, and low impact aerobics. Once exercise is begun, the number of older adults who faithfully continue to exercise is unknown. Also unclear are factors, such as painful episodes and lack of knowledge on how to progress, that deter adults from continuing physical activity. Once impediments to continuing exercise are identified, strategies for helping elders to stay on program can be developed.

Determine the mechanism(s) responsible for the age-associated loss in the maximum specific force of muscles.

The age-related loss in maximum specific force is not explained by an increase in extracellular components. Furthermore, the dry mass-wet mass ratio is not different for muscles in old compared with those in young or adult animals. Consequently, the phenomenon of muscle weakness appears to be intrinsic to at least some single muscle fibers. The hypothesis is that the decrease in maximum specific force (kN/m^2) results from a decrease in the number of cross-bridges per unit area, a decrease in the force developed by each cross-bridge, or both factors. The evaluation of force development and the underlying deficits need to be performed on permeabilized and intact single fibers.

Determine the mechanism(s) responsible for the age-related increase in the susceptibility of muscles to contraction-induced injuries.

Contraction-induced injury is more likely to occur when muscles are stretched during contractions rather than when muscles remain at the same length or shorten during contractions. After comparable protocols of lengthening contractions, muscles in old animals display a greater deficit in force and in injury to muscle fibers than muscles in younger animals. The mechanisms responsible are unknown.

Determine both the mechanism(s) by which training protects muscles from contraction-induced injury and the mechanism(s) responsible for the age-related impairment in the regeneration of skeletal muscle fibers following such injury.

Anecdotal observations and preliminary experimental data support the concept that training-related injury occurs following repeated performances of protocols of lengthening contractions. The repeated performances must allow a sufficient number of days for recovery from the injury in order to develop a trained muscle that is no longer injured by the exercise protocol. The mechanism responsible for this protection has not been identified. Following contraction-induced injury, skeletal muscles from older animals regenerate less well than those from young animals.

Identify cytokines or hormonal substances that modulate muscle mass or repair processes after injury in both young and old subjects.

Acute, and sometimes chronic responses to injury are associated with muscle protein breakdown. The adaptive value, if any, of this breakdown, and the mediators responsible, are poorly defined. Blocking the mediators of muscle protein breakdown when they are released inappropriately may delay the age-associated loss in muscle mass. Animal models of controlled muscle injury in the absence of infection need to be developed in order to determine which cytokines and hormones are induced by injury as well as where they are produced. These cytokines and hormones need to be blocked in vivo to determine their causal role for muscle protein breakdown. Efficient blocking agents might slow the progression of age-associated muscle loss.

Determine the relationship between both skeletal muscle mass and strength and the duration of time that elderly people are bedridden.

Little information exists to indicate whether sudden reduction in mobility in the elderly results in an irreversible loss of skeletal muscle and strength. Thresholds for loss of mobility due to remaining mass of skeletal muscle need to be determined to provide clinical goals for maintaining quality of life. The amount of skeletal muscle mass and strength regained by elderly people after prolonged periods of bed rest also is unknown. The extent to which countermeasures can assist the process of regaining strength is largely unexplored.

Determine the mechanism(s) responsible for the age-related denervation of muscle fibers.

The selective loss of muscle fiber number with aging in humans is presumed to be due to denervation without consequent reinnervation.

Determine the effect of combined therapy with growth hormone and insulin-like growth factor-I or other myogenic growth factors on muscle mass, strength, and endurance in elderly subjects.

In order to justify the risks and costs involved in administering these growth factors to humans, it must be demonstrated that they are actually capable of improving the ability of the elderly to perform the functions of daily living. Blinded, placebo-controlled studies should be developed that include techniques such as MRI scanning and muscle fiber area determinations in biopsies to determine precisely the degree of increased mass. Exercise tolerance test, CYBEX (or other strength) testing, and a comprehensive quality of life assessment should be performed to determine the functional impact of the growth factors.

Determine how muscle growth factors are down-regulated during maturation.

Delineation of mechanisms controlling muscle growth in development have potential application in the development of drug or gene therapy to reverse age-associated loss of muscle mass. Experiments involving transgenic mice could be used to cause conditional ablation of growth factors at various times during the life span of the animals. This methodology would allow a comparison of the local versus the systemic effects of various growth factors. Studies would evaluate the influence of receptor expression and binding proteins as well as the growth factor itself.

What are the mechanisms, both integrative for the total organism and cellular, responsible for the improvement in strength in the elderly?

Knowledge of the cellular mechanism by which strength training prevents the loss of muscle mass of the integrative factors, such as coordination, and of cellular mechanisms would promote investigations into drugs and into gene therapy that might counteract aging-associated muscle loss. Not all elderly can, or will, participate in strength training programs for their lifetime. Pharmaceutical methods are needed for those unable to undertake strength training.

Use transgenic animals in studies of aging.

A transgenic mouse overexpressing the ski oncogene has muscle hypertrophy. These animals have unusually high muscle/fat ratios, the parameter that declines in normal animals and humans during aging. Studies could be undertaken to determine (1) if these animals show the same age-related declines in muscle mass; (2) if the increased muscle mass in these animals is associated with a proportional increase in muscle strength; and (3) what muscle-related genes are expressed differently in this model.

Determine structural, matrix, and protein changes in muscles of older subjects.

Insufficient information exists on structural and functional changes in connective tissue and fascia of skeletal muscle that take place during aging, particularly during adulthood to senescence in human beings. Do these changes occur? Are they important? Does collagen and proteoglycan metabolism differ in skeletal muscles of young and old subjects in response to muscle injury? Although some information is available on muscle stiffness with aging, better correlative longitudinal studies relating changes in connective tissue of muscle to stiffness during aging is needed. Correlations of age-induced changes in connective tissue should be compared and contrasted among skeletal muscle, articular cartilage, tendons, and vertebral disks.

Determine why the proliferative and myogenic capacity of satellite cells decreases with aging.

Satellite cells are necessary for repair of muscle injury. However, muscle from older animals regenerates from injury more slowly. Also, aging-associated atrophy of skeletal muscle cannot be prevented. Studies are needed to determine if changes in satellite cells play some role in these events. Apparently the satellite cell may undergo an aging process. What is the genesis of the aging-induced changes in satellite cell proliferation and myogenic potential?

Section Four

Tendon and Ligament

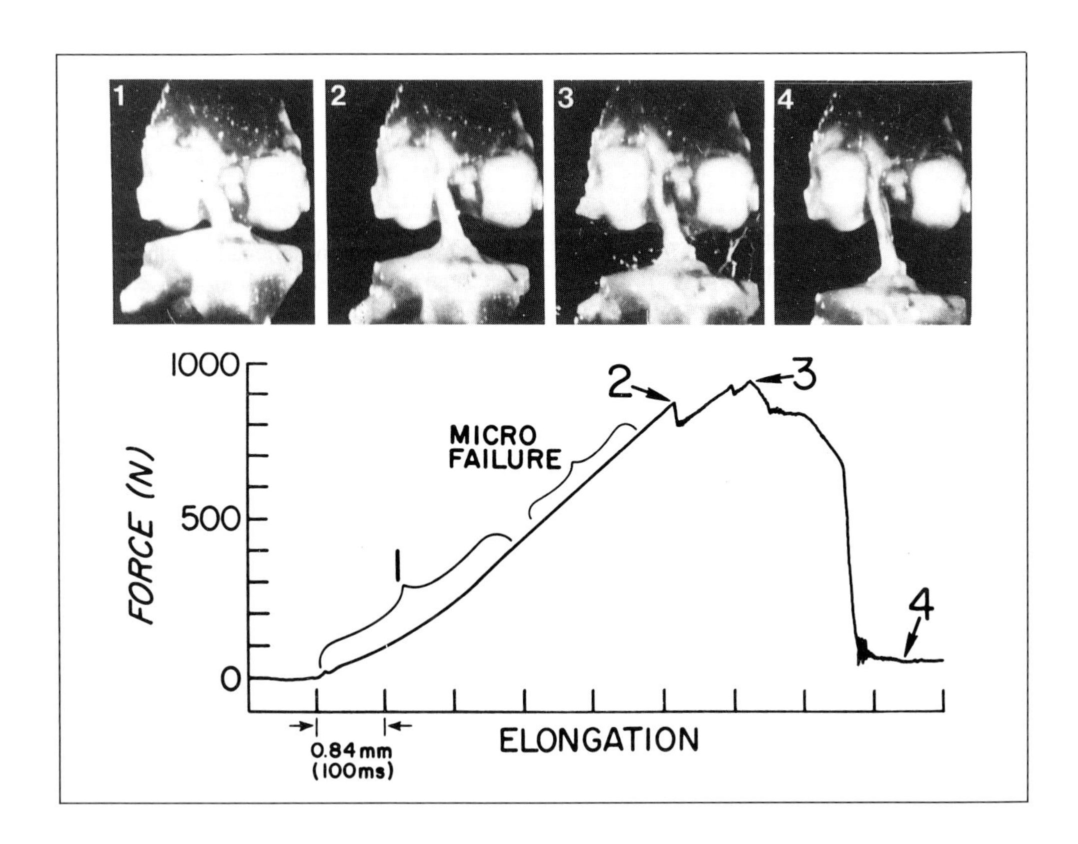

Section Leader

Savio L-Y. Woo, PhD

Section Contributors

Kai-Nan An, PhD
David L. Butler, PhD
Cyril Frank, MD
Richard H. Gelberman, MD
David A. Hart, PhD
Cato T. Laurencin, MD, PhD
John Matyas, PhD
Kathryn G. Vogel, PhD

In the tendon biomechanics chapter, the mechanical properties of tendons (stress-strain curve and viscoelastic properties) are described. The importance of considering tendon muscle coupling is emphasized. Tendon disorders involving calcifying tendinitis (with an unknown etiology), tenosynovitis between the tendon and its sheath (trigger finger), as well as the rupture of tendons due to degeneration and the effects of cumulative tendon trauma are summarized. Of particular importance is the contribution of mechanical factors to tendon disorders. The forces in the flexor tendon and the flexor digitorum profundus tendon can be quite high; on the order of 100 to 250 N (or approximately 4% strain over the in-situ strain). Therefore, it is important to consider the tendon elasticity together with muscle contraction. Furthermore, in addition to tensile loads, tendons are subjected to compressive loads. For example, when a flexor tendon is wrapped around its pulley (the angle is greater than 90%), the compressive load could be more than 1.5 times the tensile load. In addition to the high compressive load, friction occurs between the tendon and its pulley, which may lead to tendon disorders. Unfortunately, very little is known about age-related changes of human tendons. Much of the data have been based on studies using the rat-tail tendon. In general, the region for low stiffness (toe region) decreases with increasing age.

Biochemically, some age-related changes in tendons and ligaments have been described. Again, the majority of the work is based on the rat-tail tendon. Collagen fibril diameters and the reducible collagen cross-links generally decreased with age, whereas the nonreducible cross-links increase with age. The noncollagenous constituents such as GAGs decrease in length, but smaller proteoglycans (decorin) increase in length. The effects of age on other constituents, such as a fibrocartilage, lipids, and minor-type collagens, are less well known. Age also correlates with a decrease in the number of fibroblasts, with a concurrent increase of extracellular matrix (implying there is not necessarily an increase in the rate of cell death). In terms of cellular function, anaerobic glycolysis is immeasurably small in the Achilles tendon of rabbits that are over three years old. This suggests that anaerobic glycolysis is the most important energy source in older tendons. Overall, biochemical data and more advanced biochemical procedures are needed to explain the functional changes of tendons and ligaments related to aging.

One of the most profound effects on aging can be demonstrated in the tensile properties of the human anterior cruciate ligament (ACL). The structural properties of the femur-ACL-tibia complex (FATC) decrease significantly. Other data imply that the usage of the ACL decreases significantly with age, starting as early as the middle years. Therefore, measurement of in vivo forces in the ACL is of particular importance. Various strain and force transducers are being developed and implanted in the ACL in order to obtain in vivo data on elongation and forces. Recently, the in vivo force in the ACL of quadrupeds was found to be 10% of its maximum strength. In contrast, the patellar tendon experienced loads approximately 30% of its maximum strength. Ultimately, data of this nature can be useful as the basis for advanced mathematical models developed to analyze and predict the forces in the ACL for a variety of activities.

In a study on age-related changes in the tensile and viscoelastic properties of the human patellar tendon, specimens were subjected to stress-relaxation and cyclic loading tests, followed by a tensile test to failure. Statistically significant differences with respect to age could not be demonstrated for stress relaxation behavior and the modulus. It was found that age has some effect on the ultimate tensile strength and strain energy density of the human patellar tendon.

Perhaps the ligament that has received the most attention is the medial collateral ligament (MCL). This attention is partly due to its relatively simple geometry as well as its healing ability. Nevertheless, the majority of investigation on the MCL has focused on growth and development as studied in rabbits, rats, and dogs. There are well-documented gross and light microscopy changes during the growth and development phase. Also, the ligament insertion sites, particularly the tibial site, undergoes significant alterations. Ultrastructural and cellular changes as well as biochemical analysis of the changes in

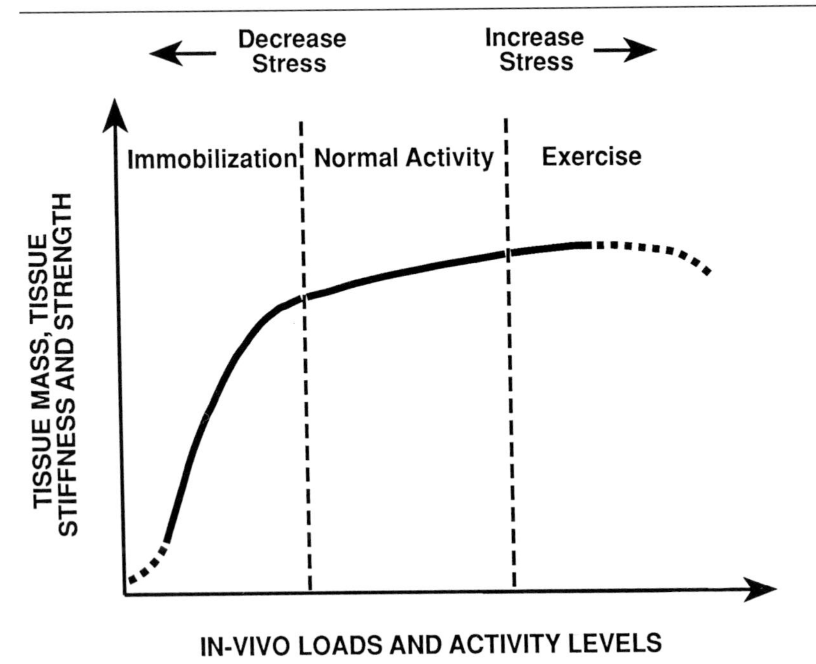

Fig. 1 *A schematic curve of the stress and motion dependent homeostatic responses for ligament and tendon.*

the MCL are also well documented. The tensile properties of the rabbit FMTC changes significantly until the time of epiphyseal closure. However, from skeletal maturity to the onset of senescence the changes are more subtle. During the aging process, there is no significant reduction in the MCL cross-sectional area. Biochemically, some loss of large-diameter collagen fibrils can be found, and the collagen cross-links shift from reducible to nonreducible. Change in the collagen organization and evidence of collagen fragmentation are potential causes for the mechanical deterioration. Data on ligament morphology and metabolic activity are inconclusive at present; hence, definitive conclusions cannot be made. Therefore, correlation of structural properties and function with aging of the MCL remains a subject to be explored.

To summarize, there are numerous clinical problems involving tendon and ligament with the aging process. Both the magnitude and cost of these problems are enormous. Unfortunately, epidemiologic data on the normal aging process and other contributing medical factors on tendon and ligament are lacking. Much of the available data that have been collected using animal models are on growth, development, and maturation, whereas relatively little information exists on aging, particularly on the truly aged animals. It is also well known that tendons and ligaments undergo stress- and motion-dependent homeostatic responses. The relationship between tissue mass and tissue properties are nonlinearly related to activity and stress levels. Exercise exhibits a possible effect on both tissue mass and mechanical properties, but the gain is relatively small. In Figure 1, it can be seen that the slope of this portion of the curve is rather shallow. However, the losses due to immobilization or lack of stress can be very large, as the slope of this portion of the curve is very steep. It should be recognized that this hypothetical curve on tissue homeostasis is based on data for normal healthy subjects. Whether aging will cause significant change of this type of nonlinear relationship is yet to be determined.

In recent years, more advanced morphologic, biochemical, and biomechanical methods have been developed to characterize the properties of these soft, parallel-fibered connective tissues, such as tendons and ligaments. For example, it is now possible to measure in vivo strain and force in tendons and ligaments. As a result, valuable data, though preliminary, are available. Continued advancement in these technological developments are necessary and need to be explored so that more sophisticated experimental methodology can be found. The field is prepared to undertake new challenges to solve the age-related problems in various tendons and ligaments. Furthermore, future investigation should also include joint capsules, which have not been covered in this section, as these tissues are very important to joint function.

Chapter 20

Overview of Disease and Treatment Related to Aging of Tendons and Ligaments

Cato T. Laurencin, MD, PhD
Richard H. Gelberman, MD

Introduction

A number of factors have been shown to influence the clinical expression of age-related diseases affecting dense regular connective tissue. Histologic and biochemical studies have demonstrated consistent age-related changes in tendons, ligaments, and capsular structures. Although data from these studies has provided insight into the aging process, a complete understanding of the manner and extent to which tendon and ligament abnormalities occur with increasing age has not been provided. It appears that a number of significant variables associated with aging, including gross anatomic influences contributing to tendon and ligament degenerative changes and the influences of associated medical conditions, significantly affect the clinical expression of disease processes.

While the scope of clinical problems resulting from age-related changes in soft tissue has not been determined fully, recent epidemiologic studies have suggested that the problem, in terms of patient morbidity and economic impact, is a major one. As the influences of age-related changes in tendon and ligament become better understood, effective preventive and therapeutic measures may be developed that will significantly improve the quality of life for the elderly.

Histologic and Biochemical Alterations in Dense Regular Connective Tissue With Increasing Age

Changes have been noted to occur with aging in both the structure and biochemical composition of tendons and ligaments. Experiments examining dense connective tissues in animals have noted several common themes. Structurally, collagen fibers increase in diameter, become varied in thickness, lose tensile strength, and become tougher with increasing age.[1-4]

Aging fibroblasts flatten, elongate, and become more slender.[1,2] At the subcellular level, there is a marked decrease in the intracytoplasmic organelles that carry out protein synthesis. Metabolically, cells progressively shift their sources of energy from aerobic glycolysis to respiration with age.[5] Overall, at the cellular level, the aging process is associated with decreasing water content, decreasing levels of glycoprotein and galactosamine in glycosaminoglycans, and decreasing biosynthetic activity.[1,2]

Experimental studies examining the properties of ligaments have further elucidated age-related histologic changes in regular dense connective tissue. Amiel and associates[1] examined the age-related properties of the medial collateral and anterior cruciate ligaments of rabbits. Morphologic studies revealed significant changes in ligamentous structure with age, including decreasing water content and collagen concentration. Biochemical analyses revealed that the concentration of reducible collagen crosslinks (associated with less mature tissue) decreased with increasing age of the animals. Conversely, nonreducible crosslinks increased, with the highest values noted in the oldest animals. Synthesis of collagen decreased with increased age, with the lowest values found in the oldest animals. The same types of morphologic and biochemical changes found in ligaments have been identified in tendons in various animal models and in human tissue.

A number of specific structural changes in human Achilles tendon serve as good clinical examples of age-related disease processes.[2] In the neonatal period, closely packed collagen fibrils, small and uniform in diameter, are arranged in undulating bundles. Cells, which are plentiful, are elongated and have rich amounts of granular endoplasmic reticulum and thin cytoplasmic processes. In the 18- to 25-year age range, collagen fibrils have widely varied diameters and regular, undulating histologic patterns. Cells are less numerous and are arranged in discontinuous rows. Characteristic spider-like cells possess thin cytoplasmic processes, which are connected by junctions that form a meshwork with woven collagen bundles. Collagen fibrils are small and uniform during the neonatal period and become larger and more variable during adolescence. With age, collagen fibrils undergo a decrease in average maximum diameter and density.[2]

Between 30 and 66 years of age, there is a reduction in the regular, undulating pattern of tendon fibers seen in the younger age groups. Collagen fibrils show heterogeneity in diameter with maximum diameter reduced. Cells, now widely separated, possess extremely elongated nuclei. In the 79- to 83-year age range, collagen fibers lose their undulating patterns and become straight. Interfibrillar areas are increased and are linked by electron dense filaments. The cell concentration of the Achilles tendon in advanced age is reduced and those cells that remain possess small cytoplasmic projections.

Factors Contributing to Age-Related Diseases of Tendon and Ligament

The changes that take place in tendon and ligament that lead to clinical disease conditions are influenced not only by basic morphologic and biochemical processes but also by a number of associated processes and factors.

Anatomic Factors

Age-related changes may be accentuated by the effects of intrinsic (vascular) and extrinsic (compressive or restrictive) tendon and ligament gross anatomic factors. Differences in the vascularity of tendons have been implicated in disease conditions of regular dense connective tissue in the rotator cuff and in the posterior tibial tendon. For example, a significant contribution to the pathogenesis of rotator cuff tears has been theorized to be traced to the microvascular pattern of the supraspinatus tendon.[6] Studies by Rathbun and Macnab[7] argue for the existence of a zone of hypovascularity in the supraspinatus tendon. Recent studies by Lohr and Uhthoff[6] agree, indicating that an area of hypovascularity exists in the supraspinatus tendon beginning 5 mm proximal to the osseous insertion of the tendon and ending at the musculotendinous junction. The articular side of the tendon was found to have the least vascularity. The authors, finding the supraspinatus tendon's area of hypovascularity in the region of supraspinatus degenerative tears, suggest that this region of hypovascularity is a significant factor in the pathogenesis of degenerative rotator cuff tears.

Fukuda and associates[8] examined rotator cuff tears, paying particular attention to the bursal side. They found that all bursal-side tears developed within 1 cm of the tendon's insertion, which is in the purported zone of hypovascularity. The distal stumps were found to be hypervascular, however, and granulation tissue was seen in that region. The authors concluded that this tissue represented a futile attempt at affecting a repair. In addition, tenocytes appeared to respond to the low oxygen environment by transforming into chondrocytes.

The relative vascularity of the rotator cuff is still a controversial subject, however. Chansky and Iannotti[9] observed that the critical zone of the rotator cuff was in fact hypervascular due to local inflammation and neovascularization. A similar concept has been put forth by Brooks and associates,[10] who failed to find any areas of hypovascularity in the rotator cuff. As the effects of regional vascularity on tendon/ligament pathology undergo further study, a resolution of this controversy should occur.

Both tendons and ligaments are subjected to compressive forces that occur extrinsically where there are bony prominences and pulleys.[11] Intermittent compression appears consistently to induce local metabolic responses. Vogel[12] has identified the accumulation of large molecular weight proteoglycans in regions of the human posterior tibial tendon posterior to the medial malleolus. Similar findings have been noted in anterior cruciate ligaments and in bovine tendons.[13] While some investigators believe that osseous or ligamentous impingement is a primary factor leading to the degenerative soft-tissue changes seen, for example, in the acromion, coracoacromial ligament, and rotator cuff, the precise roles of compression and impingement in the pathogenesis of regular dense connective tissue diseases remain to be better defined.[14]

Patterns of Use

Kannus and associates[15] examined injury patterns occurring in the elderly in a three-year prospective study focusing on exercise-related changes. A control group of younger athletes was used for comparison. The author found that 70% of exercise injuries occurring in the elderly were of the overuse type, compared with only 41% of those noted in younger athletes. In addition, injuries affecting the knee, shoulder, Achilles tendon, and posterior tibial tendon appeared to be particularly prevalent among the elderly.

Overuse trauma has been implicated as a source of injuries for older patients in a number of exercise settings. Some believe that older patients have a diminished ability to recover physiologically from the microtrauma that takes place with cumulative stress.[16]

Medical Conditions

A number of medical conditions, including diabetes mellitus, obesity, and hypertension, may influence the pathophysiology of diseases of tendons and ligaments during the process of aging. Examination of one such disease, diabetes mellitus, serves as a case in point. Hamlin and associates[17] examined the chronologic ages of human subjects by examining tendon collagen samples. They found that subjects with juvenile diabetes mellitus had, based on enzymatic digestion of collagen, higher experimentally determined ages than actual ages. Their data imply that an acceleration of the aging process of regular dense connective tissue occurs with diabetes mellitus. Leung and associates[18] found that diabetes mellitus reduced the rate of accretion of collagen in connective tissues because of an increase in the intracellular degradation of procollagen. Studies by other workers have identified a decreased ability of some mesenchymal cells, such as ligament fibroblasts, to carry out metabolic processes in diabetes mellitus.[19] Thus, diabetes mellitus appears to produce alterations at the cellular level that accelerate and/or exacerbate changes in connective tissue.

Clinical studies examining a number of pathologic processes that affect ligament and tendon have also implied that associated medical diseases play a significant role. Brotherton and Ball[20] and Stern and Harwin,[21] reporting on bilateral spontaneous, simultaneous ruptures of the quadriceps tendon, have suggested that underlying diabetes mellitus played a role in causing tendon failure. Holmes and Mann,[22] in a retrospective review of 67 patients with ruptures of the posterior tibial tendon, found that 52% suffered from diabetes mellitus, hypertension, and/or obesity. Their findings suggest that an acceleration of microvascular and macrovascular changes associated with these disorders may be responsible partially for age-related increases in posterior tibial tendon ruptures.

Adjacent Structures (Bone, Cartilage, and Muscle) Influence the Clinical Expression of Age-Related Disease Processes

The clinical expression of diseases involving dense regular connective tissue with aging is a complex interrelationship of many factors. For diseases that involve tendons and ligaments, it is important to consider concomitant changes in contiguous structures, such as bone and cartilage. For example, Brewer,[23] examining the aging process in the rotator cuff of autopsy specimens, found morphologic changes in bone, tendon, and cartilage. Osteitis of the greater tuberosity with cystic degeneration, a degenerative sulcus between the greater tuberosity and articular cartilage, disruptions in the attachment of tendon to bone via Sharpy's fibers, decreased cellularity, and fragmentation of the supraspinatus tendon were noted.

Bone and Cartilage

The immature growth centers of bone are the focus of abnormalities involving tendons in the young. Osgood-Schlatter's disease is the most common orthopaedic complaint involving tendon in the young, and has been reported to encompass 10% to 16% of sports overuse injuries.[24,25] Calcaneal apophysitis (Sever's disease) has been reported to occur in 6% to 15% of sports overuse injuries in the young.[25,26] With maturation and closure of epiphyseal plates, these orthopaedic afflictions no longer occur.

Bone mass increases through the third decade of life, at which time it reaches a plateau and subsequently decreases.[27] Women will, in the span of a lifetime, lose approximately 35% of cortical bone and 50% of trabecular bone.[27] The rate of bone loss is approximately 0.3% to 0.5% per year with an accelerated rate of 2% to 3% per year during the first 6 to 10 years after menopause.[28,29] Men overall tend to lose two thirds of the bone that women lose over a lifetime.[28]

A competition appears to exist between patterns of injury characterized by ligament/tendon wear versus bone failure over time. An example is found in the wrist, where similar mechanisms of injury cause physeal failure in the young, carpal ligament failure and resulting instability in the third and fourth decades, and distal radius fractures in the elderly. In the knee, a similar pattern can be seen.

19 million working people are estimated to be affected, with 72% of costs incurred because of diseases that affect individuals 45 years of age and over.[41,42] Soft-tissue disorders such as tendinitis, bursitis, and epicondylitis have been cited by the World Health Organization as causing considerable morbidity in the workplace.[43,44] In the United States, 115,000 injuries related to repeated soft-tissue tendon trauma at work were reported in 1988.[42]

Disorders of tendons and bursae, accounting for 59,000 hospitalizations, ranked as the 15th most common musculoskeletal condition in America in 1988.[42] Soft-tissue operations, such as repair of rotator cuff tendons, are among the 20 most frequent musculoskeletal procedures performed.[42] Most injuries involving regular dense connective tissue are not treated by hospitalization and, thus, the overall incidence of these problems is unknown. There are reasons to believe that the magnitude of the problem is increasing enormously as disorders caused by repeated trauma alone grew from 25% of reported occupational illnesses in 1983 to 45% in 1988.

References

1. Amiel D, Kuiper SD, Wallace CD, et al: Age-related properties of medial collateral ligament and anterior cruciate ligament: A morphologic and collagen maturation study in the rabbit. *J Gerontol* 1991;46:159-165.
2. Strocchi R, DePasquale V, Guizzardi S, et al: Human Achilles tendon: Morphological and morphometric variations as a function of age. *Foot Ankle* 1991;12:100-104.
3. Carpenter DG, Loynd JA: An integrated theory of aging. *J Am Geriatr Soc* 1968;16:1307-1322.
4. Houck JC, DeHesse C, Jacob R: The effect of ageing upon collagen metabolism. *Symp Soc Exp Biol* 1967;21:403-425.
5. Danielsen CC, Andreassen TT, Mosekilde L: Mechanical properties of collagen from decalcified rat femur in relation to age and in vitro maturation. *Calcif Tissue Int* 1986;39:69-73.
6. Lohr JF, Uhthoff HK: The microvascular pattern of the supraspinatus tendon. *Clin Orthop* 1990;254:35-38.
7. Rathbun JB, Macnab I: The microvascular pattern of the rotator cuff. *J Bone Joint Surg* 1970;52B:540-553.
8. Fukuda H, Hamada K, Yamanaka K: Pathology and pathogenesis of bursal-side rotator cuff tears viewed from en bloc histologic sections. *Clin Orthop* 1990;254:75-80.
9. Chansky HA, Iannotti JP: The vascularity of the rotator cuff. *Clin Sports Med* 1991;10:807-822.
10. Brooks CH, Revell WJ, Heatley FW: A quantitative histological study of the vascularity of the rotator cuff tendon. *J Bone Joint Surg* 1992;74B:151-153.
11. Amadio PC: Tendon and ligament, in Cohen IK, Diegelmann RF, Lindblad WJ (eds): *Wound Healing: Biochemical and Clinical Aspects*. Philadelphia, WB Saunders, 1992, p 384.
12. Vogel KG: Proteoglycans accumulate in a region of human tibialis posterior tendon subjected to compressive force in vitro and in ligaments, in Trans Combined Meeting Orthop Res Soc USA, Japan and Canada October 21-23, 1991, p. 58.

and Harwin,[21] reporting on bilateral spontaneous, simultaneous ruptures of the quadriceps tendon, have suggested that underlying diabetes mellitus played a role in causing tendon failure. Holmes and Mann,[22] in a retrospective review of 67 patients with ruptures of the posterior tibial tendon, found that 52% suffered from diabetes mellitus, hypertension, and/or obesity. Their findings suggest that an acceleration of microvascular and macrovascular changes associated with these disorders may be responsible partially for age-related increases in posterior tibial tendon ruptures.

Adjacent Structures (Bone, Cartilage, and Muscle) Influence the Clinical Expression of Age-Related Disease Processes

The clinical expression of diseases involving dense regular connective tissue with aging is a complex interrelationship of many factors. For diseases that involve tendons and ligaments, it is important to consider concomitant changes in contiguous structures, such as bone and cartilage. For example, Brewer,[23] examining the aging process in the rotator cuff of autopsy specimens, found morphologic changes in bone, tendon, and cartilage. Osteitis of the greater tuberosity with cystic degeneration, a degenerative sulcus between the greater tuberosity and articular cartilage, disruptions in the attachment of tendon to bone via Sharpy's fibers, decreased cellularity, and fragmentation of the supraspinatus tendon were noted.

Bone and Cartilage

The immature growth centers of bone are the focus of abnormalities involving tendons in the young. Osgood-Schlatter's disease is the most common orthopaedic complaint involving tendon in the young, and has been reported to encompass 10% to 16% of sports overuse injuries.[24,25] Calcaneal apophysitis (Sever's disease) has been reported to occur in 6% to 15% of sports overuse injuries in the young.[25,26] With maturation and closure of epiphyseal plates, these orthopaedic afflictions no longer occur.

Bone mass increases through the third decade of life, at which time it reaches a plateau and subsequently decreases.[27] Women will, in the span of a lifetime, lose approximately 35% of cortical bone and 50% of trabecular bone.[27] The rate of bone loss is approximately 0.3% to 0.5% per year with an accelerated rate of 2% to 3% per year during the first 6 to 10 years after menopause.[28,29] Men overall tend to lose two thirds of the bone that women lose over a lifetime.[28]

A competition appears to exist between patterns of injury characterized by ligament/tendon wear versus bone failure over time. An example is found in the wrist, where similar mechanisms of injury cause physeal failure in the young, carpal ligament failure and resulting instability in the third and fourth decades, and distal radius fractures in the elderly. In the knee, a similar pattern can be seen.

There, growth plate failures are seen in the very young, cruciate and collateral ligament injuries occur in more mature populations, and fractures of the tibial plateau take place in older age groups.

There have been a number of epidemiologic and basic science studies that provide support for these observations. Skak and associates[30] studied 91 consecutive metaphyseal fractures involving the knee in children aged 0 to 14 years. They found that metaphyseal fractures predominated in the youngest of the group, while teenagers experienced ligamentous rupture with low-energy trauma and physeal injury with high-energy trauma. With increased aging and closure of physes, ligamentous ruptures assumed greater importance. Osseous injuries about the knee, such as tibial plateau fractures, occurred with high-energy trauma, such as falls from heights or vehicular accidents.

Woo and associates,[31] studying the tensile properties of the human femur-ACL-tibia complex using three groups of cadaver knees, 22 to 35 years, 40 to 50 years, and 60 to 97 years, found that the structural properties (linear stiffness, ultimate load and energy absorbed) decreased significantly with increased specimen age. Woo has also studied the tensile properties of the rabbit medial collateral ligament as a function of age. With closure of the epiphysis, the mode of failure was found to undergo a transition from tibial avulsion to failure of the ligament in its mid-substance (as seen in clinical situations).

Muscle

During the aging process, muscle undergoes significant decreases in both size and strength. The extent of strength loss is difficult to quantitate, however, as the effects of disuse (which appears histologically identical to loss due to aging) must be considered also.[32] Studies in humans have shown decreased muscle size and strength with increasing age, and experiments in active animals have demonstrated that an absolute loss of muscle power occurs with the aging process alone. Brooks and Faulkner[33] examined the ability of mice of different ages to generate and sustain maximum power of the extensor digitorum longus muscles. They found that, after normalizing for muscle mass, power was 20% lower for the muscles of older mice than for those of young adult mice. With repeated contractions, both old and young adult mice demonstrated lower sustained normalized power than young mice.

Studies in humans have confirmed the observation that significant age-related changes occur in muscle.[34-37] A study of quadriceps femoris muscle cross-sectional area with ultrasound techniques found that muscles were 25% smaller in healthy men in the eighth decade of life compared with muscles in men in the third decade. The older group had 39% less isometric strength of the quadriceps muscle than their younger counterparts.[34]

Age-Related Changes in the Response of Ligament and Tendon to Injury, Repair, and Immobilization

The clinical response to injuries affecting ligament and tendon has been found to be influenced considerably by patient age. For instance, the prognosis following anterior glenohumeral dislocation has age as its most important determinate.[38] In a prospective study including 256 primary anterior dislocations of the shoulder, Hovelius[38] noted that decreasing rates of recurrence occurred with increasing age. While younger patients were prone to redislocations, older patients developed adhesive capsulitis (a syndrome in which pain and stiffness is associated with ligamentous and capsular contracture). In contrast to the population with recurrent shoulder dislocations, patients with adhesive capsulitis are usually in the fifth to seventh decades of life and frequently have associated diabetes, thyroid disease, and/or Parkinson's disease.

While there are indications that age-related changes in dense regular connective tissue influence injury and repair processes, there have been few animal studies elucidating the precise nature of these influences. A recent study examined the healing of partially and completely transected anterior cruciate ligaments in immature (epiphyses closed) and mature rabbits.[39] Mechanical testing and histologic examinations of ligaments were performed postoperatively, at two and six weeks, at three months, and at one year. Ligaments that were completely transected showed no evidence of repair. Partially transected ligaments from immature animals had one third the strength of ligaments from sham-operated knees postoperatively, while ligaments from mature animals had one quarter the strength. At three months, the stiffness of the ligaments was found to have returned to normal. By one year, the immature animals' ligaments were two thirds as strong as the sham-operated ones, and mature animals had ligaments with three quarters the strength of controls.

Preliminary investigations by a number of workers have suggested that changes in tendon and ligament with age may be modified by exercise. In a study of rat patellar tendon matrix changes with increasing age, with and without voluntary exercises, Vailas and associates[40] found decreased tendon glycosaminoglycan content in the nonexercise group. However, in comparing rats from nonexercising and voluntary exercise groups, older (28 months) rats subjected to a running regimen had levels of glycosaminoglycan content equivalent to younger (9 months) sedentary rats. These findings suggest that voluntary exercise has a positive influence on certain aging parameters.

The Magnitude of the Problem

In the United States, musculoskeletal disorders are the leading cause of disability for those in their working years, accounting for $126 billion dollars in costs to the American economy in 1988. Over

19 million working people are estimated to be affected, with 72% of costs incurred because of diseases that affect individuals 45 years of age and over.[41,42] Soft-tissue disorders such as tendinitis, bursitis, and epicondylitis have been cited by the World Health Organization as causing considerable morbidity in the workplace.[43,44] In the United States, 115,000 injuries related to repeated soft-tissue tendon trauma at work were reported in 1988.[42]

Disorders of tendons and bursae, accounting for 59,000 hospitalizations, ranked as the 15th most common musculoskeletal condition in America in 1988.[42] Soft-tissue operations, such as repair of rotator cuff tendons, are among the 20 most frequent musculoskeletal procedures performed.[42] Most injuries involving regular dense connective tissue are not treated by hospitalization and, thus, the overall incidence of these problems is unknown. There are reasons to believe that the magnitude of the problem is increasing enormously as disorders caused by repeated trauma alone grew from 25% of reported occupational illnesses in 1983 to 45% in 1988.

References

1. Amiel D, Kuiper SD, Wallace CD, et al: Age-related properties of medial collateral ligament and anterior cruciate ligament: A morphologic and collagen maturation study in the rabbit. *J Gerontol* 1991;46:159-165.
2. Strocchi R, DePasquale V, Guizzardi S, et al: Human Achilles tendon: Morphological and morphometric variations as a function of age. *Foot Ankle* 1991;12:100-104.
3. Carpenter DG, Loynd JA: An integrated theory of aging. *J Am Geriatr Soc* 1968;16:1307-1322.
4. Houck JC, DeHesse C, Jacob R: The effect of ageing upon collagen metabolism. *Symp Soc Exp Biol* 1967;21:403-425.
5. Danielsen CC, Andreassen TT, Mosekilde L: Mechanical properties of collagen from decalcified rat femur in relation to age and in vitro maturation. *Calcif Tissue Int* 1986;39:69-73.
6. Lohr JF, Uhthoff HK: The microvascular pattern of the supraspinatus tendon. *Clin Orthop* 1990;254:35-38.
7. Rathbun JB, Macnab I: The microvascular pattern of the rotator cuff. *J Bone Joint Surg* 1970;52B:540-553.
8. Fukuda H, Hamada K, Yamanaka K: Pathology and pathogenesis of bursal-side rotator cuff tears viewed from en bloc histologic sections. *Clin Orthop* 1990;254:75-80.
9. Chansky HA, Iannotti JP: The vascularity of the rotator cuff. *Clin Sports Med* 1991;10:807-822.
10. Brooks CH, Revell WJ, Heatley FW: A quantitative histological study of the vascularity of the rotator cuff tendon. *J Bone Joint Surg* 1992;74B:151-153.
11. Amadio PC: Tendon and ligament, in Cohen IK, Diegelmann RF, Lindblad WJ (eds): *Wound Healing: Biochemical and Clinical Aspects.* Philadelphia, WB Saunders, 1992, p 384.
12. Vogel KG: Proteoglycans accumulate in a region of human tibialis posterior tendon subjected to compressive force in vitro and in ligaments, in Trans Combined Meeting Orthop Res Soc USA, Japan and Canada October 21-23, 1991, p. 58.

13. Koob TJ, Vogel KG, Thurmond FA: Comprehensive loading in vitro regulates proteoglycan synthesis by fibrocartilage in tendon. *Trans Orthop Res Soc* 1991;16:49.
14. Neer CS II: Impingement lesions. *Clin Orthop* 1983;173:70-77.
15. Kannus P, Niittymäki S, Järvinen M, et al: Sports injuries in elderly athletes: A three-year prospective, controlled study. *Age Ageing* 1989;18:263-270.
16. Jobe FW, Schwab DM: Golf for the mature athlete. *Clin Sports Med* 1991;10:269-282.
17. Hamlin CR, Kohn RR, Luschin JH: Apparent accelerated aging of human collagen in diabetes mellitus. *Diabetes* 1975;24:902-904.
18. Leung MK, Folkes GA, Ramamurthy NS, et al: Diabetes stimulates procollagen degradation in rat tendon in vitro. *Biochim Biophys Acta* 1986;880:147-152.
19. Sasaki T, Ramamurthy NS, Golub LM: Insulin-deficient diabetes impairs osteoblast and periodontal ligament fibroblast metabolism but does not affect ameloblasts and odontoblasts: Response to tetracycline(s) administration. *J Biol Buccale* 1990;18:215-226.
20. Brotherton BJ, Ball J: Bilateral simultaneous rupture of the quadriceps tendons. *Br J Surg* 1975;62:918-920.
21. Stern RE, Harwin SF: Spontaneous and simultaneous rupture of both quadriceps tendons. *Clin Orthop* 1980;147:188-189.
22. Holmes GB Jr, Mann RA: Possible epidemiological factors associated with rupture of the posterior tibial tendon. *Foot Ankle* 1992;13:70-79.
23. Brewer BJ: Aging of the rotator cuff. *Am J Sports Med* 1979;7:102-110.
24. Kannus P, Nittymäki S, Järvinen M: Athletic overuse injuries in children: A 30-month prospective follow-up study at an outpatient sports clinic. *Clin Pediatr (Phila)* 1988;27:333-337.
25. Kujala UM, Aalto T, Osterman K, et al: The effect of volleyball playing on the knee extensor mechanism. *Am J Sports Med* 1989;17:766-769.
26. Orava S, Puranen J: Exertion injuries in adolescent athletes. *Br J Sports Med* 1978;12:4-10.
27. Riggs BL, Melton LJ III: Involutional osteoporosis. *N Engl J Med* 1986;314:1676-1686.
28. Riggs BL, Wahner HW, Melton LJ III, et al: Rates of bone loss in the appendicular and axial skeletons of women: Evidence of substantial vertebral bone loss before menopause. *J Clin Invest* 1986;77:1487-1491.
29. Katch FI, McArdle WD (eds): *Nutrition, Weight Control and Exercise*, ed 3. Philadelphia, Lea & Febiger, 1988, pp 240-258.
30. Skak SV, Jensen TT, Poulsen TD, et al: Epidemiology of knee injuries in children. *Acta Orthop Scand* 1987;58:78-81.
31. Woo SL, Hollis JM, Adams DJ, et al: Tensile properties of the human femur-anterior cruciate ligament-tibia complex: The effects of specimen age and orientation. *Am J Sports Med* 1991;19:217-225.
32. Wilmore JH: The aging of bone and muscle. *Clin Sports Med* 1991;10:231-244.
33. Brooks SV, Faulkner JA: Maximum and sustained power of extensor digitorum longus muscles from young, adult and old mice. *J Gerontol* 1991;46:B28-B33.
34. Young A, Stokes M, Crowe M: The size and strength of the quadriceps muscles of old and young men. *Clin Physiol* 1985;5:145-154.
35. Rice CL, Cunningham DA, Paterson DH, et al: Arm and leg composition determined by computed tomography in young and elderly men. *Clin Physiol* 1989;9:207-220.

36. Buskirk ER, Segal SS: The aging motor system: Skeletal muscle weakness, in Spirduso WW, Eckert HM (eds): *Physical Activity in Aging*. Champaign, IL, Human Kinetics Books for the American Academy of Physical Education, 1989, pp. 19-36.
37. Larsson L: Morphological and functional characteristics of the ageing skeletal muscle in man: A cross-sectional study. *Acta Physiol Scand* 1978;457(Suppl):1-36.
38. Hovelius L: Anterior dislocation of the shoulder in teen-agers and young adults: Five-year prognosis. *J Bone Joint Surg* 1987;69A:393-399.
39. Hefti FL, Kress A, Fasel J, et al: Healing of the transected anterior cruciate ligament in the rabbit. *J Bone Joint Surg* 1991;73A:373-383.
40. Vailas AC, Pedrini VA, Pedrini-Mille A, et al: Patellar tendon matrix changes associated with aging and voluntary exercise. *J Appl Physiol* 1985;58:1572-1576.
41. Association of Schools of Public Health: Proposed national strategies for the prevention of leading work-related diseases and injuries, part I. *USA, Association of Schools of Public Health*, 1986.
42. Praemer A, Furner S, Rice DP (eds): *Musculoskeletal Conditions in the United States*. Park Ridge, IL, American Academy of Orthopaedic Surgeons, 1992.
43. Raffle PAB, Lee WR, McCallum RI, et al: *Hunter's Diseases of Occupations*. Boston, Little Brown, 1987.
44. Tenth Report of the Joint ILO/WHO Committee on Occupational Health; Epidemiology of work-related diseases and accidents. Geneva, World Health Organization, 1989.

Chapter 21

Tendon Biomechanics: Age-Related Changes in Function and Forces

Kai-Nan An, PhD

Introduction

The tendon is an integral part of the musculoskeletal system. The primary function of the tendon is to transmit muscle force to the skeletal system, and to do so with a limited amount of elongation. Anatomically, the main portion of the tendon is composed of dense and parallel fibers of connective tissue. One end of the tendon is part of the tendon plate, or the aponeurosis of the muscle, and the other end inserts onto bone.

Morphologically, the tendon is a complex composite material consisting of collagen fibrils embedded in a matrix of proteoglycans.[1] The hierarchical organization of the tendon collagen has been proposed to consist of the single protein macromolecule, the tropocollagen triple helix, the microfibril of five tropocollagen units appropriately staggered lengthwise, and finally, the tetragonal lattice of microfibrils, possibly forming subfibrils and the collagen fibril.[2] Proteoglycans and glycoproteins in association with water come into play as a matrix that binds the fibrils together.

The mechanical properties of tendon greatly influence the transmission of muscle tension to skeletal parts during posture and movement. The properties of tendon depend mainly on the properties and architecture of the collagen and the interaction of the elastic fibers with each other. In addition, the interwoven proteoglycans contribute to time-dependent properties. The stress-strain relationship of the tendon follows the same characteristic curve of other collagenous tissue. It begins with an area called the toe region, in which the tendon deforms easily with slight tension. With the increase of stress and strain, the toe region is followed by a fairly linear region. The tangent modulus of this linear region has been found to be in the range of 0.6 to 1.7 GPa.[3] Within the elastic range, the elastic strain energy to recovery is 90% to 96% per cycle.[4] With a further increase in the stress, microfailure of fiber bundles and eventual gross failure of the tendon will result. The ultimate tensile strength of tendon has been found to range from 50 to 150 MPa, and the ultimate strain value, from 9% to 30%.[5]

In addition, tendon can be classified as a viscoelastic material. The rate of loading influences the mechanical response of the tendon. In general, with an increase in loading rate, the stress-strain curve shifts slightly toward the left, making the tendon stiffer. Furthermore, when the tendon is subjected to cyclic loading and unloading, the stress-strain curve shifts to the right, representing either a stress relaxation or a strain creep phenomenon.[6,7]

Dynamically, the muscle and tendon function together to provide the passive and active properties of the musculotendon actuator. The force-generating capacity of the muscle as an actuator is influenced by the length of the muscle fiber at contraction, as well as by the velocity of contraction. The tendon compliance will affect the muscle fiber length, and thus the contracting length-tension relationship. The rate of muscle contraction will affect not only the muscle velocity-tension relationship, but also the tendon viscoelastic properties. In general, the muscle, tendon, and skeletal systems constitute a coupled and multiple-input and multiple-output feedback system.[3] The storage of elastic energy in the musculotendon actuator for subsequent release often is hypothesized to be a mechanism used to perform locomotion or other motor tasks efficiently.[3]

Tendon Disorder

The loading environment sustained by the tendon can allow various types of pathologies and disorders to develop. These disorders often develop as a result of cumulative loading and degenerative processes. Therefore, aging is an important factor related to the development of such abnormalities. Tendon injuries, such as tendinitis, tenosynovitis, rupture, and dislocation, can cause significant morbidity and disability.[8] Tendinitis is the inflammation of the tendon substance itself; tenosynovitis develops secondary to mechanical and inflammatory irritation to both the tendon and the surrounding sheath. Chronic injury and degeneration of the tendon can lead to rupture of the tendon as well.

In tenosynovitis, the tendon and tendon sheath are irritated by either inflammatory or mechanical processes. Rheumatoid tenosynovitis occurs in some patients with rheumatoid arthritis. Mechanical overuse, friction, and chronic stretching also can lead to tenosynovitis. More specifically, tendinitis (inflammation of the tendon itself) is caused by trauma to the tendon, which may be caused by partial rupture of the fibers, tumors within the tendon, deposits of cholesterol, ossification within the tendon, or local rheumatoid thickening.[9]

Trigger finger and stenosing tenosynovitis commonly occur at the A1 pulley of the flexor tendon sheath of the finger. Nodular swelling of the tendon and circumferential narrowing of the pulley space make the passage of the tendon difficult. With steady pulling of the tendon, the nodule suddenly passes through the restricted pulley and cannot return to its normal position without considerable extensor force, and the finger locks in flexion.[10] On the other hand, limited flexion in the trigger finger also is possible, especially in rheumatoid disease. The etiologies of trigger finger injuries are diverse. Chronic distention of the sheath and stretching of the pulley in the rheumatoid patient frequently can cause trigger finger injuries. Repetitive trauma, such as that from pistol-gripped industrial machinery or long hours of gripping a steering wheel, also may cause trigger finger.

In addition to the finger joint, stenosing tenosynovitis can occur in the abductor pollicis longus and extensor pollicis brevis tendon within the first extensor compartment at the wrist or in the flexor digitorum longus in the tarsal tunnel beneath the medial malleolus. Biceps tendinitis at the shoulder joint also has been linked to stenosing tenosynovitis, with thickening and stenosis of both the transverse ligament and the sheath and narrowing of the tendon under the sheath.[11] Impingement and attrition of the tendon of the long head of the biceps, especially in patients with occupations that require repetitive overhead lifting and throwing, eventually lead to various disorders, including degeneration of the tendon. In the attrition type of tendinitis, local bone reaction and the formation of adhesions in the bicipital groove cause stenosis and lead to further attrition of the tendon.

Calcifying tendinitis of the rotator cuff tendon is also a common disorder of unknown etiology, in which reactive calcification is spontaneously resorbed over time, with subsequent healing of the tendon.[12] This disease can cause acute pain and discomfort and can lead to dysfunction of the shoulder joint. The pathogenesis of this disorder has yet to be determined. Degeneration of fibers of the rotator cuff tendons caused by mechanical wear and tear, or by aging, or both, has been postulated to lead to calcification.[13] With aging, there is a diminution in the vascular supply to the rotator cuff tendon, along with fiber changes.[14] Collagen bundles and fascicles that constitute the distinctive architecture of the cuff tendon show the most conspicuous age-related changes, beginning at the end of the fifth decade of life.[15] These fascicles thin, split, and fray. Calcification then occurs on these fragmented fibers and on the necrotic debris.[16,17]

A different theory outlines the process of calcification in calcifying tendinitis to be actively mediated by cells in a variable environment rather than caused by a degenerative disease.[12] The evolution of the disease has been considered to follow three distinct stages: precalcified, calcified, and postcalcified. In the precalcified stage, the site for calcification undergoes fibrocartilaginous transformation. In the ensuing calcified stage, calcium crystals are deposited in the matrix vesicles, which coalesce to form a large area of deposits.[18] At this time, the area of fibrocartilage with foci of calcification is generally devoid of vascularity. Following an inactive period of calcification, spontaneous resorption of calcium takes place with macrophage and giant cells. In this theory of calcification, however, it is still difficult to know what triggers the fibrocartilaginous transformation in the first place. Hypoxia and localized pressure have been suggested as possible etiologic factors. Calcifying tendinitis also has been commonly seen in musculotendinous units of the elbow and foot, often occurring at sites of tensile stress in areas of tendon, muscle, and periosteal damage.[19]

In individuals who are middle aged or older, a tendon may rupture spontaneously as a result of degenerative processes. Rupture frequently occurs in the tibialis posterior tendon.[9] Also, acute and insidious rupture of the Achilles tendon is often seen in older patients.[20] Rupture of the flexor carpi radialis tendon and long flexor tendon has been associated with previous carpal tunnel decompression surgery and local steroid injection.[21] Attrition of the flexor tendons on the bony spur within the carpal tunnel or on the scaphoid also has been identified as responsible for rupture of the flexor tendon in the patient with rheumatoid tenosynovitis.[22]

Biomechanically, these associated tendon disorders can be related to cumulative trauma from repetitive stretching or attrition as the tendon passes through the pulley and bony surface. The knowledge of the loading environment experienced by these tendons is important to a better understanding of the possible etiology of the disorders.

Mechanical Factors Associated With Tendon Disorders

The possible association between the mechanical loading environment encountered by the tendon and the development of disorders and adaptions is discussed using the finger joint as an example. In the past, numerous experimental and analytic methods have been used to determine the flexor tendon tensions produced in performing activities of daily living.

In general, the tensions in either flexor profundus or flexor subliminus tendons range from two to three times the applied force during tip pinch function; one to four times, during key pinch function; and three to four times, during grasp function. Translated into absolute values, during strenuous pinch and grasp functions, the tensions in the flexors can range between 100 and 250 N.[23]

One question is whether the elongation of tendon under this amount of tension has any influence on muscle action in terms of the length-tension relationship. Based on the reported tensile properties of human flexor tendon,[24] the tendon can experience 4% elongation under such an amount of tension. For a flexor tendon of normal length (from 185 to 250 mm), an elongation of 7.4 to 10 mm therefore would be possible. For the flexor muscle, using the average muscle fiber length of 60 mm, the elongation of tendon can alter the normal muscle fiber length by between 10% and 15%. Theoretically, it is therefore important to consider the elasticity of tendon along with the contractility of the muscle in the contraction process of the musculotendon actuator.

Another question is whether the tension in the tendon under strenuous pinch and grasp function would cause damage or possible rupture of the tendon. Again, according to the data reported, the tensile strength measurements of the finger flexor tendons range between 1,100 and 1,300 N.[24] In other words, during strenuous hand function, the tendon will only experience 10% to 20% of its ultimate strength. Such tendon tension therefore can be considered relatively safe against possible gross rupture. However, 20% of tension or 4% of elongation would put the tendon at the end of the toe region and at the beginning of the linear region on the force-elongation curve. Microdamage of the collagen bundles or even of the interwoven links in the linear region is possible. Cumulative trauma of such an injury and the biologic response to such an assault can lead to the inflammatory and degenerative process.

In addition to the tensile load exerted along the tendon, the transverse load as the tendon wraps around the bony structure or pulley constraint can cause compressive stress in the tendon and attrition on the surface. As the tendon changes the direction of its path, the resultant transverse load applied to the tendon is approximately equal to two times the tension in the tendon times the sine of half the angle of the tendon direction. For example, when the tendon is running straight, the angle is zero and the transverse load is equal to zero. On the other hand, when the tendon is rotated 90 degrees as the finger joint is flexed, the resultant transverse force on the tendon can be as high as 1.732 times the tension in the tendon. With this amount of transverse load, internal pressures between fibers within the tendon (up to 2 MPa), and compressive stresses approaching that in cartilage could possibly be developed.[25] The fibrocartilaginous zone developed in the tendon may represent the functional adaption to the compressive load.[26]

A tendon sliding over the curved pulley or bony surface is analogous to a belt wrapped around a pulley. The frictional force appearing at the interface is exponentially proportional to the coefficient of friction between the two surfaces and the angle of contact.[27] Under the same frictional condition, increasing the joint flexion angle will increase the contact angle, and, thus, the frictional force on the tendon surface.[28-30] The coefficient of friction also can be increased as a result of aging and inflammatory processes, which may further magnify the frictional force and possible attrition.

Stenosing tenosynovitis takes place more frequently in the ring finger than in the index finger,[10] possibly because during normal hand function, the metacarpophalangeal joint is flexed more in the ring and little fingers than in the middle and index fingers. With larger joint flexion angles, more compressive load and irritation will be encountered, and eventual nodular swelling appears in the tendon. Meanwhile, the friction and attrition on the pulley and tendon sheath also can cause inflammation in these surrounding soft tissues, further restricting the pathway of the tendon, and creating a vicious cycle.

Age-Related Change of Mechanical Properties

Intuitively, aging should have some, if not significant, influence on the mechanical properties of tendons. Unfortunately, not many reports in the literature deal with this subject. The stress-strain behavior of rat tail tendon as a function of age was examined.[2] The toe region decreases with age because of a systematic decrease in crimp angle with age.[31] The stiffness or modulus within the linear region increases up to maturity and then remains constant. Up to maturity of tendon, the linear region is followed by a single yield region, in which irreversible elongation and structural damage can take place. A near-zero apparent modulus was observed in this region. After maturity, this single yield plateau was not obvious and was replaced by two distinct yield regions. The stress and strain at failure also increased with maturation.[32,33]

In general, with increased molecular stability, there is an increase in the tensile strength of tissue. This is true only to a certain degree of structural rigidity, however.[34] Cross-linking increases with maturation and aging, thereby producing structural stability. A corresponding decrease in tensile strength is observed in this case after a certain point.[34,35] Sensitivity of tensile strength, failure strain, and failure energy density to strain rate decreases rapidly as an animal grows and matures.[33] More recently, it has been documented that the mechanical values are followed closely by the collagen content, but not at all by the elastic content, in the tendon.[36]

Tendon in different anatomic locations experiences different loading environments. The mechanical properties thus are different. The tensile strength and stiffness of digital flexor tendons in adult miniature swine were about twice those of corresponding digital extensor tendons.[37] Furthermore, the hysteresis of extensor tendons has been found to be twice as much as in the flexor tendon.[4] At birth, the digital flexor and extensor tendons have identical mechanical properties. The difference in mechanical properties of flexor and extensor tendons increases with age and maturation. This leads to the hypothesis that the increase in cross-link stabilization that normally occurs in collagen during growth and aging may take place faster in the flexor tendon than in the extensor tendon, thus providing the superior mechanical properties of the flexors at maturity. This change may be influenced by the level of stress experienced in vivo.[4] Biochemical analysis also has shown that digital flexor tendons have a much higher concentration of collagen per wet weight of tissue than do digital extensor tendons.[37]

References

1. Elliott DH: Structure and function of mammalian tendon. *Biol Rev* 1965;40:392-421.

2. Kastelic J, Baer E: Deformation in tendon collagen, in Vincent JFV, Currey JD (eds): *The Mechanical Properties of Biological Materials*. Cambridge, Cambridge University Press, 1980, pp 397-435.
3. Zajac FE: Muscle and tendon: Properties, models, scaling, and application to biomechanics and motor control. *Crit Rev Biomed Eng* 1989;17:359-411.
4. Shadwick RE: Elastic energy storage in tendons: Mechanical differences related to function and age. *J Appl Physiol* 1990; 68:1033-1040.
5. Gelberman R, Goldberg V, An KN, et al: Tendon, in Woo SL-Y, Buckwalter JA (eds): *Symposium on Injury and Repair of the Musculoskeletal Soft Tissues*. Park Ridge, IL, American Academy of Orthopaedic Surgeons, 1988, chap 1, pp 5-40.
6. Hubbard RP, Chun KJ: Mechanical responses of tendons to repeated extensions and wait periods. *J Biomech Eng* 1988;110:11-19.
7. Woo SL: Mechanical properties of tendons and ligaments: I. Quasi-static and nonlinear viscoelastic properties. *Biorheology* 1982; 19:385-396.
8. Rosenberg ZS, Cheung Y: Diagnostic imaging of the ankle and foot, in Jahss MH (ed): *Disorders of the Foot and Ankle*, ed 2. Philadelphia, WB Saunders, 1991, vol 1, chap 6, pp 109-154.
9. Jahss MH: Tendon disorders of the foot and ankle, in Jahss MH (ed): *Disorders of the Foot and Ankle*, ed 2. Philadelphia, WB Saunders, 1991, vol 2, pp 1461-1513.
10. McFarland GB: Entrapment syndromes, in Evarts CM (ed): *Surgery of the Musculoskeletal System*, ed 2. New York, Churchill Livingstone, 1990, vol 1, chap 37, pp 961-981.
11. Burkhead WZ Jr: The biceps tendon, in Rockwood CA Jr, Matsen FA III (eds): *The Shoulder*. Philadelphia, WB Saunders, 1990, vol 2, chap 20, pp 791-836.
12. Uhthoff HK, Sarkar K: Calcifying tendinitis, in Rockwood CA Jr, Matsen FA III (eds): *The Shoulder*. Philadelphia, WB Saunders, 1990, vol 2, chap 19, pp 774-790.
13. Codman, EA: *The Shoulder: Rupture of the Supraspinatus Tendon and Other Lesions in or About the Subacromial Bursa*. Boston, Thomas Todd, 1934.
14. Brewer BJ: Aging of the rotator cuff. *Am J Sports Med* 1979;7:102-110.
15. Olsson O: Degenerative changes of the shoulder joint and their connection with shoulder pain. *Acta Chir Scand* 1953;181:1-130.
16. McLaughlin HL: Lesions of the musculotendinous cuff of the shoulder: Observations on the pathology, course and treatment of calcific deposits. *Ann Surg* 1946;124:354-362.
17. Macnab I: Rotator cuff tendinitis. *Ann Royal Coll Surg Engl* 1973;53:271-287.
18. Sarkar K, Uhthoff HK: Ultrastructural localization of calcium in calcifying tendinitis. *Arch Patho Lab Med* 1978;102:266-269.
19. Dobyns JH: Musculotendinous problems at the elbow, in Evards CM (ed): *Surgery of the Musculoskeletal System*, ed 2. New York, Churchill Livingstone, 1990, vol 2, chap 58, pp 1661-1681.
20. Hattrup SJ, Johnson KA: A review of ruptures of the Achilles tendon. *Foot Ankle* 1985;6:34-38.
21. Tonkin MA, Stern HS: Spontaneous rupture of the flexor carpi radialis tendon. *J Hand Surg* 1991;16B:72-74.
22. Ertel AN, Millender LH, Nalebuff E, et al: Flexor tendon ruptures in patients with rheumatoid arthritis. *J Hand Surg* 1988;13A:860-866.

23. Chao EY, An KN, Cooney WP, et al: Biomechanics of the hand. 1989, World Scientific.
24. Pring DJ, Amis AA, Coombs RR: The mechanical properties of human flexor tendons in relation to artificial tendons. *J Hand Surg* 1985;10B:331-336.
25. Sidles JA, Clark JM, Huber JD: Large internal pressures occur in ligament grafts at bone tunnels. *Trans Orthop Res Soc* 1990;81.
26. Okuda Y; Gorski JP, An KN, et al: Biochemical, histological, and biomechanical analyses of canine tendon. *J Orthop Res* 1987;5:60-68.
27. Beer FP, Johnston ER Jr: *Vector Mechanics for Engineers: Statics and Dynamics*. New York, McGraw Hill, 1962.
28. Armstrong TJ, Chaffin DB: Some biomechanical aspects of the carpal tunnel. *J Biomech* 1979;12:567-570.
29. Goldstein SA, Armstrong TJ, Chaffin DB, et al: Analysis of cumulative strain in tendons and tendon sheaths. *J Biomech* 1987;20:1-6.
30. Horii E, Lin GT, Cooney WP, et al: Comparative flexor tendon excursion after passive mobilization: An in vitro study. *J Hand Surg* 1992;17A:559-566.
31. Torp S, Arridge RGC, Armeniades CD, et al: Structure-property relationships in tendon as a function of age, in Atkins EDT, Keller A (eds): *Structure of Fibrous Biopolymer*. London, Butterworths, 1975, pp 197-221.
32. Danielsen CC: Mechanical properties of native and reconstituted rat tail tendon collagen upon maturation in vitro. *Mech Ageing Dev* 1987; 40:9-16.
33. Haut RC: Age-dependent influence of strain rate on the tensile failure of rat tail tendon. *J Biomech Eng* 1983;105:296-299.
34. Viidik A: Biomechanical behavior of soft connective tissues, in Akkas N (ed): *Progress in Biomechanics*. Alphen aan den Rijn, Sijthoff and Noordhoff (International Publishers), 1979, pp 75-113.
35. Vogel HG: Influence of maturation and age on mechanical and biochemical parameters of connective tissue of various organs in the rat. *Connect Tissue Res* 1978;6:161-166.
36. Vogel HC: Species differences of elastic and collagenous tissue: Influence of maturation and age. *Mech Ageing Dev* 1991;57:15-24.
37. Woo SL, Gomez MA, Amiel D, et al: The effects of exercise on the biomechanical and biochemical properties of swine digital flexor tendons. *J Biomech Eng* 1981;103:51-56.

Chapter 22

Biochemical Changes Associated With Aging in Tendon and Ligament

Kathryn G. Vogel, PhD

Introduction

Compared to the major age-related musculoskeletal conditions affecting human health, such as arthritis, osteoporosis, or neoplasms, there are few problems with tendons and ligaments. The rate of dislocations and sprains even appears to decrease in older populations,[1] probably because of decreased physical activity. On the other hand, some remarkable age-related changes in tendon material properties suggest that biochemical factors could explain the changes in tendon and ligament with aging.

This chapter will primarily review measurements of the age-related composition and structure of collagen in tendon and ligament. This is appropriate because type I collagen is the major constituent of these tissues. Proteoglycan content and cellular metabolism also will be discussed, although there is little information related to age. In the end it will not be possible to relate functional deficits to specific biochemical changes, but certain directions for future research will be clarified.

Measurement of Biochemical Composition With Aging

Most studies of tendon structure and function have used rat tail tendons. Rats are a convenient vertebrate model for aging studies because they can be managed in large numbers and have a short life span (about 36 months). Rabbit tendons and ligaments also are used widely, and have some advantage over rat tendons and ligaments for studying biochemistry because the tissues are larger. The disadvantage with rabbits is that they are more expensive to house and an old rabbit must be at least 48 months of age. Bovine tissue is excellent for studies of noncollagenous tendon components because the tendons are large (making it easier to measure minor constituents) and are readily available from a slaughterhouse, however, it can be difficult to obtain old bovine tissue and to know the donor's age. Human tissue is the most relevant tissue for biomedical research.

In order to determine the quantity of a given constituent of tissue and its changes with age, some general agreement must be reached on normalizing the data. In practice, collagen content generally is expressed as a percentage of the dry weight of the tissue. As discussed by Maroudas,[2] however, it is not possible to make conclusions about an increase or decrease in the absolute quantity of a single component on the basis of a change in its percentage concentration. If collagen content increases, it could mean that there is more collagen or it could mean there is less of other constituents. It is interesting that the compositional data for human femoral head cartilage at different ages support the assumption that amounts of collagen and chondroitin sulfate remained constant throughout the aging process.[2]

Collagen

Structure

The diameter of collagen fibrils in the tendon of newborn vertebrates is generally small and homogeneous.[3] With advancing age, the fibrils show an increase in mean diameter with a broad and frequently bimodal distribution (ie, from about 50 nm in a rat tail tendon of a newborn to 400 nm in that of an adult) (Fig. 1).[4] In all cases in which fibril distribution was documented at senescence, there was a bimodal distribution.[5] In a few studies, mean fibril diameter was found to decrease during senescence.[6] It was suggested that the small fibrils found in senescent tendon could be the product of newly synthesized collagen, whereas wear and fatigue could render the oldest and largest fibrils susceptible to enzymatic degradation.[3] Virtually nothing is known, however, about the ultrastructural appearance of collagen during normal turnover.

An extraordinary histologic study of human tendon pathology following spontaneous rupture included 445 control tendons from healthy individuals.[7] Histopathologic changes were noted in 35% of the control tendons. The changes included such diagnoses as hypoxic degeneration, mucoid degeneration, and tendolipomatosis, and were more common in the older control subjects. This suggests that degenerative changes in tendons may be common in people older than 35 years, and may predispose to spontaneous rupture.

In an X-ray diffraction analysis of human toe tendons obtained post mortem from individuals aged 16 to 76 years, there was no change in the equatorial spacing of human tendon collagen with age (mean = 14.5 ± 0.1 Å).[8]

Amount of Collagen

The mean concentration of collagen was reported to be increased in old rabbit Achilles tendon[9] and rat patellar tendon[10] compared to that of young mature animals, but was not significantly changed in

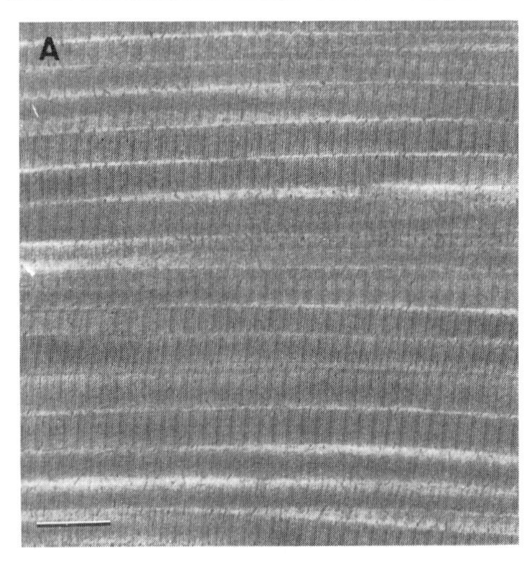

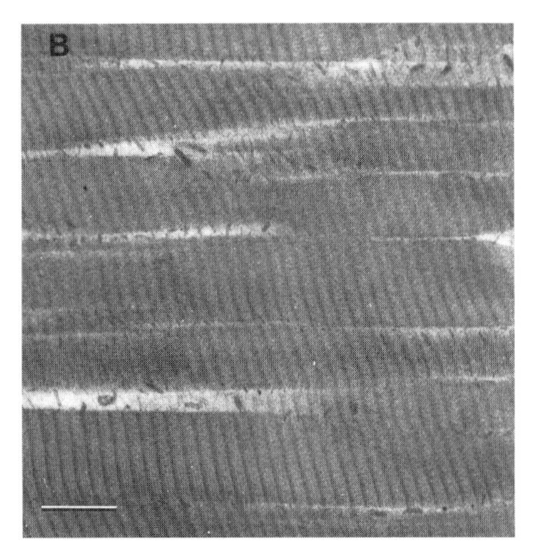

Fig. 1 *Bovine deep flexor tendon in longitudinal section. (**Left**), Fetal (7 months gestation). (**Right**), Adult. Fibril diameters were uniform in the fetal tissue (mean = 88 ± 17 nm); in adult tissue the mean fibril diameter was larger and showed a bimodal distribution (mean = 158 ± 68 nm).[39] Tissue was stained during fixation with cuprolinic blue. Sections were stained with uranyl acetate and lead citrate. Bars = 100 nm.*

old rat tail tendon.[11] In 36-month-old rabbits, a slight decrease in the concentration of collagen was found for both medial collateral ligaments (MCL) and anterior cruciate ligaments (ACL).[12] In all cases, the data were normalized to the wet or dry weight of the tissue. Vailas and associates[10] found no difference between the dry weight of the entire patellar tendon of a young rat compared with that of an old rat, and thus concluded that the increased concentration of collagen in old tendon must derive from a decrease in the noncollagenous components of the matrix. There would appear to be no convincing evidence that collagen content is significantly diminished with age.

Collagen Cross-linking

The cross-linking theory of aging gained support 20 to 30 years ago, when it was noticed that the collagen from old animals behaved differently from the collagen of young animals. The theory is that tissue becomes increasingly rigid and shows functional deficits with age because of covalent intermolecular cross-links. One changing characteristic was the sensitivity of collagen to collagenase digestion. A straight line resulted when the logarithm of time required to digest 50% of the collagen was plotted as a function of donor age (Fig. 2).[13] Using this as a standard curve, the method was used successfully to determine the age of unknown samples.[14] It was concluded that collagen is increasingly stabilized with aging, but it

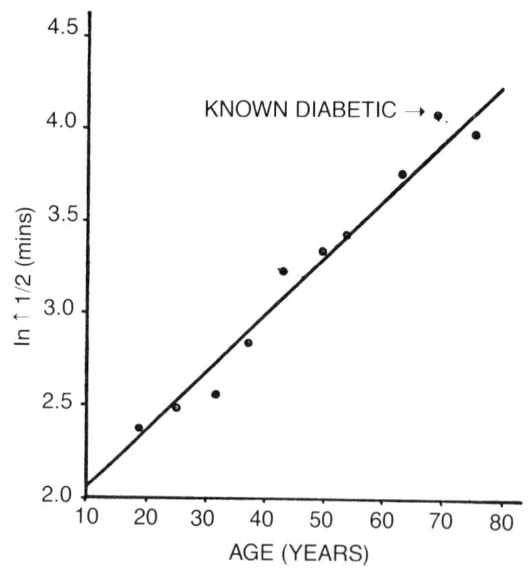

Fig. 2 *Logarithm of the time required to digest 50% of the collagen as a function of age. (Reproduced with permission from Hamlin CR, Kohn RR: Determination of human chronological age by study of a collagen sample.* Exp Gerontol *1972;7:377-379.)*

could not be determined whether the stabilizing factor was covalent cross-linking or some other physical interaction.

Other measurements used to demonstrate differences between collagen of young and old animals are decreasing solubility in neutral salt or 1% acetic acid,[15] and the breaking time of rat tail tendon collagen fibers submerged in buffered 7 M urea.[16] The relationship between breaking time (in seconds) and animal age, often referred to as "aging of collagen fibers," can be described by a straight line. Breaking time correlates with maximum life span of different rat strains[16] and even caging conditions of the animals,[17] presumably because the lower tail temperature of individually caged rats slowed the rate of collagen aging.

Knowledge of the chemistry of collagen cross-linking and the actual changes that accompany aging has developed slowly. Based on data from experiments that attempted to alter collagen aging by chemical treatment of the fibers, Bailey[18] proposed that labile bonds formed first and were gradually stabilized with increasing age to produce a permanent cross-link. Reducible aldehyde-derived cross-links in bovine skin were found to increase to a maximum at about one year of age and then to decrease and virtually disappear with increasing age.[19] This change also was found in human palmar flexor tendons.[20] In rabbit ACL and MCL, the reducible cross-links (such as hydroxylysinonorleucine) diminished relative to total collagen with age, but the pyridinoline nonreducible cross-link increased.[12]

That cross-linking changes with age in tendons and ligaments now seems to be a clearly demonstrated phenomenon. It is not clear, however, whether the cross-links that have been measured are important determinants of the tissue's physical properties, or whether the age-related changes in collagen solubilization and breaking strength are related to cross-linking. No change in the deformation characteristics of individual rat tail tendon collagen fibrils could be found as a function of age.[21] Barenberg and associates[21] suggested that the explanation for changes in mechanical properties with aging should be sought in changes of fibril size and fibril interaction with matrix proteoglycans, rather than in altered intermolecular cross-linking.

Nonenzymatic Glycation

In a number of the studies discussed above, collagen from diabetic patients appeared to age more rapidly than similar tissue from nondiabetics. The symptomatic similarities between diabetes and aging may result from the chemical interaction of proteins with free sugars, a process now called nonenzymatic glycation.[22] In one such reaction, glucose forms an unstable Schiff base with an amino group (lysine) on protein, and this is quickly rearranged to a stabler Amadori product. The Amadori product may then interact with another sugar or another Amadori product, creating a covalent cross-link. One glucose-derived cross-link for which the chemical structure is known is 2-furanyl-4(5)-(2-furanyl)-1H-imidazole, known as FFI.[23] The compound is a yellow-brown fluorescent chromophore that may be partly responsible for the color that develops during nonenzymatic browning of proteins and increased collagen fluorescence with aging. Brennan[24] concluded that nonenzymatic glycation and the subsequent formation of fluorescent cross-links did not contribute to the increased amount of acid-insoluble collagen found in tail tendons of rats with induced diabetes. Diabetes, however, did contribute to the formation of a larger fraction of nonreducible cross-links in tendon, cross-links perhaps like those found with increasing age.[25]

Another cross-link, named pentosidine, is formed when a lysine and an arginine residue are cross-linked by pentose.[26] The concentration of this cross-link increased with age in human dura mater, and its formation was particularly stimulated by the incubation of collagen with D-ribose.[26] Pentosidine was found in many tissues and was generated by fibroblasts cultured on collagen-coated dishes. The authors suggest that ribose or ribonucleotide metabolites could be pentosidine precursors in vivo. The equatorial spacing of normal human tendon collagen gradually increased when it was incubated in ribose, reaching values similar to those noted for tendon collagen from diabetic individuals.[8] These values were significantly higher than the values for normal collagen from aged individuals. The X-ray diffraction data suggest that the significant alterations in lateral packing of the molecules induced by diabetes are different from the changes associated with normal aging.

Although it is not yet clear whether the changed physical properties of collagen fibers with aging are caused by nonenzymatic glycation, evidence supporting this theory continues to accumulate. Tendon fiber breaking time was correlated directly with collagen-linked fluorescence in aging rats, with a correlation coefficient of 0.913.[27] The increased breaking time of tail tendon (ie, aging of collagen) that accompanies development of diabetes was prevented by giving the animals daily injections of aminoguanidine,[28] a compound that has been reported to inhibit the formation of nonenzymatic glycation cross-linking products.[29]

Noncollagenous Constituents

Glycosaminoglycan Content

The content of collagen, glycosaminoglycans, and water was determined in bovine heart valve tissue from fetal, young, and adult animals.[30] When expressed per gram of collagen, water content fell with increasing age and glycosaminoglycan content was rather constant. When expressed per gram of water, both collagen content and dermatan sulfate content increased. Glycosaminoglycan content was about 2% of collagen content in the old tissue. In rabbit Achilles tendon, the uronic acid content of old tendon was only about 15% of that in newborn tendon, when expressed as a percent of tissue dry weight.[9] Because collagen content was a much greater percentage of old tendon dry weight, uronic acid as a percent of collagen decreased from 2% in newborn to only 0.1% in old tendon. Cetta and associates[4] also found diminished hexosamine content in 4-year-old rabbit Achilles tendon compared to young tissue. The glucosamine-galactosamine ratio was highest in the old tendon.

Proteoglycans

The small proteoglycan now known as decorin makes up 88% by weight of the total proteoglycan extracted from adult bovine flexor tendon.[31,32] This proteoglycan migrates on sodium dodecyl sulfate/polyacrylamide gels as a diffuse band between 100 and 150 kDa and has a core protein of about 45 kDa after the enzymatic removal of the dermatan sulfate chain. Decorin from fetal bovine tendon migrated as a tighter band with lower molecular weight than in adult tissue, due to a shorter and more uniform glycosaminoglycan chain (fetal GAG chain Mr = 30,000; adult GAG chain Mr = 37,000 and polydisperse).[31,33] Although the decorin glycosaminoglycan in bovine tendon increased in length with age, a different situation was seen in rabbit flexor digitorum profundus (FDP) tendon. Decorin from one-month-old and three-month-old rabbit FDP tendon migrated as a tight band centered at about 100 kDa; decorin from one-year-old rabbit FDP tendon was more diffuse and significantly smaller (D. Hernandez, BS, and Kathryn Vogel, PhD, unpublished observation, 1990). The core protein at all ages was a 45 kDa band. This means that glycosaminoglycan chains of decorin from rabbit FDP tendon were shorter in the older animals. There are no comparable data for very old bovine or rabbit tendon.

The electrophoretic migration of decorin from human tibialis posterior tendon did not appear to change with increasing age after puberty.[34] Both decorin and biglycan small proteoglycans were found in human anterior cruciate ligament and lateral collateral ligament, with no consistent differences related to age. The number of tissues examined was too low to allow definite conclusions about age-related differences, however.

Fibrocartilage in Tendon

Fibrocartilage is found in tendons at the location where the tendon wraps around a bony pulley. Histologic analysis of age-related changes in fibrocartilage of the rat suprapatella have been reported to include the appearance of type II collagen after 12 months and increasing numbers of lipid droplets and glycogen granules in cytoplasm of the cells.[35] A significant amount of large proteoglycan accumulates in the fibrocartilaginous region of bovine deep flexor tendon.[36] Accumulation of this large proteoglycan, which is identical to aggrecan of articular cartilage,[37] means that the age-related proteoglycan changes noted in cartilage also may occur in tendon. This could be especially significant if systemic immunologic reaction to cartilage proteoglycan or type II collagen has been stimulated. In addition, mechanical properties of tendon may be altered at this site. Similar accumulation of large proteoglycan was described where the human tibialis posterior tendon passes under the medial malleolus.[34]

Lipids

Lipid deposits were identified histologically in various tendons and fascia.[7,38] Although the deposits were absent in young tendon, it was not possible to conclude that lipid accumulation increased with advancing age. A positive correlation was noted between the severity of lipid deposition in the Achilles tendon and the extent of lipid deposition in the coronary artery.[38]

Other Constituents

Few other constituents of tendons and ligaments have been analyzed. This lack of information leads to the common impression that tendons are composed of collagen and proteoglycans. Biochemical analyses indicate that many other constituents must be present, however. For example, collagen content of newborn rabbit Achilles tendon is about 40% of dry weight; glycosaminoglycan is only 1% to 2%.[4,9] However, collagen content of old tendon is up to 85% of the tissue dry weight and glycosaminoglycan is less than 1%.[9] This means that, relatively speaking, a large amount of material present in newborn tendon is absent in old tendon. Just what this material is and how it affects young and old tendon function is not clear. Enormous quantities of protein are removed from a tendon extract by ion-exchange chromatography during purification of proteoglycans (Fig. 3),[31,33] and these proteins remain virtually uncharacterized.

Cells of Tendon and Ligament

The most obvious histologic distinction between fetal and mature tendon is that fetal tissue is highly cellular, whereas adult tissues have few cells distributed throughout a dense extracellular matrix.[9,39] Based on a morphologic study of rabbit Achilles tendon at various ages, Ippolito and associates[9] suggest that the apparent decrease in number of tendon fibroblasts with aging is caused by a relative increase in amount of extracellular matrix, rather than by cell death. The cells in old tissue were longer and thinner than those in young tissue, although there was evidence that processes of these cells remained in contact with each other. The amount of endoplasmic reticulum was minimal compared to that of cells in young tendon, suggesting low metabolic activity in the old tissue.

Only a few studies have investigated cellular function as related to age. Aerobic glycolysis was high in slices of rabbit Achilles tendon up to three months of age, but was not measurable in three-year-old tendon.[40] Because anaerobic glycolysis continued at a low level in tissue of all ages, it was concluded that anaerobic glycolysis may be the most important energy source in old tendon. Incorporation of ^{3}H-proline into hydroxyproline by explant culture of three-year-old rabbit MCL and ACL was significantly lower than for one-year-old tissue.[12] Because the incorporation results were normalized to collagen content, it is not clear whether the cells exhibited a lower synthetic rate or whether there were fewer cells. When pieces of tissue from the fibrocartilaginous region of adult bovine flexor tendon are maintained in culture, they initially synthesize large proteoglycan primarily.[41] After two weeks in culture, however, a switch occurs and small proteoglycans are synthesized primarily.[41] This switch can be largely prevented by subjecting the tissue to brief, daily, uniaxial compressive loading.[42] When greater care was taken to determine donor age, it was noted that the ratio of ^{35}S-sulfate incorporated into large/small proteoglycans synthesized by disks of tendon fibrocartilage from adult animals (one year) switched from 1.54 ± 0.21 on day 1 (ie, predominantly large proteoglycan) to 0.28 ± 0.05 on day 14 (ie, predominantly small proteoglycan) (Kathryn G. Vogel, PhD, and Daniel Hernandez, BS, unpublished observation, 1992). However, tissue from older animals (five years) did not switch during the 14-day culture period (ratio = 2.65 ± 0.05 on day

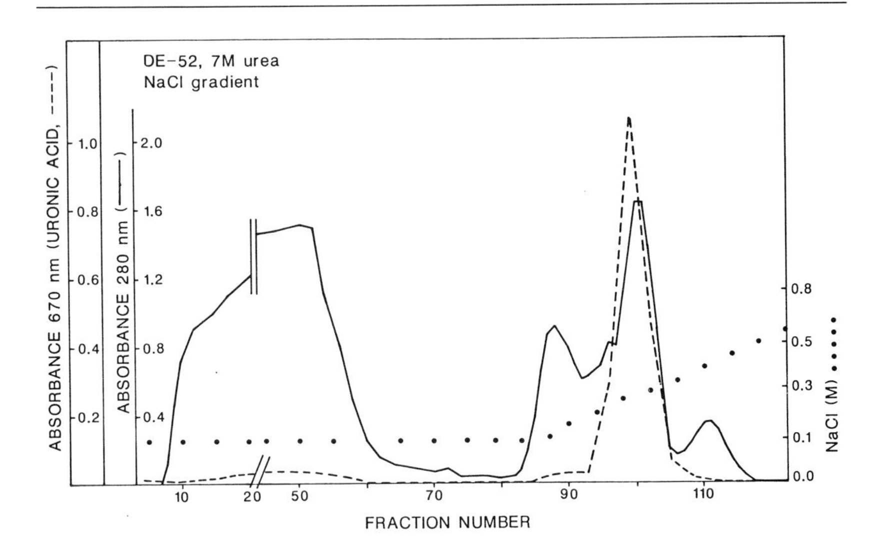

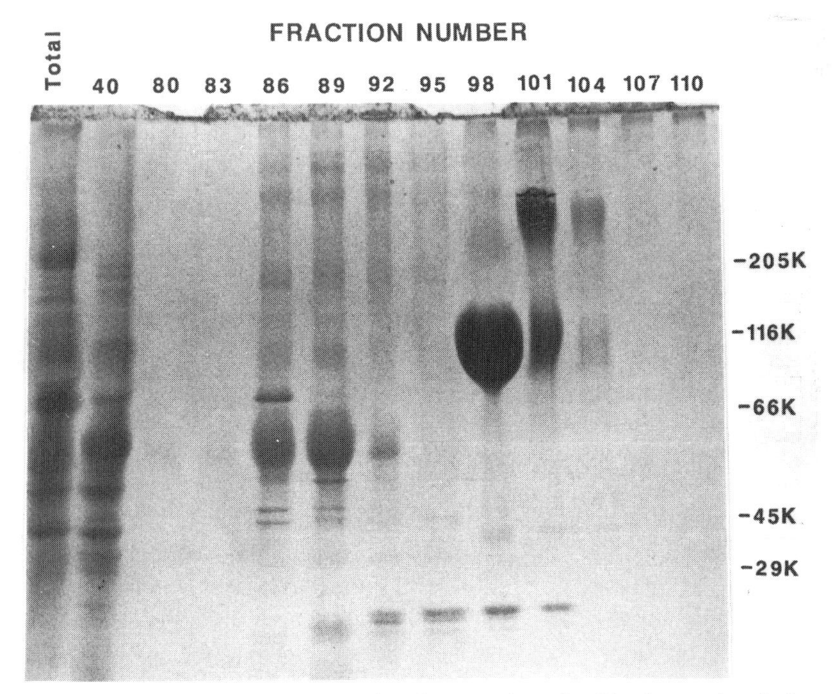

Fig. 3 *Ion-exchange chromatography of extract from fetal bovine tendon.* ***Left,*** *Proteins that do not stick to the column (fractions 10-60) represent about 90% of total extracted protein.* ***Right,*** *Sodium dodecyl sulfate/polyacrylamide gel electrophoresis of fraction 40 demonstrates the large number of different proteins in this fraction. The prominent protein in fractions 85-90 is fibromodulin. Proteoglycans were eluted in fractions 92-104. The small peak at fraction 110 is nucleic acid. (Reproduced with permission from Vogel KG, Evanko SP: Proteoglycans of fetal bovine tendon.* J Biol Chem *1987;262:13607-13613.)*

1 versus 1.99 ± 0.19 on day 14). In addition, cells from the tissue of one-year-old animals migrated to the cut disk surfaces during the two-week culture period, whereas no cells were found on the cut surfaces of disks from five-year-old animals. These examples show clear age-related differences in the behavioral and matrix biosynthetic characteristics of tendon cells in tissue explants.

Summary

Biochemical analysis has not yet been particularly helpful in explaining the functional changes of tendon and ligament related to aging. There does not appear to be a dramatic loss of collagen in old tissue. The best-documented changes are decreased amounts of reducible collagen cross-links and increased amounts of nonreducible cross-links, normalized per mole of collagen. Changes as a result of nonenzymatic glycation also may prove significant.

Collagen and proteoglycans make up only a portion of tendon and ligament composition, and the remaining constituents are largely undefined. The question of how individual proteins and proteoglycans change in structure or location with age has not been addressed significantly. Future studies should be directed toward defining currently unknown constituents of the tissue and toward ultrastructural analysis of the relationships among these constituents.

References

1. Praemer A, Furner S, Rice DP: Musculoskeletal injuries, in *Musculoskeletal Conditions in the United States*, Park Ridge, IL, American Academy of Orthopaedic Surgeons, 1992, sec 3, pp 83-124.
2. Maroudas A: Different ways of expressing concentration of cartilage constituents with special reference to the tissue's organization and functional properties, in Maroudas A, Kuettner K (eds): *Methods in Cartilage Research*, London, Academic Press Ltd, 1990, chap 53, pp 211-219.
3. Parry DAD, Craig AS: Growth and development of collagen fibrils in connective tissue, in Ruggeri A, Motta PM (eds): *Ultrastructure of the Connective Tissue Matrix*. Boston, Martinus Nijhoff, 1984, chap 2, pp 34-64.
4. Cetta G, Tenni R, Zanaboni G, et al: Biochemical and morphological modifications in rabbit Achilles tendon during maturation and ageing. *Biochem J* 1982;204:61-67.
5. Parry DA, Barnes GR, Craig AS: A comparison of the size distribution of collagen fibrils in connective tissues as a function of age and a possible relation between fibril size distribution and mechanical properties. *Proc R Soc Lond [Biol]* 1978;203:305-321.
6. Parry DA, Craig AS, Barnes GR: Tendon and ligament from the horse: an ultrastructural study of collagen fibrils and elastic fibres as a function of age. *Proc R Soc Lond [Biol]* 1978;203:293-303.
7. Kannus P, Jozsa L: Histopathological changes preceding spontaneous rupture of a tendon: A controlled study of 891 patients. *J Bone Joint Surg* 1991;73A:1507-1525.

8. James VJ, Delbridge L, McLennan SV, et al: Use of X-ray diffraction in study of human diabetic and aging collagen. *Diabetes* 1991;40:391-394.
9. Ippolito E, Natali PG, Postacchini F, et al: Morphological, immunochemical, and biochemical study of rabbit achilles tendon at various ages. *J Bone Joint Surg* 1980;62A:583-598.
10. Vailas AC, Pedrini VA, Pedrini-Mille A, et al: Patellar tendon matrix changes associated with aging and voluntary exercise. *J Appl Physiol* 1985;58:1572-1576.
11. Vogel HG: Species differences of elastic and collagenous tissue: Influence of maturation and age. *Mech Ageing Dev* 1991;57:15-24.
12. Amiel D, Kuiper SD, Wallace CD, et al: Age-related properties of medial collateral ligament and anterior cruciate ligament: A morphologic and collagen maturation study in the rabbit. *J Gerontol* 1991;46:B159-B165.
13. Hamlin CR, Kohn RR: Evidence for progressive, age-related structural changes in post-mature human collagen. *Biochim Biophys Acta* 1971;236:458-467.
14. Hamlin CR, Kohn RR: Determination of human chronological age by study of a collagen sample. *Exp Gerontol* 1972;7:377-379.
15. Everitt AV, Gal A, Steele MG: Age changes in the solubility of tail tendon collagen throughout the lifespan of the rat. *Gerontologia* 1970;16:30-40.
16. Bochantin J, Mays LL: Age-dependence of collagen tail fiber breaking strength in Sprague-Dawley and Fischer 344 rats. *Exp Gerontol* 1981;16:101-106.
17. Everitt AV, Porter BD, Steele M: Dietary, caging and temperature factors in the ageing of collagen fibres in rat tail tendon. *Gerontology* 1981;27:37-41.
18. Bailey AJ: The stabilization of the intermolecular crosslinks of collagen with ageing. *Gerontologia (Basel)* 1969;15:65-76.
19. Robins SP, Shimokomaki M, Bailey AJ: The chemistry of the collagen cross-links: Age-related changes in the reducible components of intact bovine collagen fibres. *Biochem J* 1973;131:771-780.
20. Fujii K, Tanzer ML: Age-related changes in the reducible crosslinks of human tendon collagen. *FEBS Lett* 1974;43:300-302.
21. Barenberg SA, Filisko FE, Geil PH: Ultrastructural deformation of collagen. *Connect Tissue Res* 1978;6:25-35.
22. Cerami A, Vlassara H, Brownlee M: Glucose and aging. *Sci Am* 1987;256:90-96.
23. Pongor S, Ulrich PC, Bencsath FA, et al: Aging of proteins: Isolation and identification of a fluorescent chromophore from the reaction of polypeptides with glucose. *Proc Natl Acad Sci USA* 1984;81:2684-2688.
24. Brennan M: Changes in solubility, non-enzymatic glycation, and fluorescence of collagen in tail tendons from diabetic rats. *J Biol Chem* 1989;264:20947-20952.
25. Brennan M: Changes in the cross-linking of collagen from rat tail tendons due to diabetes. *J Biol Chem* 1989;264:20953-20960.
26. Sell DR, Monnier VM: Structure elucidation of a senescence cross-link from human extracellular matrix: Implication of pentoses in the aging process. *J Biol Chem* 1989;264:21597-21602.
27. Rolandi R, Borgoglio A, Odetti P: Correlation of collagen-linked fluorescence and tendon fiber breaking time. *Gerontology* 1991; 37:240-243.
28. Oxlund H, Andreassen TT: Aminoguanidine treatment reduces the increase in collagen stability of rats with experimental diabetes mellitus. *Diabetologia* 1992;35:19-25.

29. Brownlee M, Vlassara H, Kooney A, et al: Aminoguanidine prevents diabetes-induced arterial wall protein cross-linking. *Science* 1986;232:1629-1632.
30. Meyer FA, Silberberg A: Aging and the interstitial content of loose connective tissue: A brief note. *Mech Ageing Dev* 1976;5:437-442.
31. Vogel KG, Heinegård D: Characterization of proteoglycans from adult bovine tendon. *J Biol Chem* 1985;260:9298-9306.
32. Vogel KG, Fisher LW: Comparisons of antibody reactivity and enzyme sensitivity between small proteoglycans from bovine tendon, bone, and cartilage. *J Biol Chem* 1986;261:11334-11340.
33. Vogel KG, Evanko SP: Proteoglycans of fetal bovine tendon. *J Biol Chem* 1987;262:13607-13613.
34. Vogel KG, Ordog A, Pogany G, et al: Proteoglycans in the compressed region of human tibialis posterior tendon and in ligaments. *J Orthop Res* 1993;11:68-77.
35. Benjamin M, Tyers RN, Ralphs JR: Age-related changes in tendon fibrocartilage. *J Anat* 1991;179:127-136.
36. Koob TJ, Vogel KG: Site-related variations in glycosaminoglycan content and swelling properties of bovine flexor tendon. *J Orthop Res* 1987;5:414-424.
37. Vogel KG, Sandy JD, Neame PJ, et al: Tensional and compressed regions of bovine tendon contain aggrecan. *Trans Orthop Res Soc* 1993;18:46.
38. Adams CW, Bayliss OB, Baker RW, et al: Lipid deposits in ageing human arteries, tendons and fascia. *Atherosclerosis* 1974;19:429-440.
39. Evanko SP, Vogel KG: Ultrastructure and proteoglycan composition in the developing fibrocartilaginous region of bovine tendon. *Matrix* 1990;10:420-436.
40. Floridi A, Ippolito E, Postacchini F: Age-related changes in the metabolism of tendon cells. *Connect Tissue Res* 1981;9:95-97.
41. Koob TJ, Vogel KG: Proteoglycan synthesis in organ cultures from regions of bovine tendon subjected to different mechanical forces. *Biochem J* 1987;246:589-598.
42. Koob TJ, Clark PE, Hernandez DJ, et al: Compression loading in vitro regulates proteoglycan synthesis by tendon fibrocartilage. *Arch Biochem Biophys* 1992;298:303-312.

Chapter 23

Mechanical Properties of the Anterior Cruciate Ligament (in situ) and In Vivo Forces as a Function of Age

David L. Butler, PhD

Introduction

Traditionally, musculoskeletal research on soft-tissue injuries has centered on athletes from the younger population. The rationale for this approach has been that our youth are more active, more involved in competitive team sports, and more eager to return to play after injury. Consequently, orthopaedic surgeons have more aggressively repaired and reconstructed damaged ligaments and tendons in their younger patients, while encouraging their older patients simply to modify their activity levels. This approach may have been acceptable in past decades, but such treatment programs are no longer applicable for the increasingly active middle-aged and even elderly athlete. Nor are these programs pertinent for the less active elderly patient seeking relief from soft-tissue stiffness and limited range of motion.

Sports injuries of the knee occur frequently[1,2] and often involve the anterior cruciate ligament (ACL). Balkfors[3] found that 77% of all confirmed knee ligament tears involved injury to the ACL. Positioned in the knee primarily to prevent both straight anterior translation[4,5] and coupled anterior translation and internal rotation,[6,7] the ACL is injured in 70% of patients who experience acute hemorrhage to the knee.[8,9] ACL injury or rupture is believed to occur up to 250,000 times each year in the United States alone, and the ACL is repaired and/or reconstructed between 75,000 and 125,000 times annually. Given that sports medicine surgeons are more likely to operate on younger rather than middle-aged or elderly patients, the actual number of patients who would like to remain active by having their ligament reconstructed might be much higher. In addition, if ACL-deficiency leads to increasingly abnormal motions and degenerative changes to the articular and meniscal cartilage,[10-12] clinicians and basic scientists alike may be grossly underestimating the severity and extent of the problem. Finally, if elderly patients do maintain higher activity levels, what are the associated forces that result from these activities, and to what extent do they threaten the mechanical integrity of normal and reconstructed soft tissues?

Clearly then, several questions related to soft-tissue aging need to be examined: (1) To what extent does the aging process influence the biomechanical properties of soft tissues like the anterior cruciate ligament? (2) What factors influence the relationship between age and these properties? (3) How do in vivo forces imposed on these soft tissues change with aging and how vulnerable are these tissues to injury? This manuscript considers each of these questions, using the anterior cruciate ligament as the tissue model.

How Does the Aging Process Influence the Mechanical and Material Properties of Soft Tissues Like the ACL?

This section will describe how the mechanical properties of the ACL (taken from the force-elongation curve) change with aging. Because these properties depend on both tissue quantity and quality, we will also describe age-related changes in tissue size and material properties.

Effects of Aging on the Mechanical Properties of the ACL

Mechanical properties of any tensile-bearing soft tissue like the ACL are typically determined by recording force and elongation during tensile testing. Investigators either cycle the tissue in tension to subfailure force levels or, more often, extend the tissue testing to failure (Fig. 1). The failure test is generally performed by controlling the rate of deformation. From the resulting force-deformation curve, researchers typically record the stiffness or slope in the nearly linear region, the maximum force or ultimate load, the elongation to maximum force or failure, and the energy to maximum force or failure (the energy absorbed under the force-elongation curve).

Data from several investigators have suggested how aging affects biomechanical properties. Unfortunately, the results of these studies have differed markedly. Early work by Kennedy and associates[13] showed no correlation of ligament maximum force with age. However, their study used isolated ACL specimens without bone ends, which were difficult to grip. In addition, specimens were obtained from cadavers with a mean age of 62 years, and the number of specimens obtained from younger individuals was not specified. Hence, the effects of gripping and the lack of specific donor ages make this data of limited value. Trent and associates[14] showed a wide distribution of maximum force values for bone-ACL-bone preparations from only five donors ranging in age between 29 and 59 years. While the results suggested that mechanical properties decreased with age, the variability in the data, and the small sample size over a rather broad age range prevented this study from being definitive.

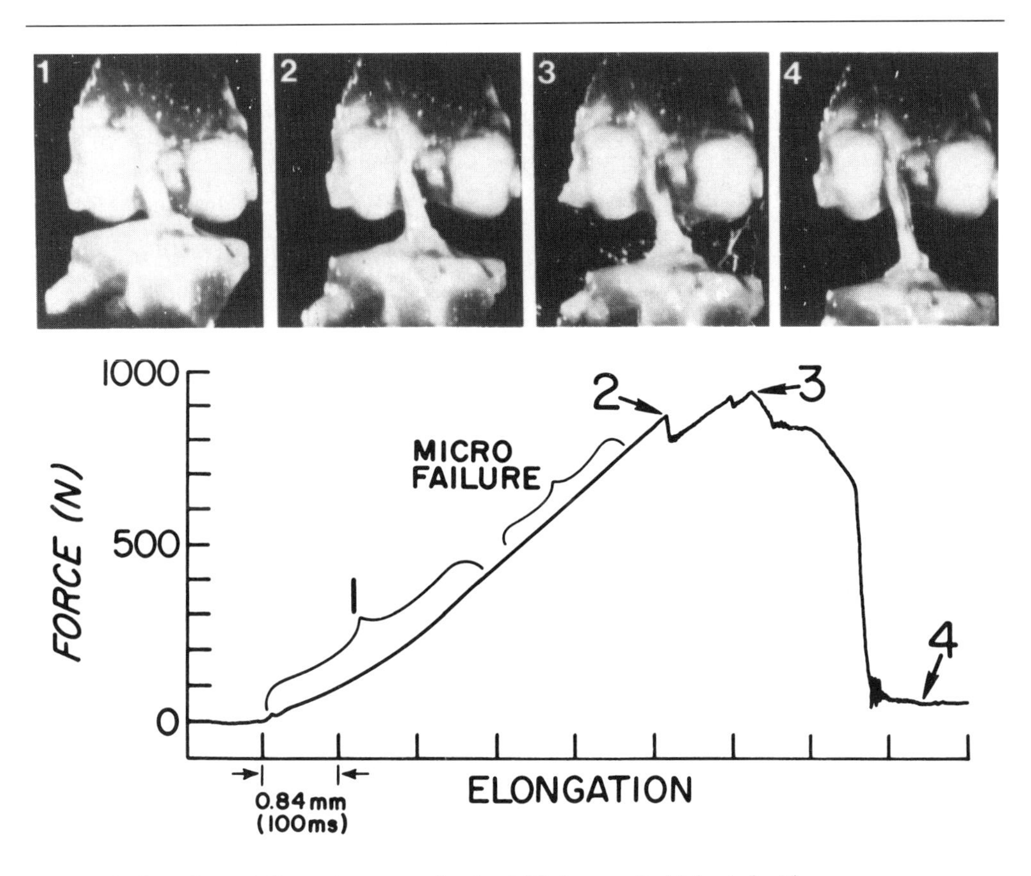

Fig. 1 *Failure force-deformation curve for the ACL-bone unit. (Adapted with permission from Noyes FR, DeLucas JL, Torvik PJ: Biomechanics of anterior cruciate ligament failure: An analysis of strain-rate sensitivity and mechanisms of failure in primates.* J Bone Joint Surg *1974;54A:236-253.)*

In 1976, Noyes and Grood[15] presented a definitive study on the effects of age on the mechanical properties of the human cadaveric ACL-bone unit. Typical force-time curves for specimens from a younger (22-year-old) and older (50-year-old) donor are shown in Figure 2. Note the much lower stiffness and maximum force in the older donor group, as well as the smaller amount of energy absorbed (area under the curve). When all data were compiled (Table 1), specimens from the younger group failed at maximum force levels 2.4 times higher than those from the older group.[15] For specimens above the age of 50 years, lower maximum forces were obtained, but these declines were not significant (see material properties description below). Of course, these mechanical property results could have been affected by the general health status of the donors, which was not specified.

Recently, Woo and associates,[16] who used a technique to more completely align the ligament axis with the load cell axis, tested a

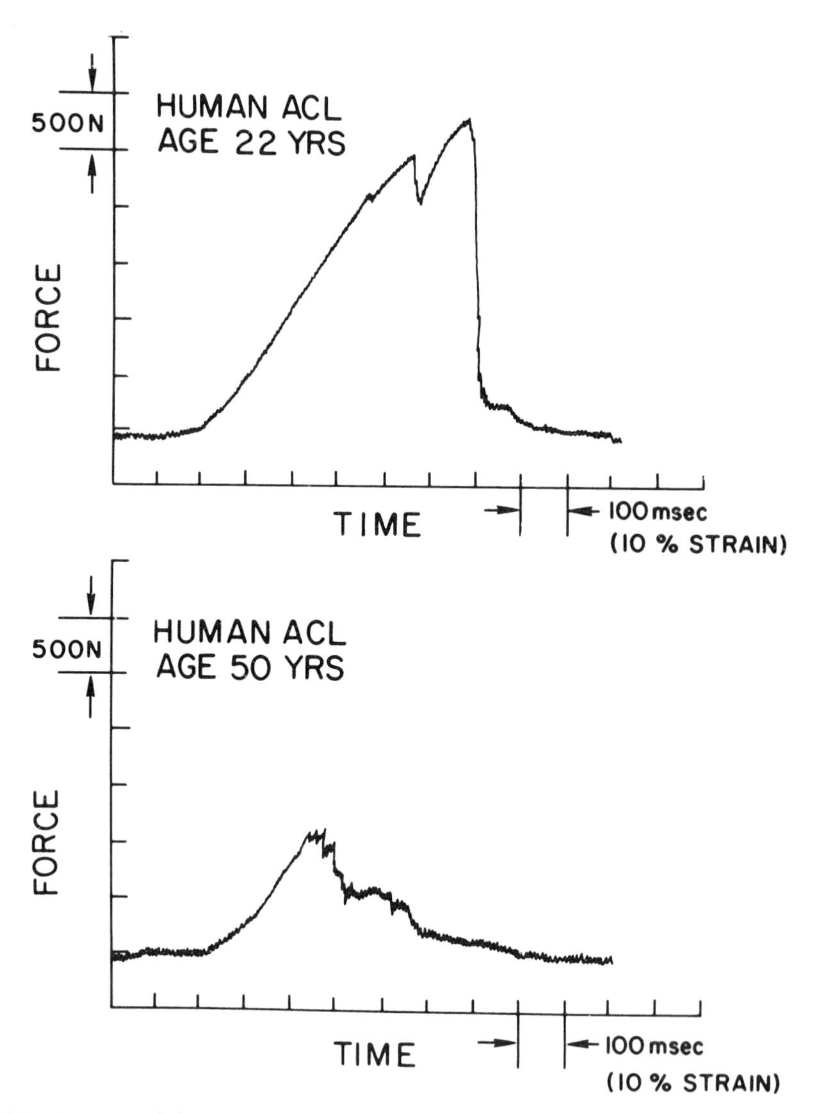

Fig. 2 *Force-deformation properties of the human ACL from young vs old donor. (Reproduced with permission from Noyes FR, Grood ES: The strength of the anterior cruciate ligament in humans and rhesus monkeys: Age-related and species-related changes.* J Bone Joint Surg *1976;58A:1074-1082.)*

large series of human cadaveric anterior cruciate ligament-bone units in tension. They, too, found a significant reduction in the material properties of this tissue between the ages of 20 and 80 years.

Regardless of concerns about the health status of the donor, most investigators would agree that age does appreciably reduce the mechanical properties of the anterior cruciate ligament-bone unit. Yet these comparisons are somewhat artificial, because properties like stiffness and maximum force depend on both tissue quantity or size

Table 1 Comparison of structural properties

	No. of Specimens	Stiffness (kN/m)	Linear Force (kN)	Maximum Force (kN)	Energy to Failure (N-m)
Older human (48-86 yrs.)	20	129±39	0.622±0.283	0.734±0.266	4.89±2.36
Younger human (16-26 yrs.)	6	182±56	1.17 ±0.75	1.73 ±0.66	12.8±5.5
Rhesus monkey	25	194±28	0.71 ±0.12	0.83 ±0.11	3.0±0.6

(Reproduced with permission from Noyes FR, Grood ES: The strength of the anterior cruciate ligament in humans and rhesus monkeys. *J Bone Joint Surg* 1976; 58A:1074-1082.)

(length and area) and tissue quality (material properties). Comparing a particularly large ligament from a young, active donor with a particularly small ligament from an older, sedentary donor accentuates the differences. Therefore, strict comparisons between groups of donors requires that material properties, rather than mechanical properties, be compared. Thus the influence of age on both tissue quantity and quality will now be examined.

Influence of Aging on Ligament Quantity or Size

How are tissue size parameters (length and cross-sectional area) influenced by the aging process? The data from Noyes and Grood[15] suggest that the cross-sectional area of the ACL of the older donors is 30% larger than that of the younger donors. However, this difference was not statistically significant. The authors examined scaling laws that might predict the size of the ligament from body mass, but they did not provide mechanisms (such as increased water content or GAG content) that would explain why area might be larger in the older donors. Similar conclusions reached in other studies indicate that aging does not appear to have a significant influence on either ACL length or cross-sectional area.

Effects of Aging on ACL Material Properties

As previously stated, material properties permit specimens from different donors to be compared, independent of the size of the tissue being tested.[17] These parameters, computed from corresponding mechanical properties and size, include: linear modulus (stiffness x length/cross-sectional area), maximum stress (maximum force/cross-sectional area or ultimate tensile strength), strain energy density (energy/cross-sectional area x length), and maximum strain (maximum elongation/length). The first three parameters are load-related, and all but maximum stress are length-related. Modulus and maximum stress are the typical material properties compared between donors.

Shown in Table 2 are comparisons of material properties from older versus younger donors.[15] Note that the linear modulus and

Table 2 Comparison of material properties*

	No. of Specimens	Elastic Modulus (MPa)	Linear Stress (MPa)	Maximum Stress (MPa)	Strain Energy to Failure (N-m/ml)
Older human (48-86 yrs.)	20	65.3±24.0*	11.3± 5.1	13.3±5.0‡	3.1±1.5†
Younger human (16-26 yrs.)	6	111±26‡	25.5±14.0†	37.8±9.3‡	10.3±3.1‡
Rhesus monkey	25	186±26	56.2±7.6	66.1±8.4	19.4±3.8

(Reproduced with permission from Noyes FR, Grood ES: The strength of the anterior cruciate ligament in humans and rhesus monkeys. *J Bone Joint Surg* 1976; 58A: 1074-1082.)

maximum stress are significantly lower for ACL-bone specimens from older donors. These data are presented graphically in Figure 3, where modulus and maximum stress demonstrate twofold and threefold decreases, respectively, between 20 and 50 years of age. Also note that the mechanism of failure changes from primarily soft-tissue, midsubstance tears in the younger group to primarily bone-avulsion failures in the older group. While the authors did not address the causes of these significant changes in material properties, clearly aging has a marked effect on tissue quality.

What Factors Influence the Relationship Between Age and Mechanical and Material Properties?

Properly interpreting the relationship between age and ACL mechanics can be difficult, because there are numerous confounding variables. These have been classified into three primary categories.

Effects of Joint Position on Mechanical Response

The relative positions of the femur and tibia strongly influence both the ligament's force-elongation curve and the mechanical and material properties derived from this curve. Joint position dictates the distance between the ends of the ACL fascicles. If this distance and the initial length of each fascicle when first loaded are known, fascicular force can be computed and total ligament force determined. With these data, a mechanical model of the ligament could be developed and the joint position that produces the most uniform loading of the bundles and the highest mechanical properties could be estimated.

Unfortunately, the ACL has a very complex fascicular structure. Fascicles vary greatly in orientation, curvature, and length. Fascicle orientation with respect to the anatomic joint axes changes greatly as knee flexion angle changes,[18,19] particularly when viewed in the sagittal plane. Bundle curvature also appears to be sensitive to joint position,[19] with large values suggesting slackened bundles unable to transmit force (for example, the posterior bundles in mid to extreme flexion). ACL fascicles also vary significantly in length.[18-22] Our

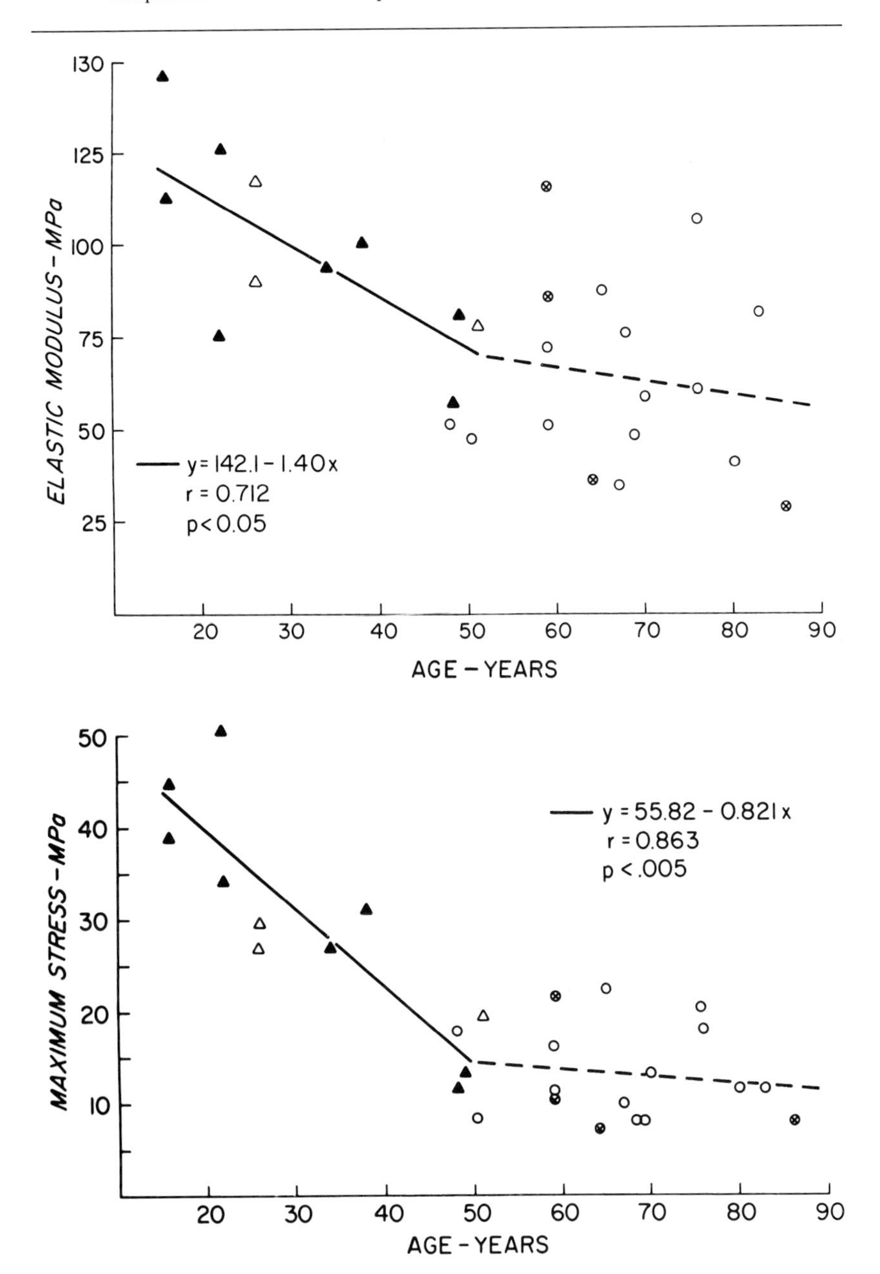

Fig. 3 *Correlation of anterior cruciate ligament strength with age. a) Elastic modulus and b) maximum stress. (Reproduced with permission from Noyes FR, Grood ES: The strength of the anterior cruciate ligament in humans and rhesus monkeys: Age-related and species-related changes.* J Bone Joint Surg *1976;58A:1074-1082.)*

group, for example, found that fascicles that attached more anteriorly and anteromedially on the tibia were two to three times longer than more posterolateral bundles.[19] These results indicate that large changes can occur in end-to-end fascicle distance, for example, during knee flexion.

Taken collectively, these studies strongly suggest that in order to assess properly the effects of aging on the ACL, it is necessary to position the bone ends so as to load the tissue consistently and reproducibly. Structurally based models that could predict fascicular load-sharing would thus be quite valuable. Add to this the biologic variability in ACL structure between donors, and one can understand how difficult it is to establish joint positions that will load fascicles of this tissue model most uniformly.

Effects of Loading Direction on Mechanical Response

Once a reasonable joint position has been selected, how does the investigator select a loading direction that will ensure nearly uniform loading and, thus, the highest structural properties of the ligament-bone unit? Selecting this direction requires aligning the tissue axis parallel to the direction of bone end displacement and, preferably, collinear with the load cell recording tissue force.

This alignment process is complicated by several issues. (1) The centroidal axis (the axis that passes through the centroid of each attachment site) is difficult to identify, primarily because the ligament has a nonuniform cross-sectional area created by nonparallel fascicles, and these nonparallel fascicles cannot be used as a guide to locating the centroidal axis. (2) The ligament attaches at acute angles to the bone, which makes it difficult to determine the precise location of the centroidal insertion sites. Short of digitizing these centroids by penetrating the tissue with a pointer, these locations can only be estimated. (3) The contours of the bones make visualization of the ligament quite difficult once the tissue is mounted. (4) If the fascicles of the tissue do not transmit force uniformly during loading, as they may not, then selecting the centroidal axis is probably not appropriate anyway. Thus the process of aligning the tissue to obtain uniform load distribution remains a difficult issue, which complicates our ability to evaluate age-related differences in mechanical and material properties.

While the process of evaluating the effects of aging on the mechanical properties remains difficult, this process is more straightforward when the material properties of component substructures or subunits of the ACL are examined.[22-24] Unfortunately, the material properties of these subunits are not entirely uniform throughout the structure.

Nonuniform Material Properties

In 1986, our group reported a technique designed to load the ACL more uniformly.[23] Multifascicular-bone units, with a more uniform length and cross-sectional area, were removed from the ligament and were mechanically failed in tension. The properties of these subunits were, on average, significantly less than those of subunits from the nearby patellar tendon-bone unit. In follow-up studies by our group, however, the anteromedial and anterolateral subunits of the ACL developed significantly larger load-related material properties than did the posterolateral band.[24] Specifically, the average modulus and maximum stress of the anterior bands were found to be two and three times the values of the posterior band. This finding is important because the loading axis for the ACL is therefore regulated not only by the geometric arrangement of the fascicles, but also by their inherent stiffnesses (moduli) and strengths (maximum stresses). For example, lower stiffness posterior fascicles would be expected to transmit smaller forces than higher stiffness anterior fascicles for equivalent end-to-end elongations.

Equally important are the spatial variations that exist in ligaments and tendons along their length. Many investigators have noted that tissues exposed to axial loading deform nonuniformly. Specifically, the tissue midsubstance tends to be stiffer than the tissue adjacent to the insertion site.[17,25-27] This variation results in higher midsubstance moduli, which have been attributed to collagen, and in lower insertion site moduli, which may be caused by the presence of fibrocartilage, less parallel fiber bundles, and stress concentrations at the soft tissue-bone interface. Recent evidence, which suggests that the strain variation along the anterior cruciate ligament subunit is less than the variation along the patellar tendon subunit,[28] will require further elucidation.

Thus, tissue quality as well as quantity must be considered in controlling the variability in recorded tissue biomechanics across different age groups. If these spatial differences in material properties are different for ligaments of different ages, accommodations must be made in the testing procedures. Such complications may make testing isolated fascicle-bone units preferable to testing whole ligament-bone preparations.

How Do In Vivo Forces Imposed on Soft Tissues Change with Aging and How Vulnerable Are These Tissues to Injury?

What are the in vivo forces transmitted by the ACL for various activities? How might aging change the level and frequency of force transmission by the ligament? Are these changes significant enough to make ligaments from younger donors more susceptible, for example, to failure than those from older donors?

These questions are difficult to answer, because in vivo forces are difficult to measure. Most in vivo recordings have been made in animals rather than in humans, and nearly all have involved indirect measurements of deformation rather than force. No studies, to our knowledge, have examined changes in in vivo ACL forces with aging.

Quadriceps contractions have been shown to impose strain on the ACL both in vitro[29,30] and in vivo.[31,32] We have reported ACL force as great as 150 pounds near full extension by pulling on the quadriceps tendon.[33] These high ACL forces during knee extension exercises have caused researchers and therapists to look for exercise modalities that do not load the ACL.

In this section, I will review those studies that have tried to measure in vivo forces in the ACL model regardless of age. I will also discuss the relationships between ACL and muscle force measurements. Finally, I will briefly review modeling efforts aimed at predicting these forces.

In Vivo ACL Force Measurements in Animals

In vitro measurements of ACL force in animals and human cadavers using buckle gages,[34-40] instrumented washers,[33,41] and an in-series, six-component load cell[42] have been an important first step to the more difficult in vivo studies. Even the indirect methods of recording tissue elongation by attaching liquid metal strain gages[43-47] and Hall effect strain transducers (HESTs)[29,30,48-50] to record bone displacement adjacent to the insertion site,[51] and to calculate tissue forces from the mechanical response and alignment of the ACL subunits[22,52] have been valuable. We have learned from these studies that in full extension, large forces are normally placed on the ACL.[33,41-43,53,54] Application of quadriceps forces loads the ACL, but only in the last 30° to 40° of extension.[33,41,53] Hyperextension is generally believed to especially load the ACL.

Few investigators have attempted to quantify knee ligament forces in vivo. Monahan and associates[55] and Fischer and associates[49] measured ligament strains caused by manually applied loads in anesthetized dogs. Several groups have also measured ligament loads in conscious animals during certain activities.[38,56,57] The first two reported minimal collateral ligament loads, lower than those of simulated clinical stress tests, which suggests that joint geometry and muscular activity largely control tibiofemoral motion in vivo. The third study suggested that even anterior drawer tests, conducted on anaesthetized goats, did not load the ACL much more than level and ramp walking activities.[57] Kain and associates[58] also noted that the primate ACL would elongate if the quadriceps was stimulated at either full extension or 45° flexion. The hamstrings counteracted the quadriceps contraction at 45° but had the opposite effect at full extension.

In Vivo ACL Force Measurements in Humans

In Vivo Deformation and Force Measurements Henning and associates[59] made in vivo ligament insertion deformation measurements in two patients during activities of daily living. To compare across subjects, an 80-lb Lachmann test was used as the standard (100%). Most activities produced less bone deformation; crutch walking (7%), squat and jump rope (21%), walking (36%). Higher values were found with jogging (64%). Only isometric quadriceps contractions with a 20-lb boot (60% to 121%) and downhill running (121%) produced larger elongations than the 80-lb Lachmann. More recently, Beynnon and associates measured ACL strains using Hall effect strain transducers (HESTs) in over 50 subjects with minimal complications (B. Beynnon, personal communication). They reported ACL strains during clinical exams with and without braces, during passive and active knee extension, and during isometric quadriceps contractions.[31,60] These results are only now appearing.[32]

Lewis and Shybut[38] used buckle gages to measure ligament loads in conscious dogs during various activities. Because minimal loads were measured in the medial collateral ligament, these investigators hypothesized that joint geometry and muscular forces normally controlled tibiofemoral motion in vivo. Ligament forces would then only develop with improper muscular coordination, as occurs in a misstep. It is not likely that these findings apply to the human ACL, however, given Henning's results and the fact that the ACL is needed to maintain equilibrium in the anterior direction.[61]

The in vivo function of the ACL and the role of muscles in controlling joint motions after ACL injury have been studied in ACL-deficient patients. Gait studies[62-64] have shown that ACL-deficient patients decrease the instability from anterior shear and internal rotation moments by avoiding full extension. These moments increase with the speed of the activity. While no significant differences exist in the vertical or sagittal shear forces during free walking, increasing the speed of gait first increases (fast walking) and then decreases (running) the vertical forces in the ACL-deficient knee.[65] Silverberg and associates[66] also noted significantly lower braking force in ACL-deficient versus contralateral control knees for both inside and outside cutting. The quadriceps was less active during these cutting maneuvers. While we have learned from these studies how important the quadriceps is in regulating anterior tibial translations near full extension, we currently have no knowledge of the actual forces in the ACL or in the patellar tendon and, hence, quadriceps, during such activities. How these forces might change as speed of activity and acceleration of the limb change is also not understood. Furthermore, knowing how these relationships between forces in the ACL and quadriceps change with aging is probably critical if ACL injury is to be avoided in the middle-aged and elderly individual.

Models Predicting ACL Forces

Several investigators have predicted that the ACL is loaded during normal walking. These forces, during locomotor activities in man, have been computed from anteroposterior shear forces determined using the inverse dynamics method.[67-70] Force estimates have ranged from 0.7 to 1.7 times body weight. Nissan[70] predicted that the ACL is loaded during the extension phase of swing and is biphasic during swing. Forces of 0.4 to 0.8 times body weight were computed. Seireg and Arvikar,[61] using modeling and optimization methods to eliminate redundancies, computed ACL force of 0.5 body weight during the early portion of stance. Andriacchi[71] has reported that ACL-deficient patients modify their gait patterns. The primary change is the avoidance of external flexion moments during stance, moments which are balanced by quadriceps contraction. For this reason the adaptation is referred to as quad avoidance gait. This altered gait may be an attempt to prevent anterior subluxations of the tibia that are normally prevented by the ACL.

Significance of Measuring In Vivo Forces

Thus, despite the extensive in vitro research and recent in vivo studies, we still do not know whether the ACL is significantly loaded in vivo. If it is, we can only speculate about which activities and conditions produce the largest forces. Does the quadriceps significantly load the ACL? Over what range of flexion angles are these loads greatest? Which activities produce the largest forces? Does walking load the ligaments more than just standing? Does the velocity of the activity increase the load? Does limb acceleration increase the load? To what extent are these factors interactive? How do these relationships change with aging?

Once a relationship between ligament and muscle forces is established, it may become possible to control these forces. We may eventually determine the force criteria (magnitude, frequency, and duration) necessary for tissue homeostasis and if these criteria change with aging. In this way, loading conditions may be found that will prevent tissue overload as well as tissue resorption. Armed with this information, we can then examine in more detail how interventions such as disuse, immobility, exercise and injury alter these load levels to affect ligament biology, structure, and mechanics.

References

1. Holbrook TL, Grazier KL, Kelsey JL, et al: *The Frequency of Occurrence, Impact and Cost of Musculoskeletal Conditions in the United States*. Chicago, IL, American Academy of Orthopaedic Surgeons, 1984.
2. Kelsey JL: *Epidemiology of Musculoskeletal Disorders*. New York, NY, Oxford University Press, 1982.
3. Balkfors B: The course of knee-ligament injuries. *Acta Orthop Scand (Suppl)* 1982;198:1-99.
4. Butler DL, Noyes FR, Grood ES: Ligamentous restraints to anterior-posterior drawer in the human knee: A biomechanical study. *J Bone Joint Surg* 1980;62A:259-270.
5. Piziali RL, Rastegar J, Nagel DA, et al: The contribution of the cruciate ligaments to the load-displacement characteristics of the human knee joint. *J Biomech Eng* 1980;102:277-283.

6. Grood ES, Noyes FR: Diagnosis of knee ligament injuries: Biomechanical precepts, in Feagin JA Jr (ed): *The Crucial Ligaments: Diagnosis and Treatment of Ligamentous Injuries About the Knee*. New York, NY, Churchill Livingstone, 1988, chap 9, pp 245-260.
7. Markolf KL, Kochan A, Amstutz HC: Measurement of knee stiffness and laxity in patients with documented absence of the anterior cruciate ligament. *J Bone Joint Surg* 1984;66A:242-252.
8. Noyes FR, Bassett RW, Grood ES, et al: Arthroscopy in acute traumatic hemarthrosis of the knee: Incidence of anterior cruciate tears and other injuries. *J Bone Joint Surg* 1980;62A:687-695.
9. DeHaven KE: Diagnosis of acute knee injuries with hemarthrosis. *Am J Sports Med* 1980;8:9-14.
10. McDaniel WJ Jr, Dameron TB Jr: Untreated ruptures of the anterior cruciate ligament: A follow-up study. *J Bone Joint Surg* 1980;62A: 696-705.
11. McDaniel WJ Jr, Dameron TB Jr: The untreated anterior cruciate ligament rupture. *Clin Orthop* 1983;172:158-163.
12. Noyes FR, Matthews DS, Mooar PA, et al: The symptomatic anterior cruciate-deficient knee: Part II. The results of rehabilitation, activity modification, and counseling on functional disability. *J Bone Joint Surg* 1983;65A:163-174.
13. Kennedy JC, Hawkins RJ, Willis RB, et al: Tension studies of human knee ligaments: Yield point, ultimate failure, and disruption of the cruciate and tibial collateral ligaments. *J Bone Joint Surg* 1976;58A: 350-355.
14. Trent PS, Walker PS, Wolf B: Ligament length patterns, strength and rotational axes of the knee joint. *Clin Orthop* 1976;117:263-270.
15. Noyes FR, Grood ES: The strength of the anterior cruciate ligament in human and rhesus monkeys: Age-related and species-related changes. *J Bone Joint Surg* 1976;58A:1074-1082.
16. Woo SL, Hollis JM, Adams DJ, et al: Tensile properties of the human femur-anterior cruciate ligament-tibia complex: The effects of specimen age and orientation. *Am J Sports Med* 1991;19:217-225.
17. Butler DL, Grood ES, Noyes FR, et al: Effects of structure and strain measurement technique on the material properties of young human tendons and fascia. *J Biomech* 1984;17:579-596.
18. Wang CJ, Walker PS: The effects of flexion and rotation on the length patterns of the ligaments of the knee. *J Biomech* 1973;6:587-596.
19. Butler DL, Martin ET, Kaiser AD, et al: The effects of flexion and tibial rotation on the 3-D orientations and lengths of human anterior cruciate ligament bundles. *Trans Orthop Res Soc* 1988;12:324-325.
20. Girgis FG, Marshall JL, Monajem A: The cruciate ligaments of the knee joint. Anatomical, functional and experimental analysis. *Clin Orthop* 1975;106:216-231.
21. Van Dijk RV: The behavior of the cruciate ligament in the human knee. Thesis. Amsterdam, 1983.
22. Hollis JM, Marcin JP, Horibe S, et al: Load determination in ACL fiber bundles under knee loading. *Trans Orthop Res Soc* 1988;13:58.
23. Butler DL, Kay MD, Stouffer DC: Comparison of material properties in fascicle-bone units from human patellar tendon and knee ligaments. *J Biomech* 1986;19:425-432.
24. Butler DL, Guan Y, Kay MD, et al: Location-dependent variations in the material properties of the anterior cruciate ligament. *J Biomech* 1992;25:511-518.
25. Noyes FR, Butler DL, Grood ES, et al: Biomechanical analysis of human ligament grafts used in knee-ligament repairs and reconstructions. *J Bone Joint Surg* 1984;66A:344-352.

26. Woo SL, Gomez MA, Seguchi Y, et al: Measurement of mechanical properties of ligament substance from a bone-ligament-bone preparation. *J Orthop Res* 1983;1:22-29.
27. Woo SL, Gomez MA, Woo Y-K, et al: Mechanical properties of tendons and ligaments. II. The relationships of immobilization and exercise on tissue remodeling. *Biorheology* 1982;19:397-408.
28. Butler DL, Sheh MY, Stouffer DC, et al: Surface strain variation in human patellar tendon and knee cruciate ligaments. *J Biomech Eng* 1990;112:38-45.
29. Arms SW, Pope MH, Johnson RJ, et al: The biomechanics of anterior cruciate ligament rehabilitation and reconstruction. *Am J Sports Med* 1984;12:8-18.
30. Renström P. Arms SW, Stanwyck TS, et al: Strain within the anterior cruciate ligament during hamstring and quadriceps activity. *Am J Sports Med* 1986;14:83-87.
31. Beynnon BD, Pope MH, Fleming BC, et al: An in-vivo study of the ACL strain biomechanics in the normal knee. *Trans Orthop Res Soc* 1989;14:324.
32. Howe JG, Johnson RJ, Kaplan MJ, et al: Anterior cruciate ligament reconstruction using quadriceps patellar tendon graft: Part I. Long-term follow up. *Am J Sports Med* 1991;19(5):447-457.
33. Paulos L, Noyes FR, Grood E, et al: Knee rehabilitation after anterior cruciate ligament reconstruction and repair. *Am Sports Med* 1981; 9:140-149.
34. Ahmed AM, Hyder A, Burke DL, et al: In-vitro ligament tension pattern in the flexed knee in passive loading. *J Orthop Res* 1987;5: 217-230.
35. Jasty M, Lew W, Lewis J: In vitro ligament forces in the normal knee using buckle transducers. *Orthop Trans* 1982;6:301.
36. Lew WD, Lewis JL: The effect of knee-prosthesis geometry on cruciate ligament mechanics during flexion. *J Bone Joint Surg* 1982;64A: 734-739.
37. Lewis JL, Lew WD, Schmidt J: A note on the application and evaluation of the buckle transducer for the knee ligament force measurement. *J Biomech Eng* 1982; 104:125-128.
38. Lewis JL, Shybut GT: In vivo forces in the collateral ligaments of canine knees. *Trans Orthop Res Soc* 1981;6:4.
39. Lewis JL, Jasty M, Schafer M, et al: Functional load directions for the two bands of the anterior cruciate ligament. *Trans Orthop Res Soc* 1980;5:307.
40. Lewis JL, Fraser GA: On the use of buckle transducers to measure knee ligament forces. ASME 1979 Biomechanics Symposium, 1979.
41. Grood ES, Suntay WJ, Noyes FR, et al: Biomechanics of the knee-extension exercise: Effect of cutting the anterior cruciate ligament. *J Bone Joint Surg* 1984;66A:725-734.
42. Markolf KL, Gorek JF, Kabo JM, et al: Direct measurement of resultant forces in the anterior cruciate ligament: An in vitro study performed with a new experimental technique. *J Bone Joint Surg* 1990;72A:557-567.
43. Draganich LF, Vahey JW: An in vitro study of anterior cruciate ligament strain induced by quadriceps and hamstrings forces. *J Orthop Res* 1990;8:57-63.
44. Edwards RG, Lafferty JF, Lange KO: Ligament strain in the human knee. *J Biomech Eng* 1970;92:131-136.
45. Kennedy JC, Hawkins RJ, Willis RB: Strain gauge analysis of knee ligaments. *Clin Orthop* 1977;129:225-229.

46. Kurosaka M, Yoshiya S, Andrish JT: A biomechanical comparison of different surgical techniques of graft fixation in anterior cruciate ligament reconstruction. *Am J Sports Med* 1987;15:225-229.
47. White AA III, Raphael IG: The effect of quadriceps loads and knee position on strain measurements of the tibial collateral ligament: An experimental study on human amputation specimens. *Acta Orthop Scand* 1972;43:176-187.
48. Arms S, Boyle J, Johnson R, et al: Strain measurement in the medial collateral ligament of the human knee: An autopsy study. *J Biomech* 1983;16:491-496.
49. Fischer RA, Arms SW, Johnson RJ, et al: The functional relationship of the posterior oblique ligament to the medial collateral ligament of the human knee. *Am J Sports Med* 1985;13:390-397.
50. Whitmer GG, Haynes DW, Hungerford DS: The effect of different axial loads upon the anterior cruciate ligament. *Trans Orthop Res Soc* 1988;13:60.
51. Paulos LE, France EP, Rosenberg TD, et al: The biomechanics of lateral knee bracing: Part I. Response of the valgus restraints to loading. *Am J Sports Med* 1987;15:419-429.
52. Hollis JM, Takai S, Adams DJ, et al: The effects of knee motion and external loading on the length of the anterior cruciate ligament (ACL): A kinematic study. *J Biomech Eng* 1991;113:208-214.
53. Draganich LF, Jaeger R, Kralj A: EMG activity of the quadriceps and hamstrings during monoarticular knee extension and flexion. *Trans Orthop Res Soc* 1987;12:283.
54. Young SK, Rigby H, Shercliff TL, et al: Antagonistic quadriceps—hamstrings action protection of the anterior cruciate ligament. *Trans Orthop Res Soc* 1988;13:197.
55. Monahan JJ, Grigg P, Pappas AM, et al: In vivo strain patterns in the four major canine knee ligaments. *J Orthop Res* 1984;2:405-418.
56. Brand RA, Rubin CT, Seeherman HJ, et al: In vivo measurements of strains in the medial collateral ligament of the sheep and horse. Proceedings Fifth Meeting of the European Society of Biomechanics. Berlin, Germany, 1986, p 78.
57. Korvick DK, Rupert MP, Holden JP, et al: Peak in vivo forces in the anterior cruciate and patellar tendon during various activities: Preliminary studies in a goat. Transactions of ASME Winter Annual Meeting, 1992.
58. Kain CC, McCarthy JJ, Arms S, et al: An in vivo study of the effect of transcutaneous electrical muscle stimulation on ACL deformation. *Trans Orthop Res Soc* 1987;12:106.
59. Henning CE, Lynch MA, Glick KR Jr: An in vivo strain gage study of elongation of the anterior cruciate ligament. *Am J Sports Med* 1985;13:22-26.
60. Howe JG, Wertheimer C, Johnson RJ, et al: Arthroscopic strain gauge measurement of the normal anterior cruciate ligament. *Arthroscopy* 1990;6:198-204.
61. Seireg A, Arvikar RJ: The prediction of muscular load sharing and joint forces in the lower extremities during walking. *J Biomech* 1975;8: 89-102.
62. Andriacchi TP, Ogle JA, Galante JO: Walking speed as a basis for normal and abnormal gait measurements. *J Biomech* 1977;10:261-268.
63. Belcher MK, Reider B, Andriacchi TP: Alterations in stairclimbing function associated with an ACL deficient knee. *Trans Orthop Res Soc* 1984;9:134.

64. Berchuck M, Andriacchi TP, Bach BR, et al: Gait adaptations by patients who have a deficient anterior cruciate ligament. *J Bone Joint Surg* 1990;72A:871- 877.
65. Tibone JE, Antich TJ, Fanton GS, et al: Functional analysis of anterior cruciate ligament instability. *Am J Sports Med* 1986;14:276-284.
66. Silverberg SA, Cherney SB, Au J, et al: Force plate and electromyographic analyses of running and cutting in patients with anterior cruciate ligament tears. *Trans Orthop Res Soc* 1987;12:200.
67. Biden F, O'Conner J, Collins JJ: Gait analysis, in Daniel DM, Akeson WH, O'Connor JJ, et al (eds): *Knee Ligaments: Structure, Function, Injury and Repair*. New York, NY, Raven Press, 1990, p 291.
68. Harrington IJ: A bioengineering analysis of force actions at the knee in normal and pathological gait. *Biomed Eng* 1976;11:167-172.
69. Morrison JB: The mechanics of the knee joint in relation to normal walking. *J Biomech* 1970;3:51-61.
70. Nissan M: Review of some basic assumptions in knee biomechanics. *J Biomech* 1980;13:375-381.
71. Andriacchi TP: Dynamics of pathological motion applied to the anterior cruciate deficient knee. *J Biomech* 1990;23(suppl 1):99-105.
72. Noyes FR, DeLucas JL, Torvik PJ: Biomechanics of anterior cruciate ligament failure: An analysis of strain-rate sensitivity and mechanisms of failure in primates. *J Bone Joint Surg* 1974;54A:236-253.

Chapter 24

Aging of the Medial Collateral Ligament: Interdisciplinary Studies

Cyril Frank, MD
John Matyas, PhD
David A. Hart, PhD

Introduction

In adult humans, the tibial or medial collateral ligament (MCL) of the knee is formed of a dense band of extracapsular connective tissue, which crosses the medial aspect of the joint. Proximally, the MCL attaches to an elliptical area on the medial femoral epicondyle. Distally, it attaches to a broad area on the medial aspect of the tibial metaphysis, immediately behind the insertions of the three pes anserinus tendons.[1] Anteriorly, it is attached to the medial patellar retinaculum.[2] Posteriorly, it merges into the posterior oblique ligament and posteromedial capsule through superior and inferior expansions to form a triangular complex (Fig. 1). A bursa separates the MCL from the underlying deep layer of the medial aspect of the knee, the so-called capsular ligaments. The capsular ligaments arise from the middle one-third of the capsule and attach to the medial meniscus—the proximal portion is called the meniscofemoral ligament, the distal portion is called the meniscotibial ligament (Fig. 2). The functional significance of each of these ligaments and the remaining structures of the medial and posteromedial aspects of the knee have been discussed extensively by Müller.[3]

The anatomy of the MCL has been described in a number of texts[4-6] and in ever-increasing detail in a series of scientific papers published over the past 60 years.[7,8] Almost without exception, the aim of these papers was to facilitate the surgical restoration of normal ligament anatomy after an injury, with no attempt made to characterize changes in form with aging. However, most textbook descriptions of MCL anatomy are likely to be from observations of fixed cadaveric material derived from elderly patients. Unfortunately, descriptions of fresh MCLs obtained during surgery are limited by the surgical approach and the extent of surgical exposure and are also without specific reference to age. Thus, the adult form of the human MCL has generally been assumed to maintain an unchanging appearance and location. However, based on evidence in

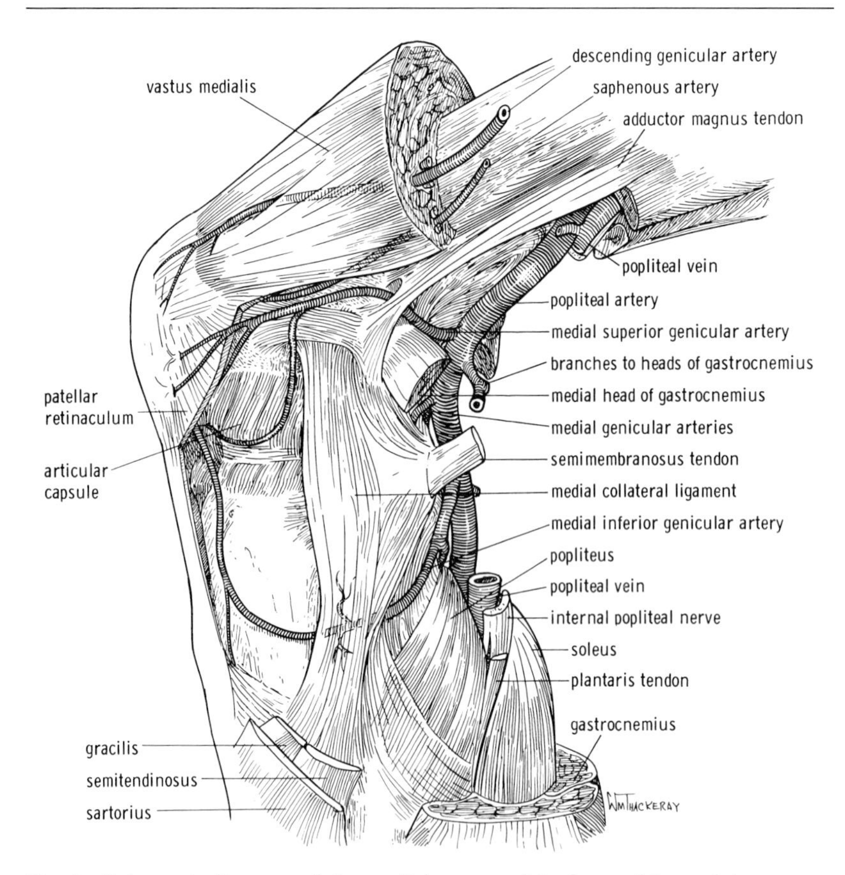

Fig. 1 *Schematic diagram of the medial aspect of the knee with special reference to the medial collateral ligament. (Reproduced with permission from Insall JN: Anatomy of the knee, in Insall JN:* Surgery of the Knee. *New York, NY, Churchill Livingstone, 1984, pp 1-20.)*

other model systems presented below, this assumption is almost certainly incorrect. The human MCL must change its form quite dramatically during at least early growth and development, and it further adjusts its gross structure with aging. As far as we are aware, however, these changes have never been defined in the human knee—a deficiency that unfortunately will not be corrected in this review. Instead, this chapter concentrates on what has been established regarding MCL growth, development, and aging in nonhuman systems, mostly in rabbits,[9-18] rats,[19-22] and dogs.[23]

In the New Zealand white rabbit it has been found that growth and development of both the animal and MCL are most rapid from birth to between six and eight months of age, with skeletal maturity (defined most relevantly as closure of the proximal tibial growth plate) being reached at about that time for males and slightly later (eight to ten months) for females.[10] Given that the life expectancy of the rabbit is between about five and eight years, then true studies of

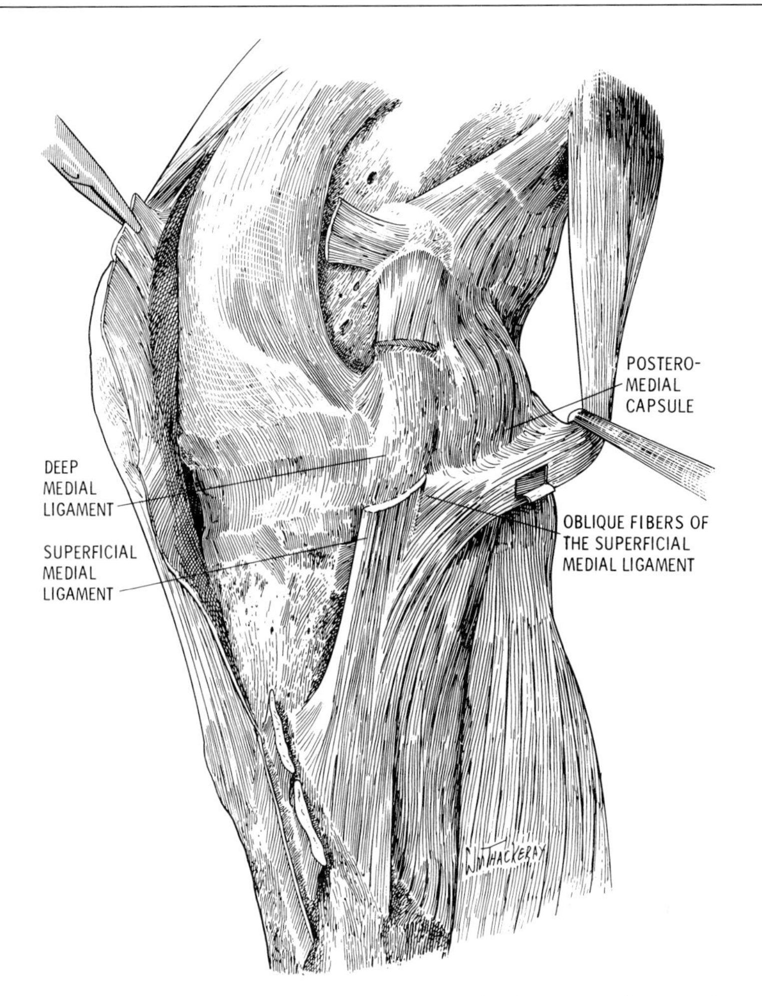

Fig. 2 *Schematic diagram of the knee with special reference to the capsular ligaments. (Reproduced with permission from Insall JN: Anatomy of the knee, in Insall JN:* Surgery of the Knee. *New York, NY, Churchill Livingstone, 1984, pp 6-7.)*

aging would have to be extended for several years after skeletal maturity is achieved. Heretofore, only a handful of studies[9,10,12-18] have documented changes in the MCL of rabbits even as old as 48 months of age. Unfortunately, these middle-aged animals are the only source of information for this review. Likewise, only relatively short-term information is available on the rat, an animal model in which skeletal maturity is actually difficult to define because the proximal tibial growth plate remains unfused for a protracted period after the lengthwise growth of the tibia has apparently ceased.

Although they are clearly not directly applicable to humans, we believe that the changes observed in the rabbit MCL during the first half of its lifespan are likely to be similar to the changes in human ligaments and other connective tissues described elsewhere in this text. Whereas the magnitude and timing of age-related changes in rabbit MCLs are undoubtedly specific to the rabbit model, and recognizing that longer term studies of truly elderly animals are still required, we believe that several important patterns of ligament maturation and aging can be seen even in shorter term studies. Furthermore, we believe that extrapolation from these animals to humans will eventually prove to be valid in a gross sense, because skeletal ligaments have similar structures and functions in all mammals. With this speculation stated clearly here, we will define aging of the MCL as completely as possible.

Growth and Development

Macroscopic Anatomy

As with other knee ligaments, the embryonic MCL forms as an early condensation of interzonal mesenchyme, which condenses on the medial side of the knee joint. During fetal life, this condensation maintains its location and takes on a more discrete form, so that it is anatomically distinct at birth.

Following birth, the MCL in the rabbit begins to change very rapidly. For the first weeks of life it is a relatively amorphous translucent band. By 3 months of age it is transformed into a discrete white-opaque structure. During this early growth period, the MCL increases remarkably in both density and size, including length, width, and thickness (Fig. 3). Although the overall relationship of the MCL to the joint is reasonably constant, closer inspection reveals that subtle changes are occurring. In particular, the dimensions of the tibial insertion begin to increase in length and width.[14] Conversely, the femoral insertion changes relatively little during this time.

The surface layer of the MCL, known as the epiligament,[24] is relatively amorphous in young animals, but during maturation it thickens, at least on the superficial (medial) portion of the MCL. On the deep (joint) side of the MCL this epiligament actually becomes thinner with maturation. Because the epiligament enters into the MCL substance at some points,[24] it is not easily separated from the deeper MCL substance at any age.

During adolescence in the rabbit (between 3 and 9 months of age), the MCL shows only minimal gross changes. Its overall dimensions change only very slightly as the underlying tibial growth plate fuses. In all other respects the MCL appears unchanged and consistently appears as the dense white band noted above.

The appearance of the MCL in adult rabbits is nearly constant. There are slight variations in size and shape between individuals, but the overall form of the ligament remains quite consistent. After skeletal maturity has been reached, there may be a subtle loss of matrix resulting in a slight decrease in MCL cross-sectional area.

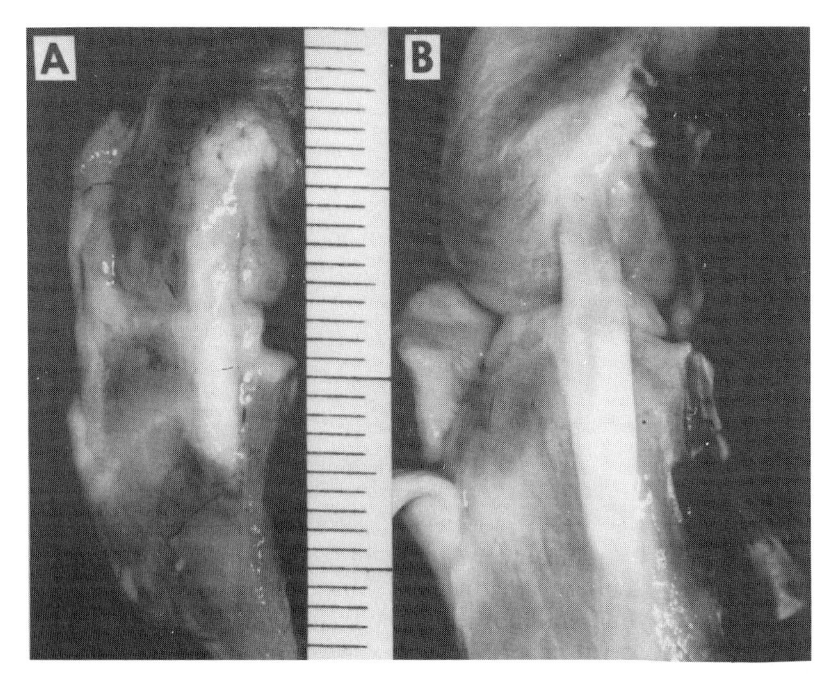

Fig. 3 *Medial macroscopic views of rabbit knee joints with MCLs.* ***Left****, a 1-month-old animal (note the translucency of the insertion distally) and* ***Right****, a 10-month-old animal (note that the growth of the tibial insertion contributes to much of the overall growth in length of the MCL, but that it retains its relative position on the tibia). (Reproduced with permission from Matyas JR, Bodie D, Andersen M, et al: The developmental morphology of a "periosteal" ligament insertion: Growth and maturation of the tibial insertion of the rabbit medial collateral ligament.* J Orthop Res *1990;8:412-424.)*

Woo and associates[10] suggest that these changes appear roughly parallel to changes in body mass and that they appear to be similar in both sexes.

Light Microscopy

Recently, several notable changes in the microscopic anatomy of the MCL have been described during aging in the MCL midsubstance and at the ligament insertions.

Ligament Midsubstance Amiel and associates[16] have reported that the pattern of collagen crimp in older animals is of smaller amplitude than in younger animals. In addition, ligament fibrocytes are transformed from rounded to spindle-shaped cells between 2 and 12 months of age, and persist as spindle-shaped cells at 36 months of age. During this transformation the cells become more organized within the collagen bundles of the ligament, although by 36 months, ligament cells lose some of their alignment. Furthermore, at 36 months of age, some regions of the MCL had decreased in cellularity and "collagen bundle fragmentation" was observed "sporadically throughout the ligament."[16] With maturation, subtle changes have also been seen in the thickness and morphology of the epiligament.[24]

Insertions The postnatal morphology of the femoral and tibial insertions of the rabbit MCL has been described to varying degrees by several investigators.[25-27] More recently, the microscopic morphology of the tibial insertion has been re-examined and described in semi-quantitative terms by Matyas and associates.[14,28,29] These studies reveal that dramatic differences in the structural organization of the tibial insertion occur during growth and maturation.

In the tibial insertion of the MCL, five different layers of tissue have been defined that change their proportions during growth and maturation.[14] In immature animals, the majority of the MCL fibers continue into the periosteum; in older animals, the majority of these collagen fibers are cemented into the tibial cortex by advancing mineral. Thus, this periosteal insertion in the immature rabbit is converted into a more direct insertion during skeletal maturation. After skeletal maturity, the adult organization was seen to persist for another year. Unfortunately, the morphology of the insertions has not been studied in older animals.

In contrast to the dynamic changes of the tibial insertion during growth and aging, the morphology of the femoral insertion appears to remain static. Even early in its postnatal development, this insertion was found to have the more classic zonal appearance seen in many other model systems; ligament tissue being continuous with zones of fibrocartilage, mineralized fibrocartilage, and finally bone. While these zones have been quantified in other systems over time,[30] they have not been quantified in the aging MCL. As with the tibial insertion, the potential changes in the femoral insertion of the MCL with true aging remain unknown.

Transmission Electron Microscopy (T.E.M.)

Cell Ultrastructure As was seen with light microscopy, T.E.M. reveals that in the immature MCL the ligament cells are ovoid, but that with progressive maturation they appear more elongated and their processes appear more closely apposed to the extracellular collagen fibers.[16] The cells from immature MCLs contain abundant rough endoplasmic reticulum (RER), Golgi apparati, and adjacent secretory vacuoles, all of which are characteristic of cells that actively synthesize extracellular products. At 36 months of age, rabbit MCLs are nearly devoid of RER and Golgi bodies, but have accumulated cytoplasmic vacuoles, dense bodies, and filament bundles. These changes, which are consistent with cell aging, suggest that the MCL cells have become less metabolically active and more degenerative.[16]

Collagen Ultrastructure As they do in many other tissues, the diameters of collagen fibrils in the MCL change during maturation.[31] A relatively unimodal population of small-diameter fibrils exists in adolescent MCLs, whereas a broader, bimodal distribution of fibril diameters occurs with maturity (Fig. 4).[32] After skeletal maturity, this distribution profile changes very little for at least one year (C.B. Frank, unpublished data). As far as we know, the fibril diameter profiles in middle-aged or elderly MCLs have not yet been described. However, based on results with other dense connective tissues, some loss of large fibrils is likely to be expected.[33]

Biochemistry

Substance During growth and maturation, a number of subtle but significant changes occur in the matrix composition of the rabbit MCL.[13] Water content varies along the length of the MCL and appears to decline somewhat during skeletal maturation. In immature and post-mature animals, water contents were slightly higher (Fig. 5). Amiel and associates,[16] on the other hand, showed a significant decline in water content with MCL maturation and subsequent aging to 36 months of age. Hexosamine content, an indication of proteoglycan content, declines with advancing age in rabbit MCLs with some slight differences between substance and insertions (Fig. 6, *left*). Collagen concentrations also vary along the length of the MCL and have been reported to vary slightly with age (Fig. 6, *right*). Amiel and associates[16] described a slight increase in collagen concentration, from 89% to 91%, during skeletal maturation but by 36 months of age the collagen concentration in the rabbit MCL has decreased dramatically, to roughly 74%. At the same time, there was also a significant change in collagen crosslinking.[16] The relative concentrations of both of the major reducible collagen crosslinks, hydroxylysinonorleucine (HLNL) and dihydroxylysinonorleucine (DHLNL), decreased (Fig. 7). Notably, there was a concomitant increase in the non-reducible crosslink pyridinoline, suggesting a possible conversion from less stable crosslinks to the more stable pyridinoline form (Fig. 8, *left*). Over the same time interval, collagen synthesis reportedly declines rather dramatically (Fig. 8, *right*), which is consistent with the concept that a matrix that is biochemically more stable turns over more slowly.

Edwards and associates[34] used both histologic and biochemical means to study cell-matrix ratios in rabbit MCL during maturation. They reported that, in the immature state, ligament cells were associated with relatively small volumes of extracellular matrix. With maturation, the cells were associated with larger volumes of matrix. This change in the ratio of cells to matrix was not caused by a decrease in the number of cells in the ligament, rather, there was more matrix between the cells. That is, as animals grew and reached skeletal maturity the overall cell density decreased, but the total

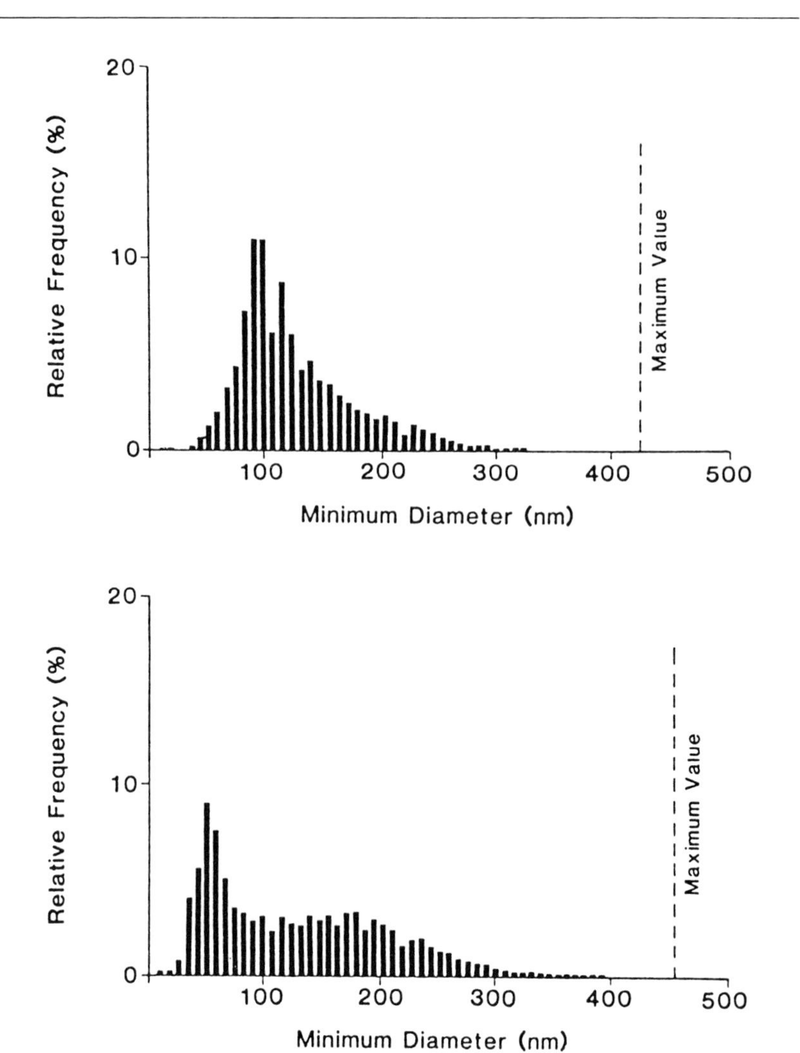

Fig. 4 *Histograms of fibril diameter distributions for* ***Top****, all 3-month-old MCLs (n = 19, 408 fibrils) and* ***Bottom****, all 10-month-old MCLs analyzed (n = 11, 911 fibrils). Note shift in mode and slightly broadened distribution of fibrils. (Reproduced with permission from Frank C, Bray D, Rademaker A, et al: Electron microscopic quantification of collagen fibril diameters in the rabbit medial collateral ligament: A baseline for comparison.* Connect Tissue Res *1989;19:11-25.)*

number of cells remained relatively constant. Whether or not this change in cell density continues during aging has not yet been studied using these methods.

Insertions While specific localization of biochemical and metabolic changes in the MCL has not been performed in detail, it is apparent that insertions, like substance, change in composition dur-

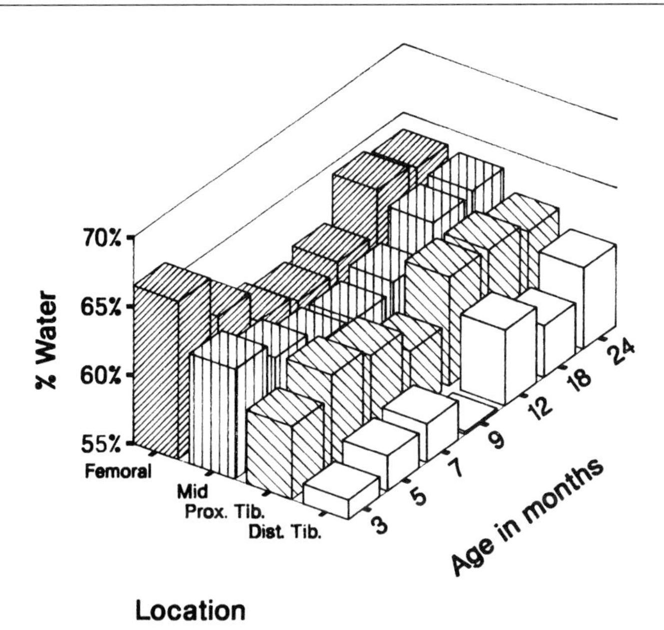

Fig. 5 *Histogram with expanded scale showing mean water contents (% of total weight) of four MCL segments for all rabbits (n=45) at various age intervals. Note the decreasing water contents from proximal to distal MCL at all ages. (Reproduced with permission from Frank C, McDonald D, Lieber R, et al: Biochemical heterogeneity within the maturing rabbit medial collateral ligament.* Clin Orthop *1988;236:279-285.)*

ing growth and development (Figs. 5 and 6). Frank and Hart[35] described the metabolic heterogeneity and changes with maturation of plasminogen activator in the MCL (Fig. 9). They found that the tibial insertion is more active metabolically than either the midsubstance or femoral insertion, at least during early growth and maturation. Whether these metabolic changes are caused by alterations in cell numbers, cell types, or changes in cell activities remains to be studied. A similar pattern of heterogeneity has been observed in human ligaments, even in tissue obtained from individuals older than 70 years of age.

Biomechanics

The biomechanical changes that take place in rabbit MCL during growth and maturation have been described by Woo and associates.[10-12,15,36] These changes will be classified below under either prefailure behaviors (laxity, cyclic-viscoelastic and load-relaxation behaviors) or failure behaviors (high load structural and material behaviors).

Prefailure Behaviors In general, immature 3-month-old MCLs have been found to be more viscous than more mature MCLs, with a

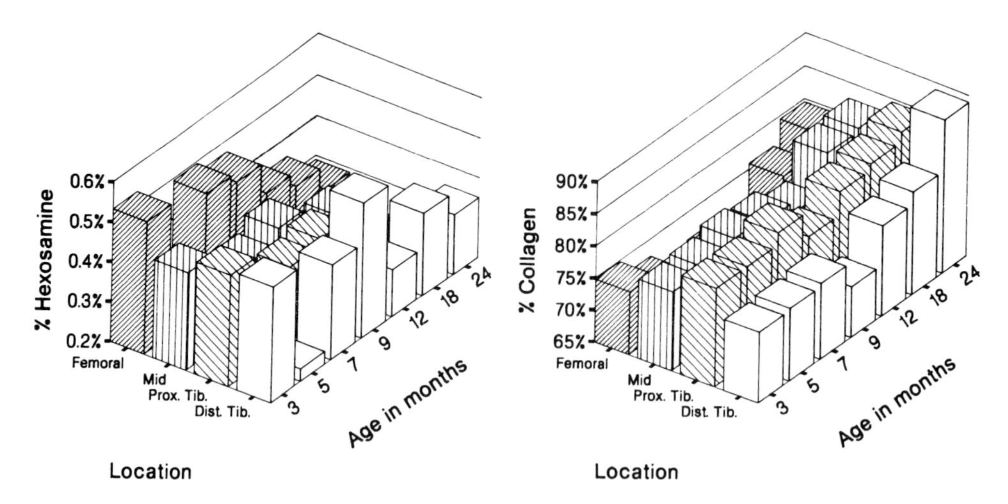

Fig. 6 *Left, Histogram with expanded scale showing mean hexosamine concentrations (% dry weight) of rabbit MCL sections at various age intervals. Insertions have highest values and a general decrease is noted with the maturation of all segments.* ***Right****, Histogram with expanded scale showing mean collagen concentration of rabbit MCL sections (% dry weight) at various age intervals. Significant differences are noted between insertions and the middle of the MCL. Also, note a subtle difference between the two insertions, as well as a significant increase in concentration of all segments after skeletal maturity. (Reproduced with permission from Frank C, McDonald D, Lieber R, et al: Biochemical heterogeneity within the maturing rabbit medial collateral ligament.* Clin Orthop *1988;236:279-285.)*

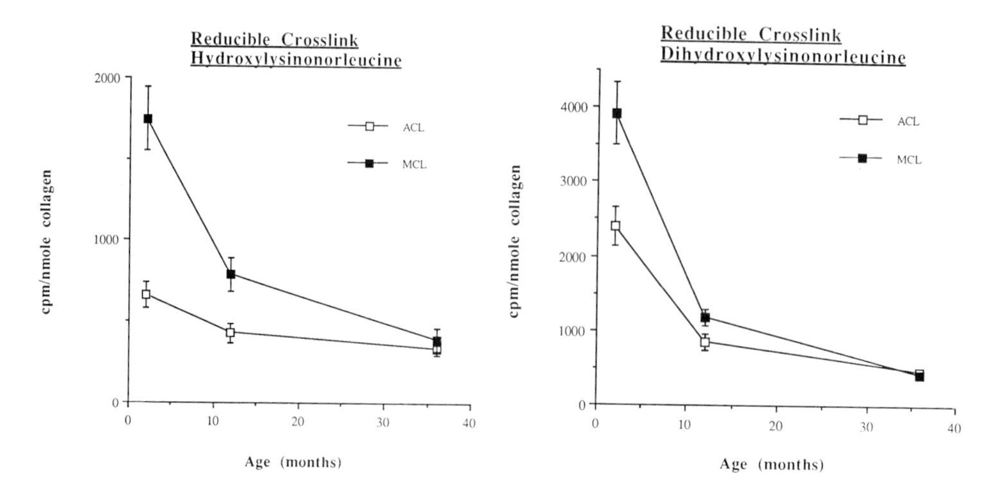

Fig. 7 *Left, Reducible crosslink hydroxylysinonorleucine content in the collagen of MCL and anterior cruciate ligament (ACL) as a function of age.* ***Right****, Reproducible crosslink dihydroxylysinonorleucine content in the collagen of MCL and ACL as a function of age. (Reproduced with permission from Amiel D, Kuiper S, Wallace CD, et al: Age-related properties of medial collateral ligament and anterior cruciate ligament: A morphologic and collagen maturation study in the rabbit.* J Gerontol *1991;46:B159-B165.)*

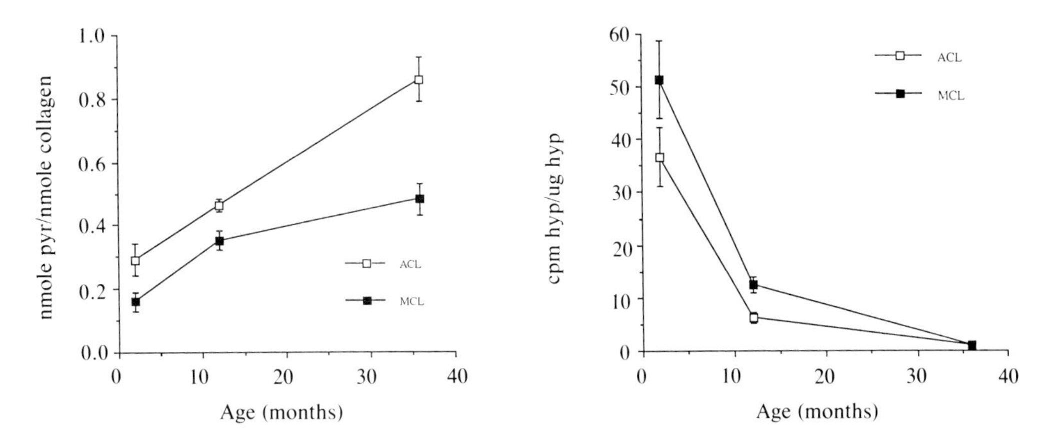

Fig. 8 ***Left,*** *Non-reducible crosslink pyridinoline content of the collagen from MCL and ACL as a function of age.* **Right**, *Relative collagen synthesis in MCL and ACL as a function of age. (Reproduced with permission from Amiel D, Kuiper S, Wallace CD, et al: Age-related properties of medial collateral ligament and anterior cruciate ligament: A morphologic and collagen maturation study in the rabbit.* J Gerontol *1991;46:B159-B165.)*

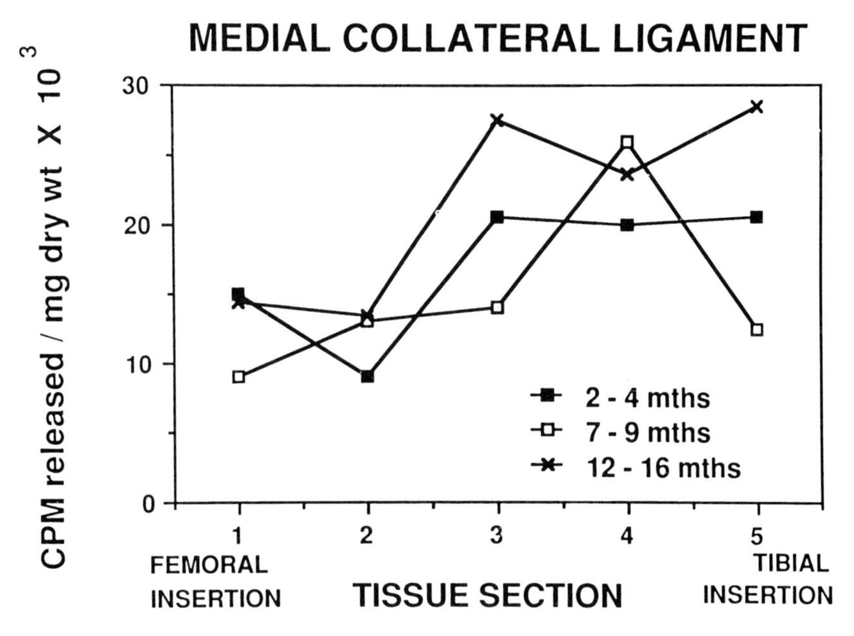

Fig. 9 *Expression of plasminogen-dependent plasminogen activator activity by medial collateral ligament tissue from rabbits of different ages. (Reproduced with permission from Frank CB, Hart DA: The biology of tendons and ligaments, in Mow VC, Ratcliffe A, Woo SL-Y (eds):* Biomechanics of Diarthrodial Joints. *New York, Springer-Verlag, 1990, pp 39-62.)*

greater static load-relaxation than their mature counterparts,[36] as well as a greater cyclic relaxation (Fig. 10, *left*). Similarly, immature ligaments are somewhat more lax, meaning that they allow slightly more deformation under low loads than older adult MCLs (Fig. 10, *right*). These results correlate well with the observations of Woo and associates,[11] that in situ stresses in the MCL increase after closure of the epiphyses (skeletal maturation).

Failure Behaviors During growth and maturation there is a very dramatic change in rabbit MCL failure behavior, which has now been very well defined[12] and characterized.[14,15,37] At the time of skeletal maturation, the MCL complex becomes dramatically stronger in a structural sense as a result of a cementing in of the deep layers of the tibial insertion by bone.[14] In immature MCLs, femur-MCL-tibia failure occurs universally at the tibial insertion. As the MCL separates from the tibia, it pulls away some of the periosteum from the distal tibia as well. MCLs from skeletally mature rabbits usually fail obliquely in their substance extending down to, but rarely involving, a bony insertion. Increases in strength of two- to four-fold (from ~ 100 N to ~ 350 N) are seen during maturation.

Structural stiffness of rabbit MCL complexes also increases with growth and maturation. The linear stiffness of male rabbit MCLs increases from 40 N/mm in 3.5-month-old animals to 64 N/mm in 1-year-olds. In females, there is an increase from 32 N/mm to 48 N/mm over that same period.[10] A slight decline in linear stiffness has been described between 12 months and 36 months of age (Fig. 11). However, based on increased elongations at failure, the energy absorbed to failure continues to increase with age. Only after 4 years did this parameter begin to decline in females.[10]

Mechanical properties of MCLs also increase significantly with growth and maturation,[10] although not as dramatically as the structural properties for the femur-MCL-tibia complex. For example, the elastic modulus of MCLs from male rabbits increased from 700 MPa in 3.5-month-old animals to 1180 MPa at 12 months. Female MCLs, over the same time period, increased from 750 MPa to 950 MPa. The moduli of both males and females subsequently declined by 25% to 50% in the ensuing two years (Fig. 12).

Strain rate had more of an effect on immature MCL properties than on those of mature MCLs.[12,21] An increase of four orders of magnitude in the extension rate increased the structural properties of immature MCLs from 50% to 150% and increased their elastic modulus by about 45%. In contrast, MCLs from skeletally mature rabbits increased their structural strength by only about 30% and their modulus by about 10% in response to the same extension rates.

Correlations of Structure and Function With Aging

As with other connective tissues, it is difficult to relate structural features and functional behaviors in any straightforward fashion.

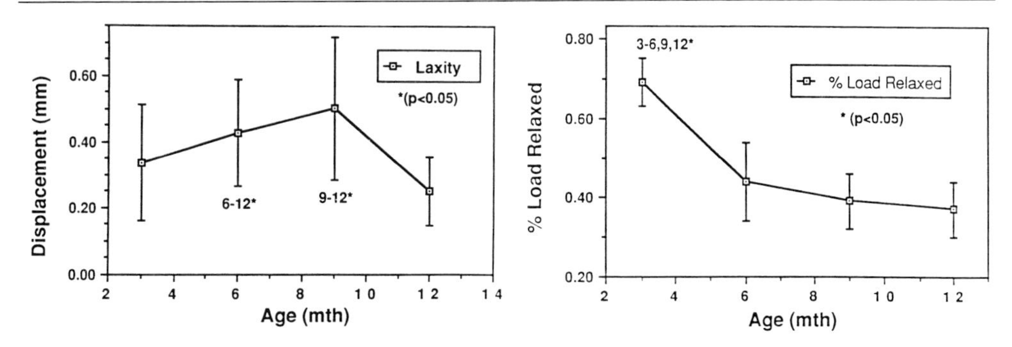

Fig. 10 *Left, Profile of percentage load relaxed for rabbit MCLs against age. **Right**, Plot of rabbit MCL laxity against age. (Reproduced with permission from Lam TC:* The Mechanical Properties of the Maturing Medial Collateral Ligament. *University of Calgary, 1988, PhD thesis.)*

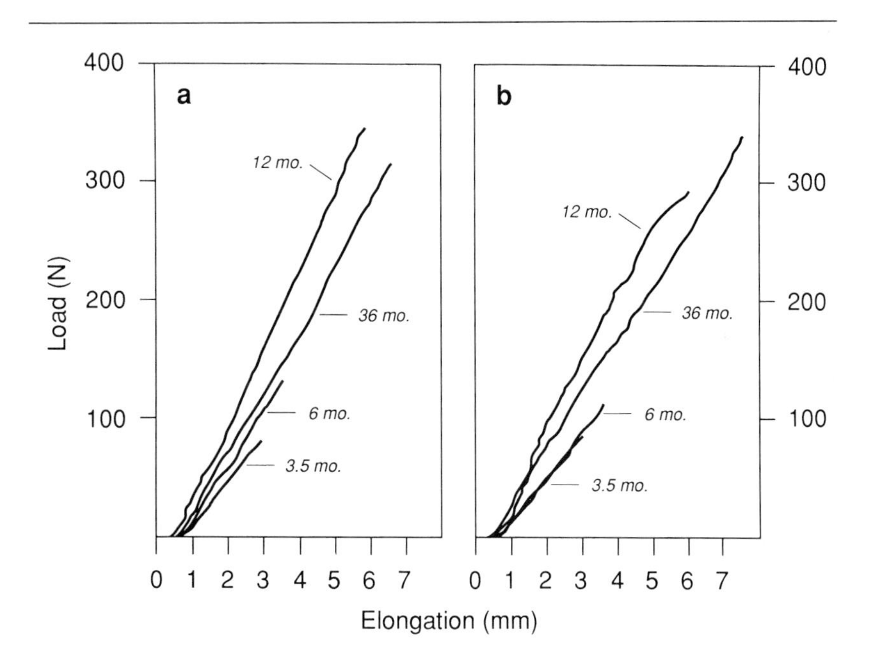

Fig. 11 *Typical load-elongation curves for the structural properties of the femur-MCL-tibia complex of both (**left**) males and (**right**) females in different age groups representing the structural properties of the femur-MCL-tibia complex. (Reproduced with permission from Woo SL-Y, Ohland KJ, Weiss JA: Aging and sex-related changes in the biomechanical properties of the rabbit medial collateral ligament.* Mech Ageing Dev *1990;56:129-142.)*

Clearly, there is a complex interaction between extracellular matrix components that translates into a mechanical product, which, in turn is measured biomechanically as MCL function. Nonetheless,

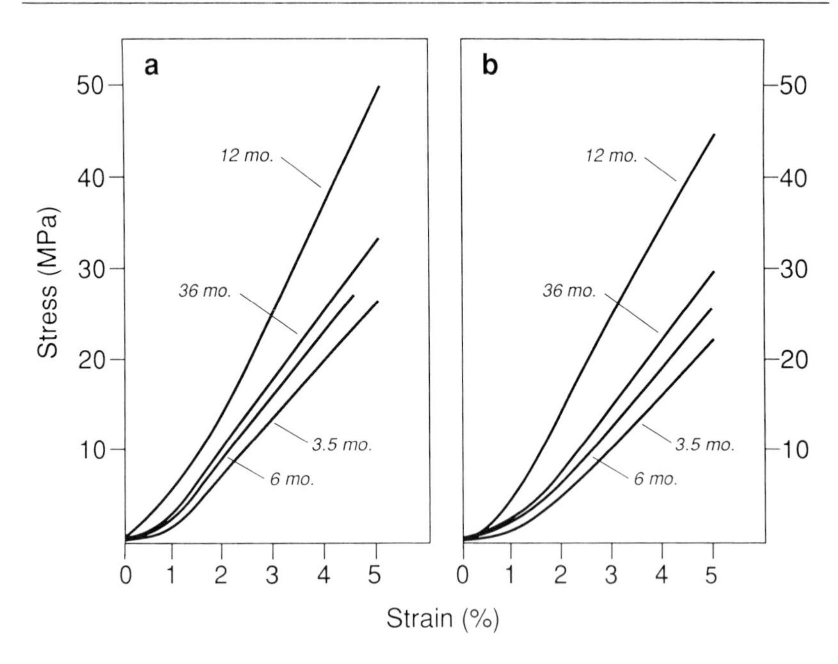

Fig. 12 *Mechanical properties of the rabbit MCL substance as represented by stress-strain curves for both (**left**) males and (**right**) females. (Reproduced with permission from Woo SL-Y, Ohland KJ, Weiss JA: Aging and sex-related changes in the biomechanical properties of the rabbit medial collateral ligament.* Mech Ageing Dev *1990;56:129-142.)*

there are some interesting and important changes in both MCL structure and function during growth, maturation, and early aging that may be related to each other and which should be discussed.

The most dramatic change in the MCL clearly occurs near the time of skeletal maturation. At this time there is a major change in the strength, stiffness, and energy-absorbing ability of the femur-MCL-tibia complex, presumably because of the cementing-in of its tibial insertion, described previously. In addition, the increased metabolic activity at the tibial insertion[35] is likely to be related to its complex cellularity[14] and to some, as-yet-undetermined local mechanisms that appear to be coupled to the growth of the tibia.[26] The conversion from a periosteal to a direct insertion at this site[14,37] is an amazing process, which involves very focal control over mineralization of ligament matrix to form the Sharpey's fibers.

While not quite as dramatic, there are more subtle changes in the extracellular matrix during this period of maturation, which can also be attributed to certain changes in its structure. Increasing quantities of collagen matrix, maturation in collagen crosslinks, changes in collagen fibril profiles, and changes in both water content and proteoglycan concentrations are all likely to contribute to stiffening and strengthening of the matrix. Decreasing viscous behavior is also likely to be connected to decreasing water content and to the progressive addition of collagen. Beyond skeletal maturity (the main interval of interest for the purpose of this workshop), changes in the mechanical properties of the MCL could be interpreted as being caused by slight deterioration of MCL structure. Femur-MCL-tibia complex stiffness and strength decline, and the MCL substance itself becomes more compliant. Decreases in collagen content, some loss of large diameter collagen fibrils, shifts in types of collagen crosslinks, changes in collagen organization (crimp), and ultrastructural evidence of collagen fragmentation could all be invoked as potential causes of this mechanical deterioration. Further changes in ligament viscoelastic behavior, with decreases in viscosity, may be related to decreases in water content and shifts in proteoglycan quantities (as measured by hexosamines). Morphologic and metabolic evidence supports the concept that ligament cells, instead of decreasing in number, are more likely to be declining in their individual abilities to regenerate normal matrix. They either fail to synthesize matrix of adequate quantity or quality, or they release degradative substances which are not being neutralized. This theory is highly speculative, however, because there are literally no data on truly aged ligaments on which to base any conclusions.

Summary

Very little is known about the growth, development, and aging of the human MCL. What is known, however, comes primarily from studies of various animal model systems, most of which have not been extended for long enough periods of time to comment on true aging.

Extrapolations from data on the rabbit MCL suggest that structure and function deteriorate slowly after adulthood, with gradual losses in the peak mechanical performances present at the time of skeletal maturation. The exact causes of mechanical and biologic deterioration in the MCL, as in other connective tissues, remain open to debate. However, a number of possibilities exist. The first possibility is genetic alteration. Subtle individual alterations may exist in the structure of the genes that encode for the molecules that make up or regulate the turnover of the extracellular matrix. In extreme cases, genetic alterations could cause premature deterioration of the MCL by coding for inferior structural molecules. More subtle variations could cause individual variations in the rate of deterioration. This concept is related to another genetic variable, namely life span. Data from animal studies are oftentimes extrapolated to humans based on proportional age, which assumes that the age at which deterioration is detected is related to the life span of the animal. However, in different species, it may be that the rate of deterioration changes the point in the life span that deterioration is detected. While studies such as those of Ozaki and Mizuno[38] support the concept of a relationship between connective tissue changes and life span, the same study also points out that other species-specific variables may influence outcome. Whether or not similar variables influence the aging process in the MCL remains to be determined.

A second important possibility is that the MCL is influenced by the environment. For example, the biomechanical environment is a factor that is likely to play a critical role in the aging process. Studies by Walsh and associates,[39] Walsh,[40] and Weir[41] indicate that immobilization (deficiency in biomechanical stimuli) of the immature rabbit MCL leads to prolonged alteration of function. Likewise, immobilization of the adult rabbit MCL leads to a detectable loss of function, but at a slower rate. Conversely, exercise is likely to alter the activity of connective tissue cells, which may lead to altered matrix metabolism with attendant changes in biomechanical properties.[42] The importance of exercise on the rate of MCL aging is unknown, but, given society's current emphasis on physical fitness, such variables should be considered. Similarly, over-use of the MCL complex may also accelerate the aging process. Indeed, if the biomechanical environment affects aging, then it is likely that all the ligaments of the knee may age at different rates, because they are likely to have different loading histories.

In addition to the mechanical environment, ligament cells and matrix may be influenced by the chemical environment.[34] As discussed earlier, MCL cells change during aging—their ultrastructure changes, they become deficient in matrix secretion, and their density declines. Whether these changes in cell function are caused by an intrinsic loss of their renewal or proliferative capacity, or by a loss in the availability of exogenous stimuli (endocrine, neuroendocrine, neurotransmitters, nutrition, etc) with aging, remains to be discovered.

A great deal of work remains to be done to even describe the changes in structure and function during aging of the MCL before being able to elucidate their correlations and the mechanisms by which they occur.

Acknowledgments

The authors gratefully acknowledge the financial support of the Alberta Heritage Foundation for Medical Research, The Canadian Arthritis Society, and the Medical Research Council of Canada. We also thank Jackie Wilson and Judy Crawford for their help in the final preparation of this manuscript.

References

1. Strobel M, Stedtfeld HW: *Diagnostic Evaluation of the Knee*. Berlin, Springer Verlag, 1990, pp 2-48.
2. Williams PL, Warwick R, Dyson M, et al (eds): *Gray's Anatomy*, ed 37. Edinburgh, Churchill Livingstone, 1989.

3. Müller W: *The Knee: Form, Function and Ligament Reconstruction.* Berlin, Springer Verlag, 1983, Part I. Kinematics, pp 8-75.
4. Gray H: *Anatomy of the Human Body*, ed 26. Philadelphia, Lea & Febiger, 1954.
5. Anson BJ: *An Atlas of Human Anatomy*, ed 2. Philadelphia, WB Saunders, 1963.
6. Crafts RC: *A Textbook of Human Anatomy*, ed 2. New York, John Wiley & Sons, 1979, pp 421-436.
7. Warren LA, Marshall JL, Girgis F: The prime static stabilizer of the medial side of the knee. *J Bone Joint Surg* 1974;56A:665-674.
8. Warren LF, Marshall JL: The supporting structures and layers on the medial side of the knee: An anatomical analysis. *J Bone Joint Surg* 1979;61A:56-62.
9. Frank C, Bodie D, Andersen M, et al: Growth of a ligament. *Trans Orthop Res Soc* 1987;12:42.
10. Woo SLY, Ohland KJ, Weiss JA: Aging and sex-related changes in the biomechanical properties of the rabbit medial collateral ligament. *Mech Ageing Dev* 1990;56:129-142.
11. Woo SLY, Weiss JA, Gomez MA, et al: Measurement of changes in ligament tension with knee motion and skeletal maturation. *J Biomech Eng* 1990;112:46-51.
12. Woo SLY, Orlando CA, Gomez MA, et al: Tensile properties of the medial collateral ligament as a function of age. *J Orthop Res* 1986;4: 133-141.
13. Frank C, McDonald D, Lieber R, et al: Biochemical heterogeneity within the maturing rabbit medial collateral ligament. *Clin Orthop* 1988;236:279-285.
14. Matyas JR, Bodie D, Andersen M, et al: The developmental morphology of a "periosteal" ligament insertion: Growth and maturation of the tibial insertion of the rabbit medial collateral ligament. *J Orthop Res* 1990;8:412-424.
15. Woo SLY, Peterson RH, Ohland KJ, et al: The effects of strain rate on the properties of the medial collateral ligament in skeletally immature and mature rabbits: A biomechanical and histological study. *J Orthop Res* 1990;8:712-721.
16. Amiel D, Kuiper SD, Wallace CD, et al: Age-related properties of medial collateral ligament and anterior cruciate ligament: A morphologic and collagen maturation study in the rabbit. *J Gerontol* 1991;46:B159-B165.
17. Wessels WE, Dahners LE: Growth of the rabbit deltoid ligament. *Clin Orthop* 1988;234:303-305.
18. Muller P, Dahners LE: A study of ligamentous growth. *Clin Orthop* 1988;229:274-277.
19. Booth FW, Tipton CM: Ligamentous strength measurements in pre-pubescent and pubescent rats. *Growth* 1970;34:177-185.
20. Tipton CM, Matthes RD, Martin RK: Influence of age and sex on the strength of bone-ligament junctions in knee joints of rats. *J Bone Joint Surg* 1978;60A:230-234.
21. Crowninshield RD, Pope MH: The strength and failure characteristics of rat medial collateral ligaments. *J Trauma* 1976;16:99-105.
22. Dahners LE, Muller P: Longitudinal ligamentous growth occurs throughout the ligament. *Trans Orthop Res Soc* 1987;12:41.
23. Cooper RR, Misol S: Tendon and ligament insertion: A light and electron microscopic study. *J Bone Joint Surg* 1970;52A:1-20.

24. Chowdhury P, Matyas JR, Frank CB: The "epiligament" of the rabbit medial collateral ligament: A quantitative morphological study. *Connect Tissue Res* 1991;27:33-50.
25. Hurov JR: Soft-tissue bone interface: How do attachments of muscles, tendons and ligaments change during growth? A light microscopic study. *J Morphol* 1986;189:313-325.
26. Videman T: An experimental study of the effects of growth on the relationship of tendons and ligaments to bone at the site of diaphyseal insertion: Part II. Determination of growth patterns and inhibition of displacement using metal markers. *Ann Chir Gynaecol* 1970;59:22-34.
27. Masoud I, Shapiro F, Moses A: Tibial epiphyseal development: A cross-sectional histologic and histomorphometric study in the New Zealand white rabbit. *J Orthop Res* 1986;4:212-220.
28. Matyas JR, Frank C: Midsubstance injury of the rabbit MCL affects the tissue architecture of the femoral insertion. *Trans Orthop Res Soc* 1990;15:34.
29. Matyas JR, Frank C: Midsubstance injury to the rabbit MCL causes changes similar to immobilization at the tibial insertion. *Trans Orthop Res Soc* 1990;15:525.
30. Matyas JR: *The Structure and Function of Tendon and Ligament Insertions into Bone*. Cornell University, New York, 1985, MSc thesis.
31. Frank C, Bray D, Rademaker A, et al: Electron microscopic quantification of collagen fibril diameters in the rabbit medial collateral ligament: A baseline for comparison. *Connect Tissue Res* 1989;19:11-25.
32. Frank C, McDonald D, Bray D, et al: Collagen fibril diameters in the healing adult rabbit medial collateral ligament. *Connect Tissue Res* 1992;27:251-263.
33. Parry DA, Barnes GR, Craig AS: A comparison of the size distribution of collagen fibrils in connective tissues as a function of age and a possible relation between fibril size distribution and mechanical properties. *Proc R Soc Lond [Biol]* 1978;203:305-321.
34. Edwards PE, Wuensche CS, Hart DA, et al: The number of cells in rabbit MCL is constant during its growth and maturation: Only matrix is added. *Proc Can Orthop Res Soc* 1990;9.
35. Frank CB, Hart DA: The biology of tendons and ligaments, in Mow VC, Ratcliffe A, Woo SL-Y (eds): *Biomechanics of Diarthrodial Joints*. New York, Springer-Verlag, 1990, vol 1, chap 2, pp 39-62.
36. Lam TC: *The Mechanical Properties of the Maturing Medial Collateral Ligament*. University of Calgary, 1988, PhD thesis.
37. Matyas JR: *The Structure and Function of the Insertions of the Rabbit Medial Collateral Ligament*. University of Calgary, 1990, PhD thesis.
38. Ozaki Y, Mizuno A: Molecular aging of lens crystallins and the life expectancy of the animal: Age-related protein structural changes studied in situ by Raman spectroscopy. *Biochim Biophys Acta* 1992;1121:245-251.
39. Walsh S, Frank C, Hart D: Immobilization alters cell metabolism in an immature ligament. *Clin Orthop* 1992;277:277-288.
40. Walsh S: *Immobilization Affects Growing Ligaments*. University of Calgary, 1989, MSc thesis.
41. Weir TMB: *Recovery of the MCL After Immobilization*. University of Calgary, 1992, MSc thesis.
42. Hansson HA, Engström AM, Holm S, et al: Somatomedin C immunoreactivity in the Achilles tendon varies in a dynamic manner with the mechanical load. *Acta Physiol Scand* 1988;134:199-208.

Future Directions

Perform epidemiological studies to determine the incidence and severity of inflammation and/or failure of these dense regular connective tissues with age.

Although clinically significant age-dependent changes in tendon and ligament occur frequently, there are no data at present that quantify the incidence and severity of these changes. These studies, which would include both midsubstance and junctional areas, should also determine the interaction of soft tissues and focus on changes in contiguous tissue, such as bone and cartilage, that occur with age.

Determine the interaction between medical factors, such as diabetes, obesity, hypertension, and smoking, and environmental factors, such as the frequency, intensity, mode, and duration of various activities, and aging.

Such studies would define more clearly the interrelationship between medical and environmental factors and aging.

Develop clinical models that allow for invasive studies to correlate anatomic and biochemical changes with aging.

No relevant clinical models currently exist for the invasive study of aging effect on dense regular connective tissue. This complex area has numerous confounding variables, and appropriate models should be sought from cadavers with no apparent preexisting disease.

Identify the age-related responses of tendon, ligament, and capsule to direct injury to soft tissue and to secondary changes that occur as a result of injury, disease, and treatment of adjacent tissues. Study the clinical effects of training, exercise, and rest periods and investigate the possible use of growth hormones and growth factors to delay aging.

The specific response of dense regular connective tissue to injury, repair, and rehabilitation as a function of age is not well understood, nor are the implications of the use of growth hormones and growth factors in older adults.

Develop specific noninvasive quantitative methodologies to study age-related changes in soft tissues (eg, biomechanical tests for joint laxity measurements, metabolic methodologies such as PET scans, MRI spectroscopy, etc).

Few noninvasive methods currently exist to assess quantitatively the tendinous, ligamentous, and capsular changes that occur with aging.

Determine the systemic and innervation effects of aging on tendon, ligament, and capsule.

Little is known about the age-related responses of tendon, ligament, and capsule to degradative changes and functional failure in blood flow and innervation. Degeneration is likely to occur with advancement of age as a result of lack of nutrition, accumulation of degradative products, or loss of proprioception. Studies are needed to define these changes. Currently available noninvasive techniques, such as biomechanical and metabolic methodologies, should be used to quantify these changes. Other technologic advances need to be made to enhance these studies.

Study the mechanical properties of tendons from various sites during the process of aging.

Studies have shown that tendon failure and/or dysfunction occur with advancing age, and some tendon sites (eg, Achilles tendon, posterior tibial tendon, and biceps tendon) are particularly prone to failure in the elderly. New studies are needed to address the following questions: (1) Is failure at these tendon sites caused by age-related weakness? (2) Do some tendons maintain their strength during aging?

Study elastic and viscoelastic properties in addition to structural and ultrastructural changes in tendon during aging to find out whether (1) changes in tendon properties and collagen fibril diameters occur with aging; and (2) if these changes translate to clinical impairment of tendon-muscle function or joint movement.

Suitable animal models must be found to perform biochemical and morphologic studies in an effort to correlate biomechanical properties to tissue structure and composition, and to collagen fibril formation and degradation.

Examine the adaptation of tendons to static and repetitive compressive loading environments.

A number of studies have implicated compressive forces around bony prominences or pulleys as responsible for alterations in the composition and mechanical properties of tendons. There is a need to determine: (1) the effect of compressive stresses on tendons; (2) if there are particular adaptive responses (synthesis/degradation) that tendons produce under compression; (3) how age is related to these adaptive responses to compression; and (4) how these adaptive responses impact on tendon mechanical performance and potential failure.

Study age-related biologic response and functional recovery of tendons subjected to static and repetitive subfailure loads.

Many of the tendon disorders experienced by the elderly are characterized by dysfunction and not complete rupture. This implies that subfailure loading may be responsible for much of the clinical impairment seen with aging tendons. Therefore, the following studies are recommended: (1) in vitro studies in which the specific sites of microscopic damage within the tendon substance are documented as tendons undergo tensile and abrasive loading; and (2) in vivo studies using suitable models for tension and abrasion to characterize the age-related reparative processes.

Study the ways in which the cumulative effects of medical disorders and aging produce tendon pathology.

Associations between disease conditions and changes in tendon that lead to dysfunction or failure have been investigated. For example, diabetes appears to have a strong association with the occurrence of tendon pathology with age. Therefore, animal studies should focus on such models as the streptozocin-induced diabetic rat or rabbit. Comparative studies of diabetic and nondiabetic animals would examine tendon composition and mechanical properties with aging. If composition and mechanical property assessment demonstrate that diabetic tendon exhibits accelerated aging, diabetic animals may be more routinely used for models of tendon aging and disease treatment.

Perform basic studies on age-related changes of morphologic, structural, and functional characteristics of ligaments

and capsule in the absence of disease processes.

Current animal models have been used to investigate growth, maturation, and early adulthood, and more studies on the effects of advanced age are required. Such studies should also be extended to human tissues.

Studies are needed to determine how structures about the knee, other than the anterior cruciate ligament (ACL), respond secondarily to: (1) reduced stiffness where significant (ie, increased knee laxity will lead to increased stresses on the meniscus and other ligaments and capsule); and (2) an ACL-deficient knee in which all loads on the ACL must be taken up by other soft tissues.

It is known that the anterior cruciate ligament (ACL) and its surgical replacement lose significant stiffness and strength with age. In vitro studies are required to understand the age-related reductions in structural and material properties of ACL. In vivo studies are also needed to document changes in functional loading on this structure with age. Modern technologies (eg, electromechanical linkage systems, implantable transducers, and robotic manipulators) are presently available to measure these quantities. Obstacles may include the development of minimally-invasive transducers to quantify force and quantitative and repetitive methods for assessing increased knee laxity in patients.

Studies should be extended to joints where clinical problems have been documented.

It is known that certain ligaments are predisposed to age-related changes, which result in stiffness, joint loss of motion, and pain, that will require physical therapy and clinical intervention. These include ankle ligaments (eg, lateral anterior talofibular ligament) that show frequent sprains that may increase with age, spinal ligaments leading to degenerative processes, digital ligaments at the base of the thumb (carpal metacarpal joint) leading to pain and instability, and capsular and ligament changes at the wrist and shoulder. Longitudinal studies need to be developed in patients to address these clinical problems with respect to age.

Studies are needed to determine why elderly people are slow to recover from injury and immobility and how this delayed recovery affects ligaments and capsule.

Morphologic, biochemical, and biomechanical studies could be used to answer these questions. Potential pharmacologic and other biomechanical regiments, such as growth hormones, cytokines, and so forth, should be explored to accelerate this recovery process.

Quantitatively evaluate the relative contributions of lysyl oxidase-mediated cross-links and nonenzymic Maillard reaction products to the mechanical properties of tendons and ligaments in aging normal subjects and animals.

Tissue from uremic and diabetic patients could be examined as a form of control.

Biomechanical tests indicate that the tensile stiffness of tendons, ligaments, and other dense connective tissues increases with development. During aging in the adult, stiffness tends to remain the same or to decline. These tissues also experience atrophy and rapidly lose normal tensile properties when deprived of ambient stress. When stress levels are restored, however, it is not known whether functional properties of aged tissues can return to normal.

Altered cross-linking of collagen (between molecules in fibrils) or altered interactions between collagen fibrils and interfibrillar macromolecules may contribute to these material property changes.

Identify animal models in which to study age-related changes in tendon and ligament metabolism, including their ability to repair and turn over the extracellular matrix.

In studies of large animals, it will be important to be able to determine the age of selected animals accurately.

Identify macromolecules other than type I collagen that change in abundance and localization with age.

These include but are not limited to other collagen types (II, III, V, XII, XIV, etc), fibronectin, fibromodulin, decorin, biglycan, and other as yet uncharacterized proteins. Although present in small amounts relative to type I collagen, such macromolecules may play important roles in organizing the tissue architecture in modifying the viscoelastic properties of the tissue.

Measure characteristics of cells in very old tendons and ligaments.

A deficiency in aerobic metabolism, for example, could severely impact tissue maintenance and healing capacity. Using both tissue explants and cells in culture, describe levels of energy metabolism, cell mobility, proliferative capacity, matrix synthesis, and catabolism. Furthermore, fibroblast mediated remodeling secondary to stress and motion deprivation as well as subsequent return to motion need to be studied as a function of age.

Define the epidemiology of injuries and disorders that occur at insertions as a function of aging.

There is evidence that with increasing age, tendon insertions to muscle or tendon and ligament insertions to bone can become sites of symptomatic change. The epidemiology of these conditions as a function of age, however, has not been defined. Studies of the incidence, prevalence, and natural histories of such insertional disorders with and without various treatments (exercise, stretching, etc) need to be defined. In particular, clinical studies are needed to define the association between insertional changes (eg, calcification, inflammation, osteophyte formation) and other diseases (diabetes, uremia, etc).

Define anatomic changes at clinically relevant insertion sites of tendons (to muscles or bones) and ligaments (to bones) with age.

Based on clinical studies aimed at defining which tendon or ligament junctions develop pathologic changes with aging, studies are required to define anatomic changes of those insertion sites in normal versus pathologic situations over time. Hypotheses regarding causes of pathologic changes could then be defined and tested clinically (eg, changing directions of force application, local microdamage, etc).

Develop animal models to study clinically relevant age-related tendon and ligament junction problems.

Based on clinical and cadaveric definitions of which junction sites develop pathology with aging, studies are required to define and validate animal models that can simulate those conditions. For example, models that would produce changes at insertions, such as inflammation, calcification, or osteophytes in aging animals, would be needed to study the natural histories of these changes and the effects of various modalities (immobilization, exercise, stretching, surgical treatments, etc) on modifying these natural histories.

Define the potential adaptability or healing of tendon and ligament junctions in aging.

It is likely that tendons and ligament junctions to bones and/or muscle are more active metabolic sites than the remainder of the structure. These sites may serve to control the length of ligament and tendon substance or heal microdamage secondary to excessive stresses or strains. Failure of these insertion sites to adapt to these influences may produce pathologic changes locally or in related joint structures. Studies should focus on understanding biological mechanisms which facilitate adaption or block it.

Section Five

Intervertebral Disk

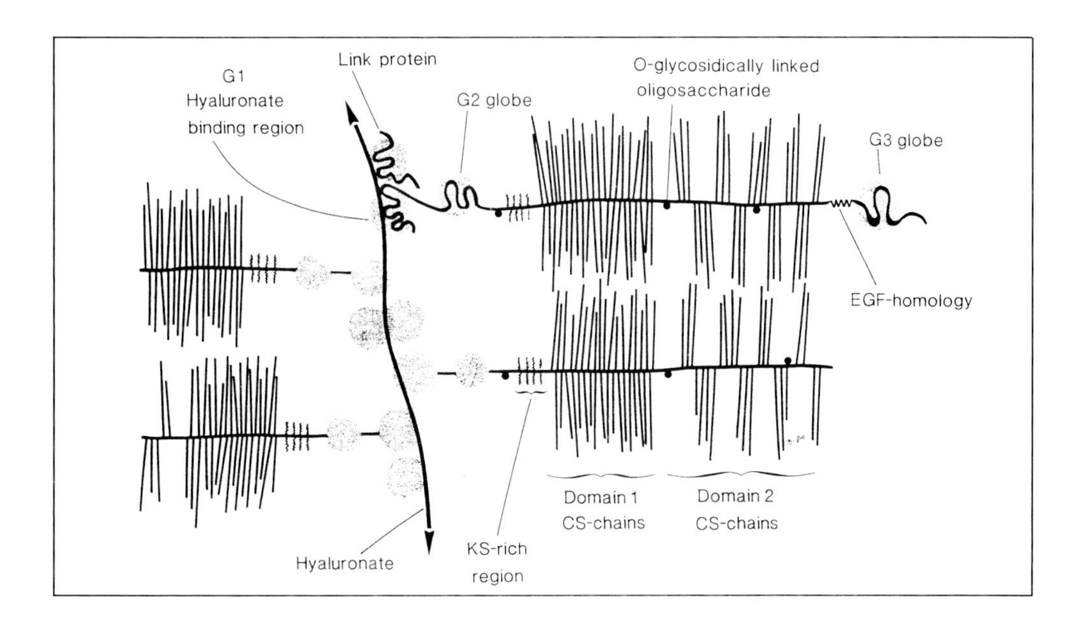

Section Leader

Ted Oegema, PhD

Section Contributors

Gunnar B.J. Andersson, MD, PhD
Dick Heinegård, MD, PhD
Pilar Lorenzo
Richard H. Pearce, PhD
Malcolm H. Pope, Dr. Med. Sc., PhD
Finn P. Reinholt, MD, PhD
Yngve Sommarin, PhD
Jill P.G. Urban, PhD

Overview

The mature intervertebral disk is the largest avascular structure in the body. Blood vessels are present only in the outer annulus fibrosus; the inner annulus fibrosus and the nucleus pulposus are bounded by calcified cartilaginous endplates. Thus, the disk depends on diffusion over long distances for nutrition and removal of waste, including matrix breakdown products. This environment may make the disk unusually vulnerable to age-related processes.

Biomechanically, the intervertebral disk with the two adjacent facet joints forms a three joint complex; that is, the motion segment of the spine that allows bending and twisting. As a result, aging changes in the disk that alter function can affect not only the disk but also the adjacent facets, possibly leading to osteoarthritis. Additionally, because motions are coupled through the vertebral bodies, mechanics of other levels can also be changed.

Grossly, there are dramatic morphologic, biomechanical, and biologic changes of the disk with age, but the exact nature and timing of the changes are unknown. More critically, the relationships of functional biomechanical alterations to the biologic changes are unexplored. Even less clear is how aging changes lead to overt clinical disease. Studies reviewed in the following five chapters clearly show several things. First, the intervertebral disk is unique, although it may have similarities to other tissue such as meniscus and hyaline cartilage, which may be useful for designing disk studies. As a result, the annulus fibrosus and nucleus pulposus have their own biology and biomechanics, and they may have their own mechanisms of aging. Because there are clear differences in the effects of aging on all levels of function and structure, studies must be appropriate for the structure and function affected. These levels range from the molecular to the most complex activities of daily living in the whole organism.

The section on the intervertebral disk is organized into topics of clinical studies, pathology, biomechanics, matrix, and cells. This division is purely for convenience, because many areas are clearly interrelated. These same areas were used as subgroups for discussion of future research. While there has been an attempt to cover the topics comprehensively, the references are selected to illustrate specific points, so the reader is referred to additional cited reviews for a more thorough discussion.

Aging plays a major role in disk degeneration, but key data for understanding these processes are frequently unavailable or were collected with inadequate documentation of the additional details needed to allow interpretation in light of subsequent studies. Unfortunately, few mechanisms of aging in the disk are understood in enough detail for the models to be predictive.

In the chapter on clinical aspects of aging, disk degeneration is shown to be so closely related to aging that it can be viewed as a consequence of aging processes. However, the rate of disk degeneration depends on other factors, such as sex, weight, work history, smoking, and, possibly, genetic background. New imaging techniques, such as magnetic resonance imaging, provide methods for noninvasively evaluating the extent of degeneration, but a need for alternatives is clear. As pointed out in "The Pathology of the Disk," decreasing proteoglycan content in the nucleus is an early sign of degeneration, which precedes the mechanical breakdown of the nucleus pulposus and the annulus fibrosus. Many of the biochemical processes that may play a role are pathways seen in other tissues, but only hints of the importance of different mechanisms have been elucidated. "Aging and Extracellular Matrix" contains a description of newly discovered matrix proteins, including several molecules that may be uniquely located within different connective tissue structures. Possible mechanisms for molecular organization of the disk's extracellular matrix are illustrated with the recently described interactions of collagen with two fibril-associated proteoglycans: decorin and fibromodulin. "Mechanical Changes That Occur With Aging" contains new studies that support the correlation of altered microstructure with mechanical changes in the disk. Theoretical models that relate changes in properties with structure are reviewed. Because micromechanical models of matrix for the disk are similar to those used in cartilage, tendon, and ligament, they are covered in more detail in those sections. The biologic response of the disk cells to the changes in the environment is covered in "The Effect of Physical Factors on Disk Cell Metabolism." Synthesis rates of macromolecules in the disk are responsive to mechanical forces but they can be moderated by available nutrition. A detailed model of the microenvironment of the disk that regulates matrix pore size and nutrient access, as well as cellular responses, is presented. Additional aspects of this model are presented in the articular cartilage section. The boundary conditions for the nutrition of the nucleus pulposus are controlled by the diffusion of nutrients from the nearest blood supply. This distance is determined by the properties of the endplates and subchondral bone. Availability of adequate nutrition is a major determinate of disk health and of its ability to repair its matrix.

The future directions for research are painted in broad strokes and identify questions that need to be addressed. The study group that formulated future research had extensive discussions on many of the problems that are unique to the disk. The disk study group also concurred with global issues raised by all the study groups. These issues included the lack of comprehensive longitudinal studies; the lack of validated animal, human, and theoretical models; the lack of detailed natural history of the aging process; and the lack of noninvasive quantitative methods. Also identified were several issues unique to the disk. One issue that stood out was the need to determine the role of the nucleus pulposus in maintaining disk homeostasis. The second unique issue was testing the hypothesis that the nucleus pulposus can serve as a "garbage can" for collecting degraded matrix macromolecules. Because reports indicate that fragments of many matrix molecules have biologic activities that differ from those of the parent molecule, the detention of these fragments could alter the ability of the cells in the disk to respond to changes. It is hoped that these brief summaries have captured the essence of the current state of the art and that the future directions will encourage both new and experienced investigators to face the challenges and rewards of studying this key component of the spine.

Chapter 25

Intervertebral Disk: Clinical Aspects

Gunnar B. J. Andersson, MD, PhD

Introduction

The intervertebral disk is perhaps that part of the spine that has received the most clinical attention. The one obvious reason for this is the historic paper by Mixter and Barr,[1] in which the importance of a disk herniation was first described—that structural failure can result in specific recognizable pathology. Ever since that time, disk herniations have been considered to be important causes of low back pain and, particularly, sciatica. At the present time, some 280,000 disk hernia operations are performed annually in the United States. A second reason for the interest in the intervertebral disk is concern about age-related degenerative changes, which in more advanced stages are easily identifiable on plain radiographs, and can be seen on almost every magnetic resonance image (MRI). These changes are not necessarily accompanied by low back pain, nor does disk degeneration have to be present in patients with pain. Nevertheless, the association between pain and degeneration has remained obvious in the mind of the public, and the fact that names such as degenerative disk disease and internal disk disruption are frequently used in the medical community, lends some clinical significance to the process.

The purposes of this paper are to: (1) review the epidemiology of disk degeneration, (2) discuss factors influencing its prevalence and severity, (3) discuss the clinical importance of disk degeneration including its role in disk herniations, and (4) identify future research directions that will enhance our understanding of the process and its relationship to low back pain and other spinal conditions.

Lumbar Disk Degeneration

Disk degeneration has been studied using radiographs, autopsy material, diskograms, and, more recently, magnetic resonance imaging techniques.[2] The problem of defining disk degeneration from radiographs is well known. Narrowing of the disk space is the usual indicator, but narrowing, when measurable, is already a sign of quite advanced disk degeneration. The fact that autopsies are typically performed on elderly individuals introduces the problem of age and disease bias. Because diskograms are usually reserved for populations with back symptoms, they may not be representative of the general population. MRI can be used for both case-control and prospective studies, but availability and cost have so far been prohibitive. These limitations should be remembered when interpreting the epidemiologic data presented.

Prevalence of Disk Degeneration

Disk degeneration occurs in all people at some time in life, frequently at relatively early age. In early cadaver studies, Schmorl and Junghanns[3] found significant disk degeneration in most spines already in the fourth decade and not infrequently as early as in the third. Diffuse degeneration was found in 100% of autopsies in persons 90 years of age. These observations were confirmed by Coventry and associates[4-6] and Hirsch and Schajowicz[7] (Fig. 1). Vernon-Roberts and Pirie[8] concluded that the degenerative process started early and progressed rapidly, so that by age 50 it was present in all spines studied. Later on, Vernon-Roberts[9,10] reported on an enlarged sample and stated that disk degeneration always exists in the lumbar spine at age 50.

There was a clear association between increasing age and progressive degeneration of the spinal structures in all of these autopsy studies. The age influence is obvious in radiographic studies as well. Kellgren and Lawrence[11] reported that 83% of 55- to 64-year-old men and 72% of women of the same age had radiographic changes of disk degeneration. This study was performed in northern England. In another study, of a Jamaican rural population, Bremner and associates[12] compared 260 males and 268 females, who were 35 to 64 years old, to an age-matched group of 225 males and 240 females studied earlier in the United Kingdom. The prevalence of disk degeneration was similar in the two populations with 65% of males and 56% of females having grade 2 to 4 degeneration on a scale that ranged from 0 = none to 4 = severe. The Jamaicans more often exhibited extensive disk changes involving more than three disks. A large number of additional radiographic studies have confirmed the influence of age on the prevalence of disk degeneration.[13-21]

Disk degeneration (at least in more advanced stages) appears to be more common in men than in women, and also to occur at an earlier age in men.[11-14,16,22,23] Miller and associates[24] reviewed published reports of cadaver disk material, which had been used for mechanical testing. They found that by age 50, 97% of all lumbar disks were reported to have degenerative changes. Male lumbar disks were found to start degenerating in the second decade of life, which was significantly earlier than the female disks, and male disks were more degenerated than age-matched female disks at any age

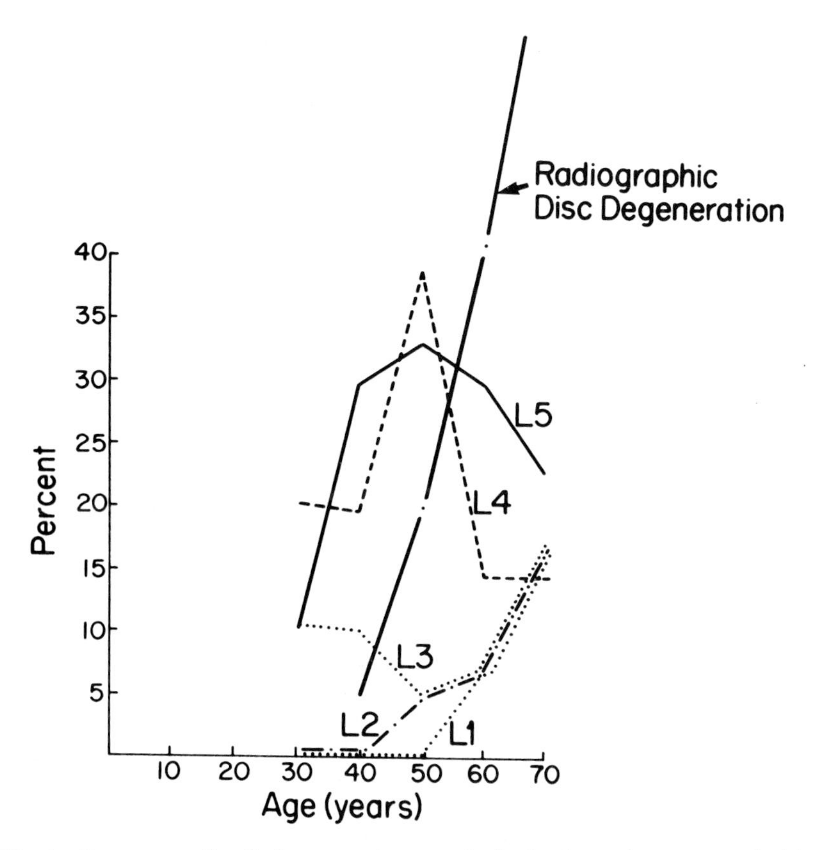

Fig. 1 *Frequency of radiating posterior tears in the lumbar spine compared with radiographic disk degeneration and prevalence of low back pain. (Adapted with permission from Hirsch C, Schajowicz F: Studies on structural changes in the lumbar annulus fibrosus.* Acta Orthop Scand *1953;22:184-231.)*

beyond age ten. Powell and associates,[25] on the other hand, studied 302 16- to 80-year-old women using magnetic resonance techniques. Disk degeneration was present by age 30 in over one third of the women and increased linearly with increasing age (Fig. 2). Most reports conclude that disk degeneration is more severe and starts earlier at the L5 and L4 levels than at the L3 level, and that L2 and L1 disks are less frequently degenerated.[11,13,14,22,25-32]

Other Factors Influencing Disk Degeneration

There are few reports on the influence of other individual factors, such as height and weight. Obesity was found to increase the prevalence of disk degeneration significantly in the study by Magora and Schwartz,[33] but not in the one by Kellgren and Lawrence.[26]

Battie and associates[34] report that greater disk degeneration was present in lumbar intervertebral disks of smokers than of non-

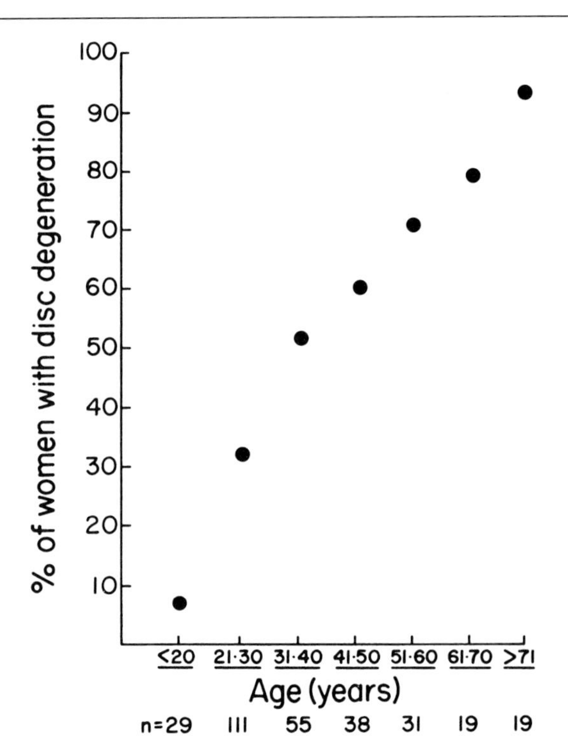

Fig. 2 *Prevalence of lumbar disk degeneration in asymptomatic women as determined by MRI. (Adapted with permission from Powell MC, Wilson M, Szypryt P, et al: Prevalence of lumbar disc degeneration observed by magnetic resonance in symptomless women.* Lancet *1986;2:1366-1367.)*

smokers. Their study, which employed MRI, was performed on identical twins who were highly discordant for cigarette smoking. The effect was present throughout the lumbar spine indicating a systemic mechanism.

Effect of Work on Disk Degeneration

Disk degeneration appears to be more frequent among workers in physically heavy jobs than in those performing light jobs.[3,11,13,14,16,22,31,35,36] Wikström and associates,[36] who reviewed the epidemiologic evidence for the effect of work on degenerative changes in the spine, concluded that work was indeed a factor influencing prevalence. Hult[13,14] reported that both low back complaints and disk degeneration were more frequent among workers in physically heavy jobs than in those in physically light jobs. In workers with heavy jobs in whom the healthy worker effect was unlikely because of a lack of job alternatives, the rate of disk degeneration was 78%, while 65% of workers with heavy work who worked in communities where other types of jobs were available had disk degeneration. By comparison, 47% of light workers had radiographically detectable degeneration. Kellgren and Lawrence[11] report that severe disk degeneration was present in 43% of 21- to 50-year-old miners compared to 7% in a group of office workers. The difference was largest in workers between 21 and 40 years of age but was consistent through all age groups. Lawrence and associates[37] found a three times higher rate in foundry workers compared to controls performing lighter work. In another study by Mach and associates,[38] lumbar disk degeneration was found in 55% of stevedores compared to 27% of an age-matched control group. Wiikeri and associates[20] classified radiographs obtained from concrete reinforcement workers into four grades of disk degeneration, from none to severe. The degree of degeneration was strongly related to age, but also to a history of back pain and sciatica and to the length of work exposure.

All studies have not reached the same conclusion, however. Caplan and associates[35] found disk degeneration in miners to be unrelated to age and duration of work experience, but, they did find it to be related to a previous work injury. Osteophytic changes, on the other hand, were related to both age and work duration, but not to a previous injury. Magora and Schwartz[33] found no clear relationship between occupation and either the occurrence of degenerative changes of the disk or of the apophyseal joints. Evans and associates[39] used MRI techniques to compare 38 ambulating and 21 sedentary employees of a company in the United States.

Disk degeneration was found to be significantly more frequent among sedentary females than among ambulating; in the men there was no difference between the groups. This study is small, but it points to an interesting new possibility of studying the influence of work on disk degeneration. Riihimäki[40] compared prevalence rates of disk degeneration in concrete reinforcement workers and house painters. Disk space narrowing occurred ten years earlier, and spondylosis five years earlier in the concrete workers. The risk ratio for disk degeneration was 1.8, while for spondylosis, it was 1.6. Earlier back accidents were found to significantly increase the risk of disk degeneration in a univariate analysis, but not in a multivariate.

Videman and associates[41] obtained careful occupational, recreational, and back pain histories from the relatives of 86 individuals who came to autopsy, all diseased before age 65. A history of back injury was related to the occurrence of symmetric disk degeneration, annular ruptures, and vertebral osteophytosis. Symmetric disk degeneration, but not annular ruptures or osteophytosis, was associated with sedentary work, and vertebral osteophytosis with heavy work.

Influence of Spinal Deformities on Disk Degeneration

In a review paper, Wiltse[42] concluded that spondylolisthesis, spinal tropism, and scoliosis contribute to an earlier and more severe development of degenerative disk changes. The influence of tropism has been confirmed by Farfan and associates[43,44] and Noren and associates.[45] Saraste[46] compared a group of patients originally treated for spondylolysis with age-matched controls who did not have lumbar spine disorders. Disk space narrowing of the disk below the spondylolytic defect was significantly more common in the spondylolysis group. These studies indicate a role of local mechanical factors in the development of disk degeneration.

The influence of neural arch defects on the prevalence of disk degeneration was studied with MRI by Szypryt and associates.[47] After the age of 25, the prevalence of disk degeneration rose significantly in patients with spondylolytic defects. In patients younger than 25, however, no such difference was observed (Fig. 3).

Lumbar Spondylosis

The term spondylosis is confusing. It should be understood to refer to vertebral osteophytosis secondary to degenerative disk disease. Osteophytes that occur at the facet joints differ from osteophytes that occur on the vertebral margins adjacent to the disks, because the facet joints are synovial joints.

Because lumbar spondylosis is related to disk degeneration by definition, the epidemiology of spondylosis closely follows that of disk degeneration discussed in the previous section. Thus, the prevalence increases markedly with age. Lumbar spondylosis appears to occur somewhat later than disk degeneration, however, and is usually not seen until age 45.[48] Vernon-Roberts and Pirie[8] report that some degree of osteophyte formation at the peripheral margins of the vertebral bodies was seen in all cases where degenerative changes occurred in the disk. Further, the more severe the degenerative changes in the disk, the more marked the osteophytes.

Effect of Disk Degeneration on the Apophyseal (Facet) Joints

Several studies have sought to clarify the relationship between disk degeneration and facet joint osteoarthritis. Ingelmark and associates[49] found that marked changes in disk structure were always accompanied by significant facet joint osteoarthritis. They suggested that disks degenerated first, and then OA developed in the facet joints. Lewin,[48] on the other hand, concluded that "disk degeneration seemed to be neither the sole nor the dominant factor predisposing to the onset and development of osteoarthritis of the lumbar synovial joints." Both of these studies were based on examination of cadaver materials. Vernon-Roberts and Pirie,[8] after dissecting more than 100 lumbar spines, came to the opposite conclusion. They felt that disk degeneration was the primary event leading to both vertebral osteophyte formation and apophyseal joint changes. This opinion was based on the fact that structural abnormalities in the disks were always accompanied by osteoarthritis in

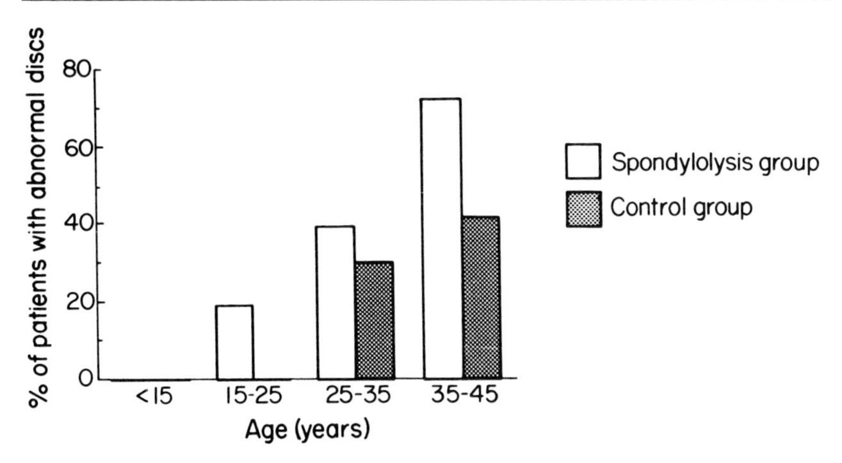

Fig. 3 *Percentage of patients with abnormal disks in the spondylolytic and control group. (Adapted with permission from Szypryt EP, Twining P, Mulholland RC, et al: The prevalence of disc degeneration associated with neural arch defects of the lumbar spine assessed by magnetic resonance imaging.* Spine *1989;14:977-981.)*

the associated facet joints, while osteoarthritis in the facet joints was absent or minimal when the disks were relatively normal or only minor degenerative changes were present. The only exceptions to this pattern were seen in patients who had structural abnormalities, such as kyphosis, scoliosis, and spondylolisthesis. This would agree with the previously discussed radiographic observations. An inverse relationship was found between the severity of osteoarthritis and the preservation of the disk structure.

Butler and associates[28] used magnetic resonance imaging to determine the degree of degeneration of disks and computed tomographic scans of the same subjects to determine the occurrence of facet joint osteoarthritis. Although disk degeneration without facet joint osteoarthritis was quite frequent, all but one level with facet joint degeneration also had disk degeneration. The one exception was in a patient with advanced Paget's disease. Although it is not possible to determine the true sequence of events in a prevalence study, this would indicate that disks degenerate before facets.

Reduction in disk thickness is part of the degenerative process and has important implications for the biomechanics of the affected spinal motion segment. Because the functional and structural integrity of the posterior intervertebral (apophyseal) joints at each level is dependent on a functionally and structurally normal disk at the same level, a reduced disk height has a substantial influence on the initiation and progression of degenerative and arthritic changes in these joints. Biomechanically, it has been shown clearly that change of disk height does influence the load on the facet joints. Dunlop and associates[50] found that the pressure across the facet joints increased significantly with narrowing of the disk space.

The Role of Disk Degeneration in the Development of Disk Herniations

Disks do not fail or herniate, in the clinical sense, from compressive loading alone. This is true even with advanced degeneration. Neither does the combination of flexion and compression seem to result in disk herniations, except in vitro, when the posterior elements are removed and flexion is contrived beyond the normal 6° to 8° up to 15°. Torsion has been found to cause annular tears, in vitro, but those tears are circumferential and do not cause the nucleus to herniate. The only consistent method by which clinically observed herniations have been produced in the laboratory is by flexion and lateral bending of motion segments in which the posterior elements were removed.[51] Moderately degenerated disks at L4 and L5 were those that herniated most consistently. This finding is consistent with the clinical appearance of disk herniations, which most frequently occur at L4 and L5 and affect people when their disks are moderately degenerated, usually between the ages of 35 and 50.

More frequently, disk herniations are probably caused by repetitive loading and crack propagation.[52] These types of failure are difficult to reproduce in the laboratory. Fatigue failure can result from stresses that are relatively low compared to those required for a static failure.

Sciatica caused by disk herniations often resolves with conservative treatment, but it can lead to hospitalization and operation. Epidemiologic data on disk herniations are often based on operative samples. Generalizations from these studies are difficult, because multiple factors other than disease severity influence the decision to perform surgery. Information from these studies does, however, shed some light on the influence of the degenerative process on herniation.

Spangfort[53] reviewed 15,235 operations derived from multiple published reports. 46.9% involved the L5/S1 level, 49.8% the L4/L5 level, and 3.3% were performed at higher levels. In his own case material, which included 2,504 operations, the distribution was 50.5% at the L5/S1 level, 47.4% at L4/L5 and 2.1% at higher lumbar levels. The mean age at surgery for both women and men was just above 40 (Fig. 4). An increase in mean age at operation with the level of herniation in the cranial direction was noted (Figs. 5 and 6). This mimics the development of disk degeneration, which tends to start at L5 and progress to L4 and higher levels, and has led to the hypothesis that there is a period in the process of degeneration when disk herniations can occur. Without degeneration, disk herniations are rare, and with advanced degeneration there is also little risk of herniation.

Heliövaara and associates[54] studied hospital records in a group of 57,000 Finnish women and men who had participated in screening examinations over a period of 11 years (1966-1977). Using various social and health registers, they attempted to identify factors pre-

Fig. 4 *Percent distribution of disk hernia operations by sex and age at operation. (Adapted with permission from Spangfort EV: The lumbar disc herniation: A computer-aided analysis of 2,504 operations.* Acta Orthop Scand Suppl *1972;142:1-95.)*

dicting herniated disks and sciatica. For each case accepted into the study, four control subjects matched for sex, age, and place of residence were selected. A total of 1,537 subjects were hospitalized because of back pain during the 558,074 person years of follow-up. The discharge diagnosis was herniated nucleus pulposus (HNP) in 30%, sciatica in 24%, and other back disease in 46%. Men had a 1.6-fold increased risk of HNP compared to women; other back diseases were equally distributed between sexes. Odds ratios were calculated for different predictive factors.[54-57] The risk of HNP was higher in tall people of both sexes, in obese men, but not women, in industrial workers and motor vehicle drivers in men; in women who did "strenuous work", were smokers, or had multiple pregnancies; and in subjects with symptoms indicating psychological distress. Marital status and leisure-time physical activities were not risk factors. The relative risk of HNP and sciatica was lower in rural than in urban areas. These researchers concluded that the main influences on risk were sex, occupation, workload, and body height; the other factors were of lesser predictive importance.

In the United States, 125 persons per 100,000 population were discharged from acute care hospitals with a first-listed diagnosis of herniated lumbar intervertebral disk in 1983.[58] The overall rates for men and women were 151.7 and 99.9 per 100,000, respectively. The

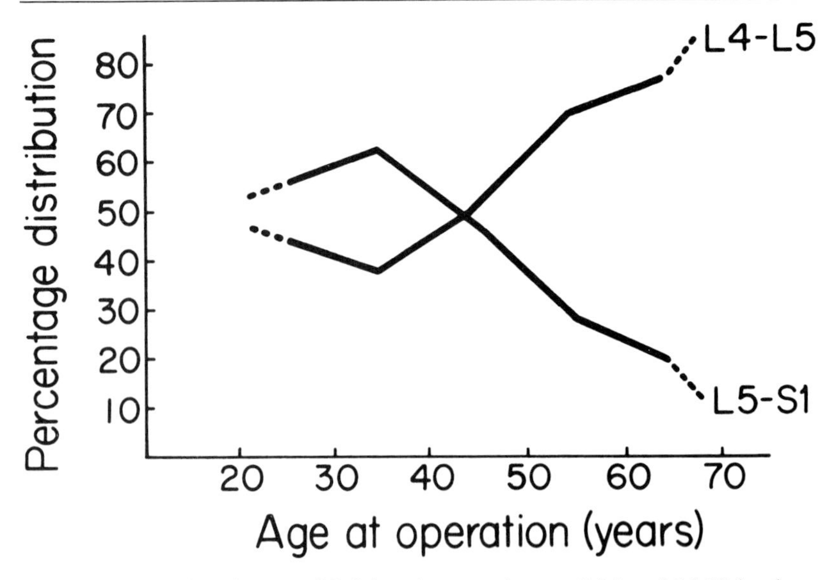

Fig. 5 *Percent distribution of disk hernia operations at L4-5 and L5-S1 levels as a function of age at operation. (Adapted with permission from Spangfort EV: The lumbar disc herniation: A computer-aided analysis of 2,504 operations.* Acta Orthop Scand Suppl *1972;142:1-95.)*

Fig. 6 *Higher level herniations (above L4) as a function of age. (Adapted with permission from Spangfort EV: The lumbar disc herniation: A computer-aided analysis of 2,504 operations.* Acta Orthop Scand Suppl *1972;142:1-95.)*

rate among persons aged 15 to 44 was 157.2; in the 45 to 64 year age group, it was 213.4 per 100,000.

A few studies include not only hospitalized but other patients with clinically determined disk herniations as well. Kelsey[59-63] sampled 20 64-year-old women and men residing in the New Haven (Connecticut) area who had lumbar radiographs taken over a two-year period for suspected herniated nucleus pulposus. A case-control design was used to determine risk factors for disk herniation. Associations were found between disk herniations and sedentary occupations, driving of motor vehicles, chronic cough and chronic bronchitis, lack of physical exercise, participation in baseball, golf, and bowling, suburban residence, and pregnancy. Jobs involving lifting, pushing, and pulling were not found to be associated with increased risk of HNP.[59-61,64,65]

Kelsey and associates[66] later performed another case-control study in Connecticut in 1979-1981 with minor methodologic modifications. The study population was 20 64-year-old women and men. A control group of non-back patients admitted for in-hospital services was matched for sex and age. Frequent lifting of heavy objects and twisting were both found to be significant risk factors.[67] Lifting while twisting the body with the knees almost straight increased the risk to particularly high levels (odd ratio 6:1). The number of hours spent in a motor vehicle and smoking were also associated with an increased risk, while pregnancy, height, weight, and participation in sports were not.[66,67]

Heliövaara[55-57] determined the prevalence rate of sciatica and its impact on society based on a sample of 8,000 persons representative of the Finnish population aged 30 or over. He carefully defined the diagnosis of lumbar disk syndrome based on medical history, symptom history, and a standardized physical examination, and found it present in 5.3% of men and 3.7% of women. In both sexes, the prevalence rates were higher in the 45- to 64-year age group. The prevalence rates for definite herniated disks were 1.9% for men and 1.3% for women.

The Role of Disk Degeneration in Spinal Stenosis

Based on autopsy studies, Kirkaldy-Willis and associates[68] suggested a process referred to as the "degenerative cascade," which uses degenerative changes to explain the development of spinal instability, degenerative spondylolisthesis, and spinal stenosis. Briefly, at some point in the degenerative process, he suggested that the stiffness of the component structures of the spinal motion segment decreases, which results in "segmental instability" (Fig. 7). During this process, back pain may be present continuously or intermittently; sciatica usually is not. Degenerative spondylolisthesis can develop at this stage. Over time, the degenerative process leads to increased stiffness of the motion segment which, coupled with osteophyte formation, leads to restabilization. Unfortunately, spi-

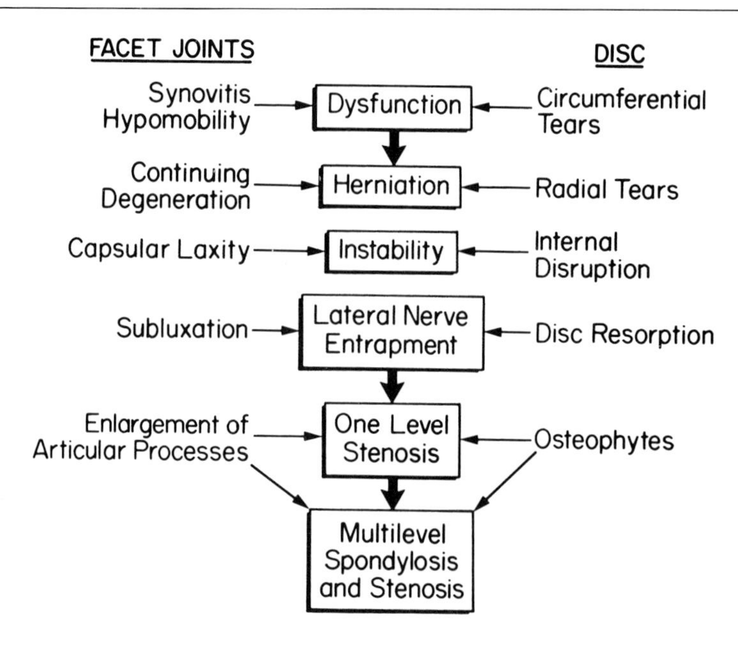

Fig. 7 *The stages of degeneration. (Adapted with permission from Kirkaldy-Willis WH, Wedge JH, Yong-Hing K, et al: Pathology and pathogenesis of lumbar spondylosis and stenosis.* Spine *1978;3:319-328.)*

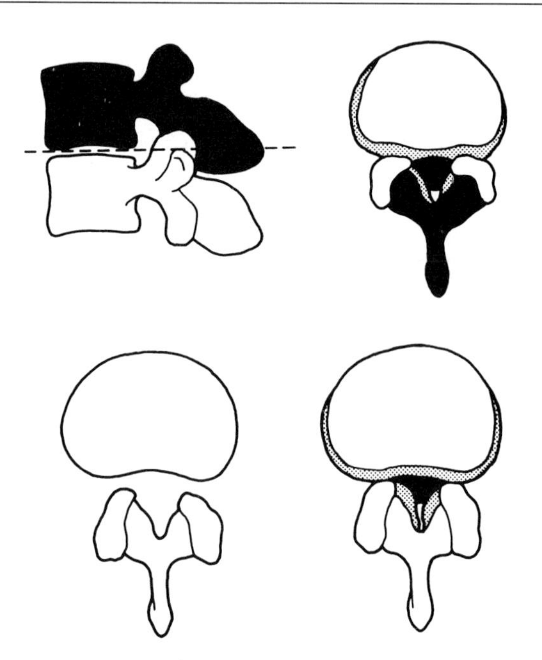

Fig. 8 *The structures surrounding the cauda equina at the disk level. (Adapted with permission from Hansson T, Schönström N: The three joint complex, in Andersson GBJ, McNeill TW (eds):* Lumbar Spinal Stenosis. *St. Louis, MO, CV Mosby Year Book, 1992, chap 10, pp 121-128.)*

nal stenosis may result from this process, as the vertebral canal and root canals are narrowed by both bony and soft tissue.

The risk for degenerative spinal stenosis is greatest at the disk level,[69] where the spinal canal is outlined by the disk, the superior facets from the inferior vertebra, the joint capsules of the facet joints, the ligamentum flavum, and the caudal part of the lamina of the superior vertebra (Fig. 8). Bulging of the disk alone may be sufficient to cause stenosis if the canal is narrowed. Thickening of the ligamentum flavum caused by buckling, and a relative hypertrophy as the space between adjacent lamina is reduced is another cause of narrowing.[70] Facet hypertrophy and osteophytes can also contribute, as can degenerative spondylolisthesis. Farfan and associates[71,72] have suggested that torsional overloads are particularly responsible for spinal stenosis. Their theory is that repetitive torsional overloads result in "crumpling" of the neural arch and asymmetric facet degeneration. A rotational deformity results causing stenosis. Thus, there are several different theories about the development of degenerative spinal stenosis.[73] All theories suggest that disk degeneration is a contributing factor, but it is not clear at present why some individuals develop clinical symptoms of stenosis while others are spared.

Summary

This chapter reviews epidemiologic data on disk aging and degeneration and relates those to structural and pathologic features of spinal conditions. Aging plays a central role in this development, as does mechanical loading, whether caused by sports, work, or deformity. Although aging of intervertebral disks should be considered normal, aging does have direct and indirect clinical effects. Factors that accelerate the time of onset and the progression of disk degeneration are not yet well understood. Nor do we understand the relationship of disk degeneration to pain.

References

1. Mixter WJ, Barr JS: Rupture of the intervertebral disc with involvement of the spinal canal. *N Engl J Med* 1934;211:210-215.
2. Andersson GBJ: The epidemiology of spinal disorders, in Frymoyer JW, Ducker TB, Hadler NM, et al (eds): *The Adult Spine: Principles and Practice*. New York, NY, Raven Press, 1991, vol 1, chap 8, pp. 107-146.
3. Schmorl G, Junghanns H: *The Human Spine in Health and Disease*, 2nd American edition (trans. EF Besemann). New York, NY, Grune and Stratton, 1971.
4. Coventry MB, Ghormley RK, Kernohan JW: The intervertebral disc: Its microscopic anatomy and pathology; anatomy, development, and physiology. Part I. *J Bone Joint Surg* 1945;27:105-112.
5. Coventry MB, Ghormley RK, Kernohan JW: The intervertebral disc: Its microscopic anatomy and pathology; changes in intervertebral disc concomitant with age. *J Bone Joint Surg* 1945;27:233-247.

6. Coventry MB, Ghormley RK, Kernohan JW: The intervertebral disc: Its microscopic anatomy and pathology; pathological changes in the intervertebral disc. *J Bone Joint Surg* 1945;27:460-474.
7. Hirsch C, Schajowicz F: Studies on structural changes in the lumbar annulus fibrosus. *Acta Orthop Scand* 1953;22:184-231.
8. Vernon-Roberts B, Pirie CJ: Degenerative changes in the intervertebral discs of the lumbar spine and their sequelae. *Rheumatol Rehabil* 1977;16:13-21.
9. Vernon-Roberts B: Disc pathology and disease states, in Ghosh P (ed): *The Biology of the Intervertebral Disc*. Boca Raton, FL, CRC Press, 1988, vol 2, chap 11, pp 73-119.
10. Vernon-Roberts B: The normal aging of the spine: Degeneration and arthritis, in Andersson GBJ, McNeill TW (eds): *Lumbar Spinal Stenosis*. St. Louis, MO, Mosby Year Book, 1992, chap 6, pp 57-75.
11. Kellgren JH, Lawrence JS: Rheumatism in miners: X-ray study. *Br J Ind Med* 1952;9:197-207.
12. Bremner JM, Lawrence JS, Miall WE: Degenerative joint disease in a Jamaican rural population. *Ann Rheum Dis* 1968;27:326-332.
13. Hult L: The Munkfors investigation. *Acta Orthop Scand* 1954; 16:S1-S76.
14. Hult L: Cervical, dorsal, and lumbar spinal syndromes. *Acta Orthop Scand* 1954;16:S1-S102.
15. Lawrence JS: Rheumatism in cotton operatives. *Br J Ind Med* 1961;18:270-276.
16. Lawrence JS: Disc degeneration: Its frequency and relationship to symptoms. *Ann Rheum Dis* 1969;28:121-138.
17. Lawrence JS, Bremner JM, Bier F: Osteo-arthrosis: Prevalence in the population and relationship between symptoms and X-ray changes. *Ann Rheum Dis* 1966;25:1-24.
18. Lawrence JS, Molyneux MK, Dingwall-Fordyce I: Rheumatism in foundry workers. *Br J Ind Med* 1966;23:42-52.
19. Torgerson WR, Dotter WE: Comparative roentgenographic study of the asymptomatic and symptomatic lumbar spine. *J Bone Joint Surg* 1976;58A:850-853.
20. Wiikeri M, Nummi J, Riihimäki H, et al: Radiologically detectable lumbar disc degeneration in concrete reinforcement workers. *Scand J Work Environ Health* 1978;4:S47-S53.
21. Riihimäki H, Wickström G, Hänninen K, et al: Predictors of sciatic pain among concrete reinforcement workers and house painters: A five year follow-up. *Scand J Work Environ Health* 1989;15:415-423.
22. Lawrence JS: Rheumatism in coal miners: Occupational factors. *Br J Ind Med* 1955;12:249-261.
23. Lawrence JS: *Rheumatism in Populations*. London, Heinemann Medical, 1977.
24. Miller JA, Schmatz C, Schultz AB: Lumbar disc degeneration: Correlation with age, sex, and spine level in 600 autopsy specimens. *Spine* 1988;13:173-178.
25. Powell MC, Wilson M, Szypryt P, et al: Prevalence of lumbar disc degeneration observed by magnetic resonance in symptomless women. *Lancet* 1986;2:1366-1367.
26. Kellgren JH, Lawrence JS: Osteo-arthrosis and disc degeneration in an urban population. *Ann Rheum Dis* 1958;17:388-397.
27. Nathan H: Osteophytes of the vertebral column: An anatomical study of their development according to age, race, and sex with considerations as to their etiology and significance. *J Bone Joint Surg* 1962;44A:243-268.

28. Butler D, Trafimow JH, Andersson GB, et al: Discs degenerate before facets. *Spine* 1990;15:111-113.
29. Kelsey JL, White AA III: Epidemiology and impact of low back pain. *Spine* 1980;5:133-142.
30. Frymoyer JW, Newberg A, Pope MH, et al: Spine radiographs in patients with low back pain: An epidemiological study in men. *J Bone Joint Surg* 1984;66A:1048-1055.
31. Biering-Sorensen F, Hansen FR, Schroll M, et al: The relation of spinal X-ray to low back pain and physical activity among 60-year-old men and women. *Spine* 1985;10:445-451.
32. Riihimäki H: Back disorders in relation to heavy physical work. Thesis, Institute of Occupational Health, Helsinki, Finland, 1989, pp 1-72.
33. Magora A, Schwartz A: Relation between the low back pain syndrome and X-ray findings. I. Degenerative osteoarthritis. *Scand J Rehabil Med* 1976;8:115-125.
34. Battie MC, Videman T, Gill K, et al: 1991 Volvo Award in clinical sciences: Smoking and lumbar intervertebral disc degeneration: An MRI study of identical twins. *Spine* 1991;16:1015-1021.
35. Caplan PS, Freedman LM, Connelly TP: Degenerative joint disease of the lumbar spine in coal miners: A clinical and X-ray study. *Arthritis Rheum* 1966;9:693-702.
36. Wickström G, Hänninen K, Lehtinen M, et al: Previous back syndromes and present back symptoms in concrete reinforcement workers. *Scand J Work Environ Health* 1978;14:S20-S29.
37. Lawrence JS, Graft R, de Laine VAI: Degenerative joint diseases in random samples and occupational groups, in Kellgren JH, et al (eds): *The Epidemiology of Chronic Rheumatism*. Oxford, England, Blackwell Scientific Publ, 1983, pp 98-119.
38. Mach J, Heitner H, Ziller R: Die Bedeutung der beruflichen Belastung fur die Entstehung degenerativer Wirbelsaulenveranderungen. *Z Hygiene Grenzgebite* 1976;22:352-354.
39. Evans W, Jobe W, Seibert C: A cross-sectional prevalence study of lumbar disc degeneration in a working population. *Spine* 1989;14:60-64.
40. Riihimäki H: Back pain and heavy physical work: A comparative study of concrete reinforcement workers and maintenance house painters. *Br J Ind Med* 1985;42:226-232.
41. Videman T, Nurminen M, Troup JD: 1990 Volvo award in clinical sciences: Lumbar spinal pathology in cadaveric material in relation to history of back pain, occupation, and physical loading. *Spine* 1990;15:728-740.
42. Wiltse LL: The effect of the common anomalies of the lumbar spine upon disc degeneration and low back pain. *Orthop Clin North Am* 1971;2:569-582.
43. Farfan HF, Cossette JW, Robertson GH, et al: The effects of torsion on the lumbar intervertebral joints: The role of torsion in the production of disc degeneration. *J Bone Joint Surg* 1970;52A:468-497.
44. Farfan HF, Huberdeau RM, Dubow HI: Lumbar intervertebral disc degeneration: The influence of geometrical features on the pattern of disc degeneration: A post mortem study. *J Bone Joint Surg* 1972;54A:492-510.
45. Noren R, Trafimow J, Andersson GB, et al: The role of facet joint tropism and facet angle in disc degeneration. *Spine* 1991;16:530-532.
46. Saraste G: Long-term clinical and radiological follow-up of spondylolysis and spondylolisthesis. *J Pediatr Orthop* 1987;7:631-638.

47. Szypryt EP, Twining P, Mulholland RC, et al: The prevalence of disc degeneration associated with neural arch defects of the lumbar spine assessed by magnetic resonance imaging. *Spine* 1989;14:977-981.
48. Lewin T: Osteoarthritis in lumbar synovial joints: A morphologic study. *Acta Orthop Scand* 1964;73:S1-S112.
49. Ingelmark BE, Moller-Christensen V, Brinch O: Spinal joint changes and dental infections. *Acta Anat* 1959;38:S36.
50. Dunlop RB, Adams MA, Hutton WC: Disc space narrowing and the lumbar facet joints. *J Bone Joint Surg* 1984;66B:706-710.
51. Adams MA, Hutton WC: Gradual disc prolapse. *Spine* 1985;10:524-531.
52. Andersson GB: Intervertebral disc, in Wright V, Radin EL (eds): *Mechanics of Human Joints: Physiology, pathophysiology, and treatment*. New York, NY, Marcel Dekker, 1993, pp 293-311.
53. Spangfort EV: The lumbar disc herniation: A computer-aided analysis of 2,504 operations. *Acta Orthop Scand Suppl* 1972;142:1-95.
54. Heliövaara M, Knekt P, Aromaa A: Incidence and risk factors of herniated lumbar intervertebral disc or sciatica leading to hospitalization. *J Chron Dis* 1987;40:251-285.
55. Heliövaara M: Occupation and risk of herniated lumbar intervertebral disc or sciatica leading to hospitalization. *J Chron Dis* 1987;40:259-264.
56. Heliövaara M: Body height, obesity, and risk of herniated lumbar intervertebral disc. *Spine* 1987;12:469-472.
57. Heliövaara M: Epidemiology of sciatica and herniated lumbar intervertebral disc. The Research Institute for Social Security, Helsinki, Finland, pp 1-147.
58. Kozak LJ, Moien M: Detailed diagnoses and surgical procedures for patients discharged from short-stay hospitals. United States, 1983. Vital and Health Statistics, Series B: Data from the National Survey, No 82.
59. Kelsey JL: An epidemiological study of acute herniated lumbar intervertebral discs. *Rheumatol Rehabil* 1975;14:144-159.
60. Kelsey JL: An epidemiological study of the relationship between occupations and acute herniated lumbar intervertebral discs. *Int J Epidemiol* 1975;4:197-205.
61. Kelsey JL: Epidemiology of radiculopathies. *Adv Neurol* 1978; 19:385-398.
62. Kelsey JL: Idiopathic low back pain: Magnitude of the problem, in White AA III, Gordon SL (eds): *American Academy of Orthopaedic Surgeons Symposium on Idiopathic Low Back Pain*. St. Louis, MO, CV Mosby, 1982, chap 1, pp 5-8.
63. Kelsey JL, Ostfeld AM: Demographic characteristics of persons with acute herniated lumbar intervertebral disc. *J Chron Dis* 1975;28:37-50.
64. Kelsey JL, Hardy RJ: Driving of motor vehicles as a risk factor for acute herniated lumbar intervertebral disc. *Am J Epidemiol* 1975;102:63-73.
65. Kelsey JL: Epidemiology of radiculopathies. *Adv Neurol* 1978; 19:385-398.
66. Kelsey JL, Githens PB, O'Connor T, et al: Acute prolapsed lumbar intervertebral disc: An epidemiologic study with special reference to driving automobiles and cigarette smoking. *Spine* 1984;9:608-613.
67. Kelsey JL, Githens PB, White AA III, et al: An epidemiologic study of lifting and twisting on the job and risk for acute prolapsed lumbar intervertebral disc. *J Orthop Res* 1984;2:61-66.
68. Kirkaldy-Willis WH, Wedge JH, Yong-Hing K, et al: Pathology and pathogenesis of lumbar spondylosis and stenosis. *Spine* 1978;3:319-328.

69. Schönström NS, Bolender N-F, Spengler DM: The pathomorphology of spinal stenosis as seen on CT scans of the lumbar spine. *Spine* 1985;10:806-811.
70. Hansson T, Schönström N: The three joint complex, in Andersson GBJ, McNeill TW (eds): *Lumbar Spinal Stenosis*. St. Louis, MO, Mosby Year Book, 1992, chap 10, pp 121-128.
71. Farfan HF, Osteria V, Lamy C: The mechanical etiology of spondylolysis and spondylolisthesis. *Clin Orthop* 1976;117:40-55.
72. Farfan HF: The pathological anatomy of degenerative spondylolisthesis: A cadaver study. *Spine* 1980;5:412-418.
73. Frymoyer JW: Degenerative Spondylolisthesis, in Andersson GBJ, McNeill TW (eds): *Lumbar Spinal Stenosis*. St. Louis, MO, Mosby Year Book, 1992, chap 11, pp 129-152.

Chapter 26

Aging and the Extracellular Matrix

Dick Heinegård, MD, PhD
Pilar Lorenzo
Finn P. Reinholt, MD, PhD
Yngve Sommarin, PhD

Connective tissues are made up of a group of molecules that are of similar character in different tissues, albeit their gene products differ in fine structure. A typical example is the fibril forming collagens. Collagen I, found in most fibrous tissues, provides tensile strength. In cartilage, however, the slightly different collagen II fulfills this function. In their overall composition, the structures of the intervertebral disk are similar to those of cartilage. Many of the molecules found primarily in cartilage, such as aggrecan, collagen II and cartilage oligomeric matrix protein (COMP), are also present in the intervertebral disk. There are, however, other proteins that are found only in the disk structures. The fact that these structures are similar in composition allows us to learn more about the components of the disk from the known components in cartilage. Below is given an account of some of the major components in the tissues.

The Collagen Network

Collagen fibers are a major constituent of connective tissues. These fibers are assembled from collagen molecules, each of which contains three α-chains of some 100 kDa, which together form a typical triple helical, rodlike molecule with a diameter of 1.5 nm and a length of some 300 nm. The molecules are secreted from the cells as procollagen consisting of three pro-α-chains. Once extracellular, the N- and C-terminal extensions are cleaved off, and the collagen molecule is formed. A large number of these molecules combine in a very specific manner to form the aggregates seen as collagen fibers. This process is tightly regulated such that the fibers are arranged in parallel with identical dimensions in a given layer in cornea, in parallel with much larger dimensions in tendon, and in different directions in skin. In articular cartilage the fibers lie parallel with the surface in the superficial layer and perpendicular in the deeper layer, with a change in direction in the intermediate layer. In the annulus of the intervertebral disk, the fibers are arranged in bundles parallel to the circumference of the disk.

It is not clear how the cells accomplish this apparently very tightly regulated assembly, but it is likely that one factor involved is represented by molecules that bind to the surface of the collagen fibers. Examples of such molecules are the group of related small proteoglycans, fibromodulin, decorin, and biglycan,[1] which all bind to collagen,[2] albeit with different specificity. These molecules are depicted schematically in Figure 1. They all contain an anionic domain in their N-terminal parts, either represented by bound glycosaminoglycan chains or dermatan sulfate or chondroitin sulfate depending on the tissue or in the case of fibromodulin represented by a tyrosine sulfate repeat domain with some five to seven residues.[3] Fibromodulin contains additional anionic groups in the form of keratan sulfate chains.[1] A fourth member of this proteoglycan family, lumican,[4] also binds to collagen.[5] Other molecules that bind to collagen fibers are represented by a specific group of collagens, referred to as FACIT collagens.[6] These are collagen IX, bound along fibers of collagen II in cartilage and in the vitreous, and collagens XII and XIV, primarily found in noncartilage tissue and, therefore, mostly found along fibers of collagen I. These collagens have several short triple helical sequences interrupted by nontriple helical domains. Collagen IX, which has been most extensively studied, appears to present the N-terminal globular domain of the α1-chain as a cationic unit protruding from the collagen II fibers. The other collagens showing overall homology may well have similar functions. It is presently not clear whether collagen XIII, which has an overall structure similar to that of the other three known members of this family, also binds to collagen fibers.

Thus, the network of the collagens is a composite structure with the central collagen fiber made up of the fibril-forming collagens. The surface of the fiber is extensively modified by bound molecules represented by the FACIT collagens and proteoglycans, schematically indicated in Figure 2. Also other molecules, such as fibronectin and thrombospondin, may bind collagen. It is not known if these molecules are all found along a given fiber or if they are primarily found in a given territory in a given tissue. One consequence of this coating of the collagen fibers with molecules having different charge may be that the molecules interact, thereby promoting the tensile properties of the collagen fibers.

Major Proteoglycans

A major constituent of cartilage is aggrecan, the large proteoglycan that forms specific aggregates with hyaluronic acid. The central core protein of some 200 kDa is substituted with glycosaminoglycan side chains and oligosaccharides to form a molecule with an M_r of some 3×10^7. This extremely large molecule is secreted from the cells and is then assembled into aggregates that contain one long

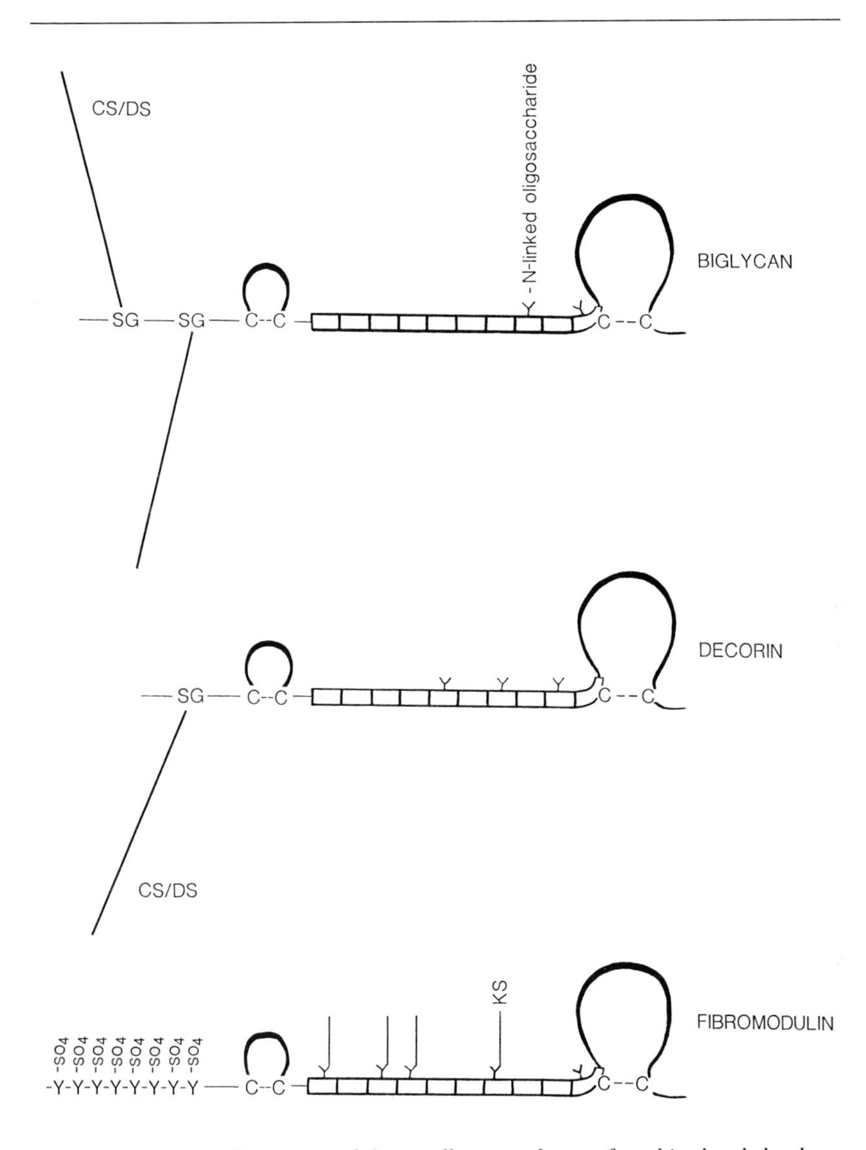

Fig. 1 *Schematic illustration of the small proteoglycans found in the skeletal tissues. CS is chondroitin sulfate, DS is dermatan sulfate, KS is keratan sulfate and Y is a tyrosin residue carrying a sulfate group.*

strand of the hyaluronate, with several (up to 100) bound proteoglycan molecules. The bond between the proteoglycan and the hyaluronate is stabilized by the link protein, which is structurally homologous to the hyaluronate-binding G1-domain of the proteoglycan.[7-9] This stable complex formation is a prerequisite for retaining the proteoglycans in the tissue, where they (by virtue of their extreme number of negatively charged glycosaminoglycans, primarily the some 100 chondroitin sulfate chains) form domains with very

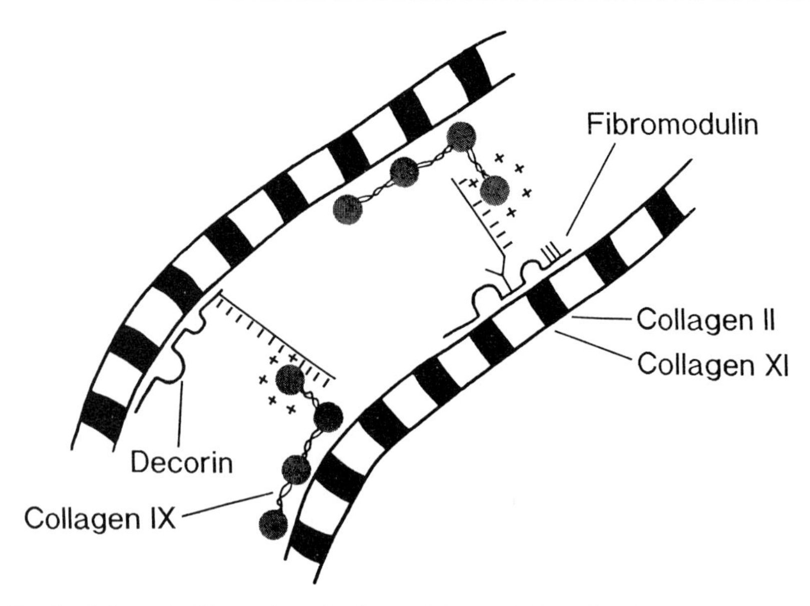

Fig. 2 *Schematic illustration of collagen fibers with bound molecules of collagen IX, decorin and fibromodulin. Note the anionic side chains of dermatan (chondroitin) sulfate on decorin and of keratan sulfate on fibromodulin and the cationic globular domain of collagen IX protruding from the fiber.*

high osmotic pressure and, therefore, swelling pressure.[10] Nevertheless, the tissue is prevented from swelling by the network of the collagen fibers. Thus the proteoglycans are the key contributor to the resilience of the cartilage, a function that also depends on the integrity of the collagen network.

Other domains of aggrecan are a C-terminal G3-globular domain showing homology with lectins,[11-13] EGF-repeat domains that may be spliced out[14] and a G2-globular domain that is close to the hyaluronate-binding G1-domain, but which does not itself bind to hyaluronate. The keratan sulfate-rich domain is interspaced between the G2-domain and the chondroitin sulfate-rich domain indicated in Figure 3.[15] This should contribute stiffness to the molecule by carrying tightly spaced keratan sulfate chains. This domain, as is discussed below, shows major alterations with aging. As is indicated in Figure 3,[14,15] the proteoglycans bound in a given aggregate vary in the length of their protein core, such that some have lost parts of their C-terminal structure.[16] This appears to happen via a slow proteolysis of the proteoglycans in the matrix. The fragment, which usually represents the major part of the proteoglycan, remains bound to the hyaluronate and is retained in the tissue; the C-terminal fragment is released and is lost from the tissue. The regulation of this process is currently unknown.

The portion of the proteoglycan containing G1, G2, and the keratan sulfate-rich region closer to the central hyaluronate chain in the

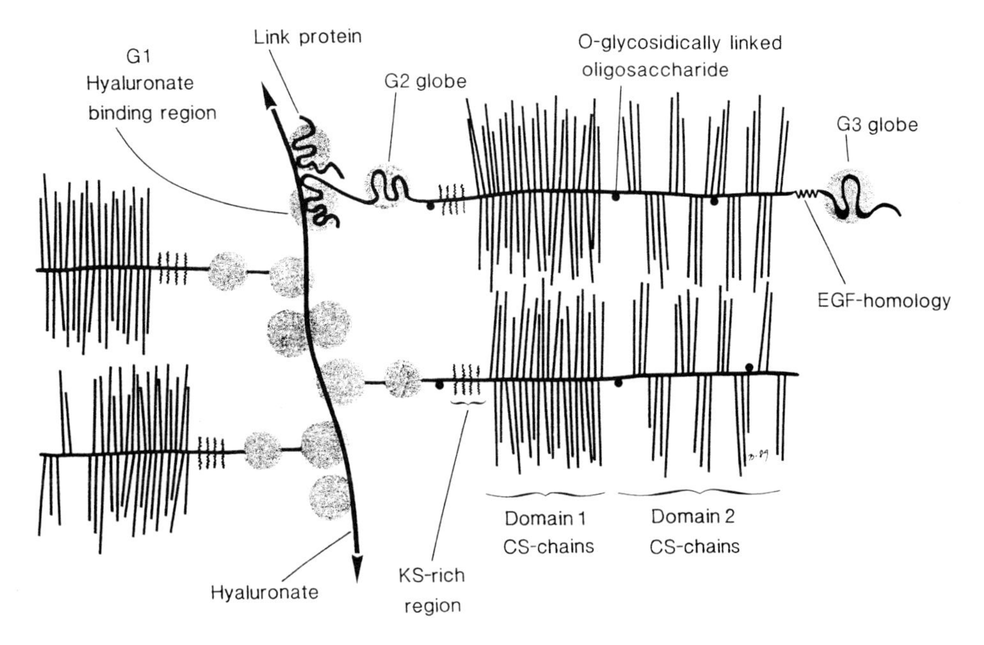

Fig. 3 *Schematic illustration of an aggrecan aggregate with hyaluronate. The partially cleaved proteoglycans, from the C-terminal G3 end, cleaved are indicated. KS is keratan sulfate, CS is chondroitin sulfate. EGF is epidermal growth factor.*

aggregate contains fewer, less bulky substituents. This region may form a space that allows diffusion and interactions of other matrix constituents. Interestingly, the keratan sulfate chains found in this domain vary considerably with age and between tissues.

Aggrecan in Aging

Aggrecan undergoes major alterations with aging. Thus, in fetal life and in the newborn, its content of keratan sulfate is very low.[17] Still, the protein core is the same and the only spliced variants that have been detected are those with differences close to the C-terminus where the EGF-repeats are located.[13,14] Thus, fetal aggrecan also contains the keratan sulfate-rich region with its potential attachment sites.[15] These sites are also substituted in the young, but with the O-glycosidically linked oligosaccharides, which have a structure identical to that of the linkage region of the keratan sulfate.[17] With increasing age, aggrecan contains an increasing number of keratan sulfate chains, and these chains are also longer in older individuals. It appears that the sum of O-glycosidically linked oligosaccharides and keratan sulfate is rather constant with age, although their ratio decreases.[17] It is noteworthy that in osteoarthritis, which is primarily a disease of old age, proteoglycans of the fetal type, which contain less keratan sulfate, are reexpressed.[18] This may follow from the increased synthetic rate of these proteoglycans.

The functional consequences of these alterations in the domain nearest to the hyaluronate central filament in the aggregate is not clear. The physical environment in this domain, however, can be expected to show major alterations, both with regard to charge and with regard to interactions as a consequence of the altered presence of keratan sulfate.

Aggrecan in Joint Disease

Processes in the articular cartilage result in release of those fragments of aggrecan that are not retained as a result of interactions with hyaluronate. Thus, elevated levels of aggrecan in synovial fluid occur early in osteoarthritis, rheumatoid arthritis, and reactive arthritis (secondary to bacterial infection, often with chlamydia, yersinia or salmonella). This occurs before alterations can be seen on arthroscopy or radiograph.[19-21] In late disease, when the organization of the tissue is failing, fragments that are retained in early disease are also released into the synovial fluid. For example, levels of the G1-hyaluronate binding domain in rheumatoid arthritis are elevated with more advanced joint destruction, when the levels of the fragments from the more C-terminal chondroitin sulfate-rich domain have decreased.[22]

Other tissues contain proteoglycans that have a similarly sized core protein, which represents a distinct gene product, one that carries fewer glycosaminoglycan side chain substituents. Versican contains a hyaluronate-binding domain in one end and a lectin-like domain in the other end and carries some 30 glycosaminoglycan side chains.[23-25] The molecule forms link stabilized aggregates with hyaluronate.[24] Fibroblasts apparently also make a second type of proteoglycan. Because it contains the three globular domains, this core protein is similar in appearance to aggrecan, but is apparently distinct, as it has fewer side chain substituents and different epitopes seen in immune reactions.[24,25]

Matrix Proteins

Cartilage

Cartilage Matrix Protein The tissue contains several proteins that are primarily found only in cartilage. The one first described is CMP. This trimeric protein, with three identical subunits of M_r 50,000,[26] has been cloned and sequenced. It contains two EGF-repeat domains and a central portion that is homologous with a number of other connective tissue proteins that are capable of binding to collagen.[27]

The protein is found only in cartilage. It is abundant in tracheal cartilage, but is not present in articular cartilage, or in the structures of the intervertebral disk.[28] Interestingly, the amount of this protein in the tissue increases markedly, from low levels during neonatal life to high concentrations (up to 6% to 7% of the wet weight of the tracheal cartilage) in old age.[29] At this time one pool of the protein resists extraction even with strongly chaotropic solvents like 4M guanidine-HC1. It thus appears as if a portion of the protein becomes more firmly retained in the matrix, perhaps via crosslink formation.

The function of the protein is an enigma. It is present in low concentrations throughout the fetal skeleton, but appears to be down regulated in the developing articular cartilage.

Cartilage Oligomeric Matrix Protein (COMP) Another oligomeric matrix protein is COMP,[30] which on rotary shadowing looks like a bouquet of five tulips.[31] Its five subunits of M_r (each 100,000) are connected in their N-terminal end to form the pentameric 527 kDa (sedimentation equilibrium centrifugation) protein.[30,31] The protein is detected in cartilage only by immunoassay.[30] Interestingly it is homologous with thrombospondin, having an overall some 60% identical amino acids (Oldberg, Antonsson, Lindblom and Heinegård, unpublished data). Although its function is not clear, it may be involved in the repair and remodeling of articular cartilage.

COMP is fragmented and released from articular cartilage in joint disease, and elevated levels can be detected in synovial fluid.[32] Interestingly, an elevated level in serum in early rheumatoid arthritis appears to be prognostic for very aggressive disease, with extensive joint destruction within few years. Patients with serum levels no higher than those of blood donors did not develop joint destruction within the next few years. In both groups of patients, serum levels of aggrecan fragments were the same as those in blood donors (Saxne, personal communication).

The 36 kDa Protein This protein is primarily found in cartilage.[33] It is somewhat basic, and two forms with somewhat different sizes are found in articular cartilage. It is not clear how the two forms differ, but the larger molecule is somewhat more basic.[34] This protein can bind chondrocytes in vitro, but a receptor has not been identified.

Other Matrix Proteins More ubiquitous proteins are fibronectin and thrombospondin, which represent minor components in normal cartilage. Fibronectin, however, is much increased in osteoarthritis.[35]

Normally, the tissue content of plasma proteins is low, but when proteoglycans are being removed from the tissue in a disease process, permeability increases, with an ensuing increase in the content of plasma proteins.

Bone

Bone Sialoprotein (BSP) SP is detected only in bone.[36,37] The protein, which is very acidic,[38] appears as a rod on electron microscopy of rotary shadowed preparations.[37] The acidic residues include a number of oligosaccharides, primarily O-glycosidically linked. Other acidic residues are the some 30% of the serine residues carrying a phosphate group[39] and a number of tyrosine-O-sulfate residues,[40] primarily in the C-terminal part of the molecule. There are also stretches of glutamic acid residues, the longest of which contains 10 consecutive such residues. The protein contains an arginyl glycyl glutamyl or RGD-cell binding sequence,[38] which promotes binding of some cell types in vitro via a vitronectin receptor.[41] BSP is synthesized by osteoblasts, and its most prominent expression is found at cartilage-bone interphases. Interestingly, although the protein is found throughout the osteoid, it is initially laid down to form a distinct interphase between the very early osteoid and the mineralized cartilage.

Future studies on the role of the protein in maintaining this interphase intact should help us understand the important interactions between cartilage and bone. This interphase appears to play a part in joint disease, because the BSP immune reactivity can be found in the synovial fluid in osteoarthritis as well as in rheumatoid arthritis.

In rheumatoid arthritis, the protein is found in the general circulation, which indicates a more generalized involvement of the skeleton. The levels in females aged 50 to 65 years are higher than those of females below the age of 50 and also than in males of the same age group. This may be taken to demonstrate an altered bone metabolism in this group of females, which includes postmenopausal females.[42]

Osteopontin This protein is synthesized by osteoblast and is found in bone. It is, however, also synthesized by a number of other cells, including tubular cells in the kidney. Also, a number of transformed cells start synthesizing the protein.[7] In bone, one function of the protein appears to be related to osteoclastic bone remodeling (Fig. 4). The synthesis of the protein in osteoblasts is markedly stimulated by 1,25-dihydroxyvitamin-D3.[43] Osteopontin, which binds osteoclasts in vitro via an RGD-sequence,[44] is laid down at the mineral interphase as a result of its interaction with hydroxyapatite. Osteoblasts or precursor cells can then bind to the protein. Studies of its immunolocalization at the ultrastructural level show the protein to be much enriched at the clear zone attachment area of osteoclasts, where also the vitronectin receptor for the protein is selectively found at the osteoclast plasma membrane.[45] This is one of the first instances in which a specific locale in the tissue of a cell-binding protein and its receptor have been demonstrated. Supporting information by in situ hybridization shows that the osteopontin is ex-

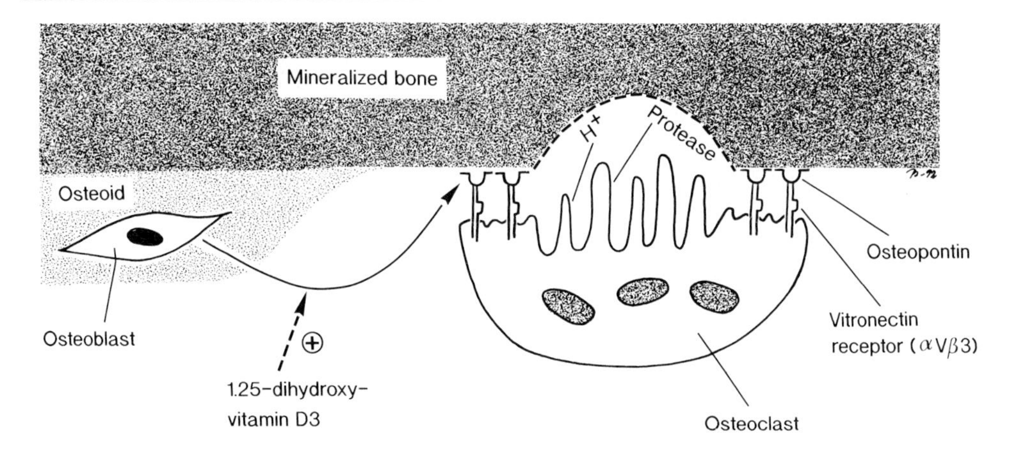

Fig. 4 *Schematic illustration of a bone surface and roles of osteoblasts and osteoclasts in bone resorption.*

pressed in a region of the metaphysis where osteoclastic activity is high and osteoclasts can be seen surrounded by osteoblasts that contain mRNA for the protein.

Osteocalcin This low molecular mass protein is made only by osteoblasts, its synthesis being stimulated by 1,25-dihydroxyvitamin-D3.[46] The protein contains three gamma-carboxyglutamic acid residues, which contribute binding to calcium and probably to hydroxyapatite. The function of this protein has evaded detection despite intense study. It is, however, released from bone in increased amounts in bone disease, where it appears primarily to be an indicator of bone apposition.

Osteonectin (SPARC) A predominant protein in the bone matrix, osteonectin is also present in many other tissues.[7] Osteonectin seems to be involved in the regulation of cell growth.

Other Bone Matrix Proteins Bone also contains proteins like thrombospondin and fibronectin, that are made by the osteoblasts.[7] An acidic protein, BAG-75, is structurally similar to BSP.

A major protein in bone is made in liver but apparently rather selectively bound into bone. This protein is referred to as α2-HS and its role in bone is not clear.

Conclusions

Connective tissues represent a complex association of a relatively small number of constituents. These constituents vary with age, an indicator of the dynamic nature of these tissues, which remodel constantly throughout life. Today, our understanding of the factors that regulate the assembly of this matrix, which occurs entirely in the extracellular space, is very limited. The situation is made more complex by the fact that connective tissues show microheterogeneity in their composition. For example, cartilage matrix closest to the cells is in a very different composition from that found further away. Articular cartilage matrix shows rather prominent differences with distance from the articular surface.

At another level there are clear differences between tissues with overall very similar composition. Thus, tracheal cartilage contains a matrix protein, CMP, which is neither present in articular cartilage nor in intervertebral disk. On the other hand, there are proteins that appear to be detectable only in the intervertebral disk structures.[34] To understand remodeling and repair we need to improve our knowledge of the composition and organization of the matrix in different compartments as well as identify key factors in the regulation of matrix assembly. We would learn more about the individual tissues by including those proteins that are present in different concentrations or that are unique to a given tissue. Ongoing work with characterization of the matrix in the structures of the intervertebral disk will allow more detailed comparisons with weightbearing articular cartilage and nonweightbearing nasal cartilage. This should promote a better understanding of key features in tissue organization and how the specific functional demands correspond to unique tissue features.

Acknowledgments

Grants were obtained from the Swedish Medical Research Council and Axel and Margareta Axson Johnsons stiftelse.

References

1. Oldberg Å, Antonsson P, Lindblom K, et al: A collagen-binding 59-kd protein (fibromodulin) is structurally related to the small interstitial proteoglycans PG-S1 and PG-S2 (decorin). *EMBO J* 1989;8:2601-2604.
2. Hedbom E, Heinegård D: Interaction of a 59-kDa connective tissue matrix protein with collagen I and collagen II. *J Biol Chem* 1989;264:6898-6905.
3. Antonsson P, Heinegård D, Oldberg Å: Posttranslational modifications of fibromodulin. *J Biol Chem* 1991;266:16859-16861.
4. Blochberger TC, Vergnes J-P, Hempel J, et al: cDNA to chick Lumican (corneal keratan sulfate proteoglycan) reveals homology to the small interstitial proteoglycan gene family and expression in muscle and intestine. *J Biol Chem* 1992;267:347-353.
5. Rada JA, Cornuet PK, Hassell JR: Regulation of corneal collagen fibrillogenesis in vitro by corneal proteoglycan (lumican and decorin) core proteins. *Exp Eye Res* 1993;56:635-648.
6. van der Rest M, Garrone R: Collagen family of proteins. *FASEB J* 1991;5:2814-2823.
7. Heinegård D, Oldberg Å: Structure and biology of cartilage and bone matrix noncollagenous macromolecules. *FASEB J* 1989;3:2042-2051.

8. Wight T, Heinegård D, Hascall V: Proteoglycans. Structure and function, in Hay ED (ed): *Cell Biology of Extracellular Matrix*, 2nd edition. New York, Plenum press, 1991, pp 45-78.
9. Hascall V, Heinegård D, Wight T: Proteoglycans. Metabolism and pathology, in Hay ED (ed): *Cell Biology of Extracellular Matrix*, 2nd edition. New York, Plenum press, 1991, pp 149-175.
10. Maroudas A: Physicochemical properties of articular cartilage, in Freeman MAR (ed): *Adult Articular Cartilage*, 2nd edition. Kent, UK, Pitman Medical, 1979, chap 4, pp 215-290.
11. Doege K, Sasaki M, Horigan E, et al: Complete primary structure of the rat cartilage proteoglycan core protein deduced from cDNA clones. *J Biol Chem* 1987;262:17757-17767.
12. Oldberg Å, Antonsson P, Heinegård D: The partial amino acid sequence of bovine cartilage proteoglycan, deduced from a cDNA clone, contains numerous Ser-Gly sequences arranged in homologous repeats. *Biochem J* 1987;243:255-259.
13. Doege KJ, Sasaki M, Kimura T, et al: Complete coding sequence and deduced primary structure of the human cartilage large aggregating proteoglycan, aggrecan. Human-specific repeats, and additional alternatively spliced forms. *J Biol Chem* 1991;266:894-902.
14. Baldwin CT, Reginato AM, Prockop DJ: A new epidermal growth factor-like domain in the human core protein for the large cartilage-specific proteoglycan. Evidence for alternative splicing of the domain. *J Biol Chem* 1989;264:15747-15750.
15. Antonsson P, Heinegård D, Oldberg Å: The keratan sulfate-enriched region of bovine cartilage proteoglycan consists of a consecutively repeated hexapeptide motif. *J Biol Chem* 1989;264:16170-16173.
16. Paulsson M, Mörgelin M, Wiedemann H, et al: Extended and globular protein domains in cartilage proteoglycans. *Biochem J* 1987;245: 763-772.
17. Inerot S, Heinegård D: Bovine tracheal cartilage proteoglycans. Variations in structure and composition with age. *Collagen Rel Res* 1983;3:245-262.
18. Carney SL, Billingham ME, Muir H, et al: Structure of newly synthesised (^{35}S)-proteoglycans and (^{35}S)-proteoglycan turnover products of cartilage explant cultures from dogs with experimental osteoarthritis. *J Orthop Res* 1985;3:104-147.
19. Saxne T, Heinegård D, Wollheim FA, et al: Difference in cartilage proteoglycan level in synovial fluid in early rheumatoid arthritis and reactive arthritis. *The Lancet* 1985;2:127-128.
20. Saxne T, Wollheim FA, Pettersson H, et al: Proteoglycan concentration in synovial fluid: Predictor of future cartilage destruction in rheumatoid arthritis? *Brit Med J (Clin Res)* 1987;295:1447-1448.
21. Lohmander LS, Dahlberg L, Ryd L, et al: Increased levels of proteoglycan fragments in knee joint fluid after injury. *Arthritis Rheum* 1989;32:1434-1442.
22. Saxne T, Heinegård D: Synovial fluid analysis of two groups of proteoglycan epitopes distinguished early and late cartilage lesions. *Arthritis Rheum* 1992;35:385-390.
23. Zimmermann DR, Ruoslahti E: Multiple domains of the large fibroblast proteoglycan, versican. *Embo J* 1989;8:2975-2981.
24. Mörgelin M, Paulsson M, Malmström A, et al: Shared and distinct structural features of interstitial proteoglycans from different bovine tissues revealed by electron microscopy. *J Biol Chem* 1989;264: 12080-12090.

25. Heinegård D, Björne-Persson A, Cöster L, et al: The core proteins of large and small interstitial proteoglycans from various connective tissues form distinct subgroups. *Biochem J* 1985;230:181-194.
26. Paulsson M, Heinegård D: Purification and structural characterization of a cartilage matrix protein. *Biochem J* 1981;197:367-375.
27. Kiss I, Deák F, Holloway RG Jr, et al: Structure of the gene for cartilage matrix protein, a modular protein of the extracellular matrix. Exon/intron organization, unusual splice sites, and relation to alpha chains of beta 2 integrins, von Willebrand factor, complement factors B and C2, and epidermal growth factor. *J Biol Chem* 1089;264:8126-8134.
28. Paulsson M, Heinegård D: Radioimmunoassay of the 148-kilodalton cartilage protein. Distribution of the protein among bovine tissues. *Biochem J* 1982;207:207-213.
29. Paulsson M, Inerot S, Heinegård D: Variation in quantity and extractability of the 148-kilodalton cartilage protein with age. *Biochem J* 1984;221:623-630.
30. Hedbom E, Antonsson P, Hjerpe A, et al: Cartilage matrix proteins: An acidic oligomeric protein (COMP) detected only in cartilage. *J Biol Chem* 1992;267:6132-6136.
31. Mörgelin M, Heinegård D, Engel J, et al: Electron microscopy of native cartilage oligomeric matrix protein purified from the Swarm rat chondrosarcoma reveals a five-armed structure. *J Biol Chem* 1992;267:6137-6141.
32. Saxne T, Heinegård D: Cartilage oligomeric matrix protein: A novel marker of cartilage turnover detectable in synovial fluid and blood. *Brit J Rheumatol* 1992;31:583-591.
33. Larsson T, Sommarin Y, Paulsson M, et al: Cartilage matrix proteins: A basic 36-kDa protein with a restricted distribution to cartilage and bone. *J Biol Chem* 1991;266:20428-20433.
34. Heinegård DK, Pimentel E: Cartilage matrix proteins, in Kuettner KE, Schleyerbach R, Peyron JG, et al (eds): *Articular Cartilage and Osteoarthritis*. New York, NY, Raven Press, 1992, chap 7, pp 95-111.
35. Burton-Wurster, Lust G: Molecular and immunologic differences in canine fibronectins from articular cartilage and plasma. *Arch Biochem Biophys* 1989;269:32-45.
36. Fisher LW, Hawkins GR, Tuross N, et al: Purification and partial characterization of small proteoglycans I and II, bone sialoproteins I and II, osteonectin from the mineral compartment of developing human bone. *J Biol Chem* 1987;262:9702-9708.
37. Franzén A, Heinegård D: Isolation and characterization of two sialoproteins present only in bone calcified matrix. *Biochem J* 1985;232:715-724.
38. Oldberg A, Franzén A, Heinegård D: Cloning and sequence analysis of rat bone sialoprotein (osteopontin) cDNA reveals an Arg-Gly-Asp cell-binding sequence. *Proc Natl Acad Sci USA* 1986;83:8819-8823.
39. Heinegård D, Hultenby K, Oldberg A, et al: Macromolecules in bone matrix. *Connect Tissue Res* 1989;21:3-14.
40. Ecarot-Charrier B, Bouchard F, Delloye C: Bone sialoprotein II synthesized by cultured osteoblasts contains tyrosine sulfate. *J Biol Chem* 1989;264:20049-20053.
41. Oldberg Å, Franzén A, Heinegård D, et al: Identification of a bone sialoprotein receptor in osteosarcoma cells. *J Biol Chem* 1988;263:19433-19436.
42. Saxne T, Heinegård D: Increased synovial fluid content of bone sialoprotein relfects tissue destruction in rheumatoid arthritis. Trans 39th Annual Meeting Orth. Res. Soc., 1993, p. 216.

43. Oldberg Å, Jirskog-Hed B, Axelsson S, et al: Regulation of bone sialoprotein mRNA by steroid hormones. *J Cell Biol* 1989;109: 3183-3186.
44. Flores ME, Norgård M, Heinegård D: RGD-directed attachment of isolated rat osteoclasts to osteopontin, bone sialoprotein and fibronectin. *Exp Cell Res* 1992;201:526-530.
45. Reinholt FP, Hultenby K, Oldberg Å, et al: Osteopontin: A possible anchor of osteoclasts to bone. *Proc Natl Acad Sci USA* 1990;87: 4473-4475.
46. Price P: Vitamin K-dependent bone protein, in Cohn DV, Martin TJ, Meunier PJ (eds): *Calcium Regulation and Bone Metabolism: Basic and Clinical Aspects*. Amsterdam, Elsevier Sci Pub Co, 1987, vol 9, pp 419-425.

Chapter 27

Morphologic and Chemical Aspects of Aging

Richard H. Pearce, PhD

Introduction

The morphology of the human intervertebral disk changes strikingly with age. The functional significance of these changes, which are regarded by some as the most profound of any tissue in the body,[1] and the mechanisms by which they occur are poorly understood.

These morphologic changes have the essential attributes of aging[2]; they are universal, progressive, irreversible, regressive, and occur after maturity. Much of the literature describes changes that occur before maturity, which should be described as maturation or development. Sometimes, morphologic changes occur at a younger age than usual or are confined to one disk or region of the spine; in such instances, they might be considered as degeneration, that is, as a disease process, not aging. Although pathologists have defined the changes that occur with advancing age as degenerative spondylosis,[3] the usual progression of morphologic change does not represent a disease. In the following review, the anatomic and biochemical features of the aging human disk will be considered.

Gross Morphology

Morphologic Features

Schmorl and Junghanns[4] based their classic description of the sequence of morphologic changes associated with the aging of the intervertebral disk on the study of over 10,000 spines postmortem. Subsequent research has been reviewed thoroughly by Vernon-Roberts[3,5,6] and is summarized below.

The human intervertebral disk reaches maturity at 21 to 25 years. At this time the central nucleus pulposus (NP) in about 40% of men and 60% of women is a colorless, translucent gel that swells on sectioning; in the remaining subjects it is a moist, white, homogeneous fibrous pad.[7] The NP is surrounded in the horizontal plane by the annulus fibrosus (AF), which is made of closely apposed concentric rings of coarse collagenous fibers with their longitudinal axes oriented at approximately 30° to the axis of the spine. Alternate layers are oriented at opposing angles to the spinal axis so that, when viewed from the side, a two-dimensional structural meshwork is seen. The posterior AF is thinner and narrower than the anterior and lateral portions. The NP is bounded above and below by the vertebral end-plates (EP), each of which is made up of a layer of compact bone overlaid by a layer of hyaline cartilage approximately 0.6 mm thick, which is adjacent to the NP and to the inner regions of the AF.[8] The EP is thinner centrally than peripherally.[8] The outer fibers of the AF are inserted directly into compact bone.

With the passage of time, the NP is converted to a fibrous pad in all individuals (97% by age 40 years[7]), and the tissue develops a drier rubbery texture with coarse soft fibers. Later, fine crevices appear between the fibers and, subsequently, clefts appear parallel to the EP. These clefts usually are above and below the central NP (48% by age 60 years[7]), but may extend with time to the AF, completely isolating the central NP. In the end, the central NP may disappear completely. The NP, normally colorless or white in the third decade, progressively develops a brown pigmentation with age. These pigments result, at least in part, from a fluorophore (possibly formed by nonspecific glycation) attached to the lysine and/or hydroxylysine residues in the collagen.[9]

After maturity, the spaces between the laminae of the AF become infiltrated with mucinous material. Later, islands of cartilaginous tissue may appear (chondroid degeneration). The boundary between the NP and AF becomes diffuse until the two tissues can be distinguished only with difficulty. Three types of tears, which are more frequent in individuals over 35 years, have been recognized in the AF[10]: (1) rim lesions, tears at the periphery of the AF adjacent to and in the same plane as the EP, (2) concentric tears, circumferential ruptures of the AF between the collagenous lamellae, and (3) radial tears, tears that are roughly perpendicular to the EP and divide the collagenous lamellae. The rim lesions occurred predominantly anteriorly in L1-L2 through L4-L5, but they were equally abundant anteriorly and posteriorly in L5-S1. The observation of vascular ingrowth and granulation tissue in these lesions suggested that they may have been associated with diskogenic pain. Circumferential tears were associated frequently with rim lesions, which may promote circumferential tear development. Radial tears occurred almost exclusively in the posterior annulus and were associated with morphologic changes in the NP.

The cartilaginous EP either becomes calcified or is replaced by bone as the person ages.[11,12] Focal regions may thin after maturity and progress to become full thickness defects. These discontinuities coalesce in the elderly to form EP devoid of cartilage. Sclerosis of the underlying bone is associated frequently with loss of the cartilaginous EP. In all individuals, osteophytes form on the anterior rims of the vertebrae after middle age; posterior osteophytes are found less frequently. NP material prolapses vertically into the vertebral body (Schmorl's nodes) in about three spines in four. Prolapses through the AF posterolaterally into the spinal canal are found postmortem in about one subject in six. Osteoarthritis of the posterior apophyseal joints is also associated with morphologic changes in the disk. Thus, aging is associated with changes in the morphology of all components of the spinal motion segment.

Grading Schemes

Although the morphologic changes described above occur in all individuals, the rate at which they develop differs. Therefore, age alone does not define the state of the disk, and many scientists undertaking research involving the intervertebral disk have adopted grading schemes to categorize the morphology of their specimens.

The most widely used scheme[13] was based on the appearance of a horizontal section of the disk midway between the vertebral bodies. The four categories depended on the fibrous appearance of the NP, the ability to distinguish NP from AF, and the presence of fissures and cavities in both the AF and NP. The distribution of these categories in 600 disks from 273 cadavers was reviewed recently.[7] The mean grade increased steadily with age, as did the proportion of the disks in the higher grades. Women lagged approximately one decade behind men in the appearance of advanced grades. L3-L4 and L4-L5 disks degenerated more frequently than other disks, including L5-S1.

The Nachemson grading scheme[13] is limited, because it does not consider the EP and vertebral body, structures not seen in a horizontal section, in the evaluation of disk morphology, and because of its emphasis on the early stages of degeneration. To overcome these limitations, a scheme based on the appearance of midsagittal sections has been proposed.[14] This scheme has been evaluated for intra- and inter-observer reproducibility, ability to distinguish grades, and applicability to all disks. Its validity was supported by an increase in the average grade with age.

Histologic Features

Sections of intervertebral disk prepared using conventional histologic methods are difficult to interpret because the tissues are relatively poor in cells and the detailed structure of the abundant collagenous fibers is obscured by the amorphous components of the extracellular matrix. The recent use of various forms of electron microscopy has clarified many aspects of the tissue structure and organization. The currently accepted concepts of disk histology have been reviewed.[1,15]

The average number of cells in the hyaline laminae of the EP is comparable to that in articular cartilage, 15,000/mm^3; that in the AF averages 9,000/mm^3; and that in the NP, 4,000/mm^3.[15] The notochordal cells found in the young NP are rarely seen after the age of 10 years.[16] The predominant cells of the NP and inner AF resemble chondrocytes and, with age, develop a dense territorial matrix.[17] The mature cells are found in chondrons.[11,15] Up to one half the cells of the adult NP are necrotic on examination by electron microscopy.[17] The cells of the outer AF resemble the fibrocytes of tendon and ligament.[15]

The ultrastructure of the collagenous fibers of the disk shows the 60-nm crossbanding and the detailed intraband structure characteristic of structural collagens.[18] A structure with cross bands at 100 nm intervals has been seen in both NP and AF[18,19] and may represent type VI collagen.[20] Occasional elastic fibers are also found.[21]

Chemical Composition

The changes in the morphology of disk tissues with aging are accompanied by changes in chemical composition. The compositional changes between conception and maturity are more striking than those between maturity and death. In considering aging of the disk, the magnitude and significance of the trends occurring after maturity will be emphasized. Earlier work has been reviewed.[1,22,23]

Water

The classic data describing the water contents of the NP and AF were published by Püschel[24] in 1930. Her study of 28 spines collected postmortem showed that the water of the NP fell from close to 9 g/g dry weight in utero to about 4 g/g dry weight at 18 years, then further to 2.3 g/g dry weight at 84 years; the changes in the AF were smaller: about 3.6 g/g dry weight in utero, 2.6 g/g dry weight at 18 years, and 2.1 g/g dry weight at 84 years. These data have been confirmed.[25] In individual disks, the water content increases from the outer to the inner AF to a value that is maintained across the width of the NP.[26] Wide individual differences in water content obscured a small but significant decrease of whole disk water from an average value of 2.8 g/g dry weight in a healthy mature disk to 2.6 g/g dry weight in a degenerate aged disk.[27] Healthy and degenerate disks from the same spine had similar water contents; these data suggest that the trend was attributable to age, not degeneration. After correction for the gel exclusion properties of collagen, the equilibrium swelling pressure of the NP follows a simple relationship to the fixed charge density, which corresponds to the dependence of osmotic pressure on the concentration of proteoglycans (PG) in solution.[28] Thus, the water content of the NP will reflect the fixed charge density of the disk PG, the gel exclusion properties of the collagenous fibers, and the axial pressure on the disk resulting from body weight as well as from muscular and ligamentous forces. Because the wide individual differences in water content of the NP probably result from all these factors, the concentration of components in disk tissue are best expressed relative to the dry weight. In disks collected postmortem, the hydration of L5-S1 was low and that of L1-L2 slightly low relative to L2-L3 through L4-L5.[29] Relative to disks collected postmortem, disks collected surgically had a lower fluid content in the inner AF and a higher fluid content in the outer AF. Measurements demonstrated that postmortem swelling pressures were constant across the width of the AF but increased from the outside inward in surgical specimens. The latter, allowed to relax overnight at 4°C, developed a profile similar to postmortem disks as a result of internal fluid shifts.

The average value of the water content of the endplate cartilage was close to 1.2 g/g dry weight, which is substantially less than that of other disk tissues. The water content was higher over the NP than over the AF, and higher in the layer near the NP than in that close to the bone.[8]

Collagen

Collagen is the most abundant nonaqueous component of the intervertebral disk. The literature has been summarized.[1,22,23,30]

In the young disk, collagen, as estimated by tissue hydroxyproline content, makes up two thirds of the dry weight of the AF and one quarter that of the NP. In the AF, its abundance increases from the inner to the outer layers and from the thoracic through the lower lumbar disks.[31] No relationship of collagen content to disk level or age was noted in the NP.[28] In a study of 75 disks, a slight increase in whole-disk collagen with morphologic grade was noted, but individual values ranged widely within each grade.[27]

The disk contains a complex mixture of collagens: the NP contains types II, VI, IX, and XI; and the AF, types I, II, III, V, VI, and IX.[1,23,30] A collagen related to type XII has been reported as well.[32] Types I and II comprise 80% of the total disk collagen. The outer AF is primarily type I and the inner primarily type II, with a decreasing ratio of type I to type II from the outer to inner layers.[33,34] In adolescents and young adults, the collagen of the AF was distributed equally in the four quadrants of the disk: anterior, posterior, left, and right. However, in mature adults, the concentration increased in the posterior relative to the anterior quadrant, had a higher ratio of type I to type II in the outer AF, and had a lower ratio in the inner AF. These differences were viewed as a response of the tissue to mechanical forces.[35] Immunohistochemical stains demonstrated types III and VI collagens to be distributed pericellularly in a chondron-like structure.[36,37] In the calf, type VI collagen is 20% of the total collagen in the NP and 5% of the total in the AF and is believed to form the fibrils with a 110-nm repeat pattern seen on electron microscopy.[20]

Collagen is more concentrated in the EP than in either the NP or the AF; the concentration is higher adjacent to the AF than to the

NP.[8] Immunohistochemical staining showed type II collagen throughout the EP and types II, III, and VI concentrated pericellularly.[36]

Proteoglycans

The constitution of the disk proteoglycans (PG) and the effects on them of aging have been the subject of several reviews.[1,22,23,38]

In contrast to water and collagen, the NP of the human disk clearly loses PG on aging. In a study of 50 donors, the mean hexosamine content of the NP fell from 14% at 11 to 20 years to 11% at 21 to 30 years, then declined steadily to 6% at 81 to 90 years.[39] A comparison of disks from a 5 year old and from a 65 year old showed the latter to have a lower hexuronate content, particularly in the NP, but also in several layers of the AF where the hexuronate decreased from the NP outwards.[31] Three healthy disks from individuals 23 to 25 years old had a higher hexuronate content than three degenerate 39- to 57-year-old specimens.[40] In a study of 75 disks from 15 donors, a clear decline in tissue hexuronate was observed as the morphologic grade increased.[27] The values for the various morphologic grades overlapped except for the high values in the disks with gel-like NP. Analysis of the data suggested that loss of PG from all disks of a spine usually occurred before the morphologic features of degeneration in any disk became apparent. In a study of eight spines, the fixed charge density of the NP, a measure of PG concentration, decreased with age, particularly after 40 years.[28] In a study in which the dimethylmethylene blue assay was used to analyze 40 to 85 tissues, the PG content of the NP averaged about 55 chondroitin sulfate equivalents, the AF about 32, the EP adjacent to the AF about 16, and the EP adjacent to the NP about 20.[8]

Disk PG contains a much lower proportion of aggregate than does bovine nasal cartilage.[38] The proportion depends upon both the tissue and the age of the donor. Few investigators have studied a substantial number of specimens, but the data from several laboratories are qualitatively quite consistent (Table 1). In healthy adult disks, the proportion of aggregate observed in PG from NP has been consistently less than that from AF. In the newborn, the proportion is higher than in adults and similar in both tissues. The adult pattern emerges by the age of 6 months. The distinction between NP and AF may diminish with degeneration. The proportion in the CEP is similar to that in AF.

The effect of aging on the relative proportions of the glycosaminoglycans (GAG) of disk PG is controversial because different analytic methods give differing results. The principal GAG of the disk PG are chondroitin 6-sulfate (Ch6S) and keratan sulfate (KS). The proportion of KS relative to chondroitin sulfate (ChS) in the disk is much higher than in other tissues.[38] The ratio of glucosamine (glcN) to galactosamine (galN), determined by ion-exchange chromatography, has been used widely to assess KS/ChS. In a study of 55 speci-

Table 1 The effects of tissue and age on the proportion of aggregate in disk proteoglycans

Age	No.	Proportion of aggregate (%) in NP	AF	CEP	Reference
8y	1	26	38	—	41
16y	1	24	33	—	
44y	1	12	24	—	
15y	1	23	56	—	42
30y	1	39	44	—	
1-10d	?	52	—	—	43
9m-75y	?	28±6	49±5	—	
N 23,23,25y	3	20	41	—	40
D 39,50,57y	3	36	38	—	
61,64,68y	3	25,11,30	46,36,44	—	44
16y	?	—	—	42±4	45
1-10d	?	52	59	45	46
6-8m	?	28	47	40	
11,60,105d	3	53	53	55	47
25,27,28y	3	32	40	41	

d - days; m - months; y - years; N - normal; D - degenerate; NP - nucleus pulposus; AF - annulus fibrosus; CEP - cartilaginous end plate

mens of whole disk tissue, glcN/galN of NP increased steadily from about 0.2 at birth to 1.5 at 90 years; that for 16 specimens of AF increased from 0.3 at birth to 1.0 at 20 years, and increased little thereafter.[48] For purified PG from AF the ratio increased only slightly between the ages of 14 and 52 years.[49] GlcN/galN in adult NP PG is less than that in AF PG.[40,42,45,50] GlcN/galN measured by gas-liquid chromatography was used to analyze PG subfractions from disk. The values were very different from the ratio of immunologically assayed KS to hexuronate, another measure of KS/ChS; this difference suggests that ratio data should be interpreted with caution.[51] When KS/ChS was determined by analysis of partially purified GAG for hexose and hexuronate respectively, the ratio was found to lie between 0.8 and 1.2 across the width of a 15-year-old disk.[52] When anion-exchanger bound hexose was used as a measure of KS, the KS content in both the posterior AF and the NP was lower in degenerate disks than in nondegenerate; the non-KS hexose averaged 63% of the total in the nondegenerate and 43% in the degenerate disks.[53] Considered in toto, these data suggest that both disk tissue and PG may contain a significant source of glcN other than KS, perhaps other glycoproteins or N- or O-linked oligosaccharides. Specific assays should be used for studies of the GAG composition of disk PG.

Hyaluronan, another glycosaminoglycan present in the intervertebral disk, participates in the formation of PG aggregates. Hyaluronan represents 1% of the total GAG in a 5-year-old NP and 3% of the total GAG in the AF. In a 65-year-old disk, hyaluronan represents 4% of the total GAG in both tissues.[50] Values in the same range were obtained for partially purified GAG from a 15-year-old disk.[52]

Limited studies of the biosynthesis of disk PG suggest that it is synthesized as a large aggregate. Radioactive sulfate was incorporated into NP PG in vitro, and the PG was isolated from the tissue after chasing with nonradioactive sulfate for intervals of 0.5 to 18 hours. Although only 17% of the total tissue PG (measured by hexuronate) was aggregate, most of the newly synthesized PG (measured by radioactivity) was incorporated initially into the aggregate. The newly synthesized PG monomer was much larger than the total tissue PG. The chain sizes of the labeled and unlabeled GAG were similar, indicating the change in monomer size resulted from differences in the core protein, not in the size of its GAG.[54] The incorporation of sulfate into an aggregating PG with a monomer larger than that of the total tissue PG also was demonstrated for dog NP and AF in vivo.[55] The rate of sulfate incorporation into disk PG depends strongly on the hydration of the tissue and is optimal at close to physiologic conditions.[56] When the osmotic pressure in the medium is adjusted to attain optimal tissue hydration in vitro, the incorporation of sulfate into NP PG in fetal and infant tissue is 2.5-fold greater than into AF PG and its incorporation into adult AF is one fifth that in fetal and infant AF.[57] Autoradiographic studies demonstrated that the maximal synthetic rate in the fetus was in the inner AF; whereas, that in the adult was in the middle AF.[57]

The heterogeneity of disk PG may arise from the accumulation within the tissue of degradation products of the initially synthesized PG aggregate. The aggregating PG of the disk, when examined by composite agarose-polyacrylamide gel electrophoresis, has two components of distinctive composition, and the nonaggregating PG has three.[51,58] Small low-density fragments were more abundant in both a 53 year old and a scoliotic disk than in a 19-year-old disk.[59] Further studies are needed to establish with certainty the effect of aging on heterogeneity.

Link protein, which stabilizes the aggregate formed by the interaction of PG monomers and hyaluronan, is present in PG aggregates from the human intervertebral disk.[44] However, in contrast to aggregates from articular cartilage, link-protein-containing disk aggregates are not resistant to dissociation by hyaluronan oligosaccharides. In infant disks, the high-molecular-weight link protein, LP1, predominates; whereas, in 20- to 30-year-old donors, a lower-molecular-weight link protein, LP3, is most abundant, and low-molecular-weight fragments appear.[47] These findings were confirmed independently by Pearce and associates,[60] who also showed that the contents of link protein in AF and NP were much less than in hip and knee articular cartilage and were less abundant in extracts of NP than of AF. Thus, in adult disks, the link protein is both less abundant and more degraded than in cartilage.

Degradative Enzymes and Inhibitors

Enzymes capable of degrading the components of the extracellular matrix of the disk have been found in NP, AF, and CEP.[61] Collagenase, gelatinase, and elastase activities were found in extracts of NP and AF. The first was more active in NP than in AF and was activated several-fold by treatment with N-ethyl maleimide, an activator of collagenase.[62] Neutral proteinase activity was concentrated from AF and NP and was found to be plasmin-like in its response to a variety of class-specific inhibitors.[63] The enzyme was activated by trypsin affinity chromatography and was capable of degrading disk PG. However, unlike plasmin, calcium was required for activity. In another study, antibodies to prostromelysin were used to demonstrate this latent enzyme in explants of disk tissue.[64] The activity was higher in cultures of NP than AF and was activated by para-amino phenylmercuric acetate and inhibited by EDTA; both are characteristics of connective tissue metalloproteinases. A comparison of the terminal amino acid sequences indicated that the conversion of LP1 to LP3 involved cleavage at a stromelysin-sensitive site in LP1. Interleukin-1 enhanced the release of the proenzyme from some specimens. Procollagenase was much less abundant than prostromelysin.

The disk contains protease inhibitors. Low-molecular-weight protein fractions from NP and AF are capable of inhibiting the action of trypsin, chymotrypsin, and polymorphonuclear leukocyte extract on cowhide power substrate. The degradation of disk PG by these enzymes was inhibited by these fractions as well; their activity was in the outer AF, progressively increasing in samples taken from the inner AF and NP.[65] The low-molecular-mass serine proteinase inhibitors of the disk were purified to homogeneity and shown by amino acid sequence analysis to be identical with the mucus proteinase inhibitors found in many connective and mucus-producing tissues.[66] These inhibitors were depleted in degenerate disks, and an α-1-proteinase inhibitor was more abundant. A portion of the latter molecule was not in an active form.[67]

Noncollagenous Proteins

These molecules, a major component of the intercellular matrix of the disk, increase in concentration with age from 0.2 g/g dry weight in 3- to 19-year-old NP to 0.45 g/g dry weight in 65- to 89-year-old NP and from 0.05 g/g dry weight in 3- to 19-year-old AF to 0.25 g/g dry weight in 65- to 89-year-old AF.[68] Despite their substantial concentration in the tissues, these proteins have been poorly characterized. Early data have been reviewed.[22,61]

A major component is the collagen-bound glycoprotein or structural glycoprotein containing mannose, fucose, and sialic acid. Extracts of NP with 4M guanidinium chloride, followed by separation of the protein components from PG by CsCl density gradient ultracentrifugation, yielded several proteins separable by SDS-PAGE electrophoresis including lysozyme.[69] The lysozyme content of the disk increases with increasing age and morphologic grade and may affect the stability of PG aggregates in the disk.[70] Thrombospondin is present as well in the NP.[71]

Other components are serum proteins including albumin and α-1-proteinase inhibitor.[61,72] As occurs in many other connective tissues, plasma proteins may gain access to the disk by diffusion from the vasculature once the loss of PG has created accessible space in the disk.[73]

Animal Models of Disk Aging

The conclusions that can be drawn from the study of human intervertebral disks are necessarily limited because of both the wide genetic diversity of humans and the limited availability of serial samples from human donors. In contrast, serial samples from experimental animals of homogeneous genetic background can be used to provide a description of the processes of aging and degeneration. Such data may have limited relevance because of the postural differences between humans and the common laboratory animals and because of species differences in tissue structure and composition. However, some animal models of disk aging and degeneration have provided suggestive insights into the aging processes in humans.

The production of nuclear herniation in the rabbit by transverse incision of the AF produces, over a period of 200 days, a series of gross and microscopic morphologic changes. The end-result of these changes resembles in some respects a degenerate human disk[74-76] with conversion of the NP to a pad of fibrocartilage that is poorly distinguishable from the AF and with the formation of osteophytes.[77] After the loss of NP, the fibrocytes of the inner AF transform into chondrocytes, and the nuclear space fills with fibrocartilage. In the AF, all but the dorsal portion reorganizes into fibrocartilage. The extruded nucleus becomes fibrosed, then cartilaginous, then bony, forming an osteophyte. The water and PG content of the tissue fall initially, then rise to a peak at 7 days and decline gradually over 200 days. The proportion of aggregate falls to a minimum at 14 days, peaks at 28 days, then declines slowly over 200 days. The tissue hyaluronan falls over the course of the experiment. These data suggest an active response of the disk to the loss of the NP, leading to the formation of a fibrocartilage linking the vertebrae. This model may reflect the response to disk herniation more closely than it does the response to aging.

A similar model has been proposed in which sheep are used and only the outer two thirds of the AF is incised close to the end-plate.[78] With this model, early nuclear herniation is rare, and the progressive changes in morphology and composition occur more slowly. The model was developed to test the hypothesis that rim lesions, discrete peripheral tears in the AF, lead to secondary degeneration. A collagenous scar forms in the outer third of the AF during the first 4 months, and the remaining adjacent collagenous lamellae become distorted. However, the defect in the middle third of the AF does not heal, and circumferential clefts develop in the inner AF after, usually, 4 to 12 months. The NP is displaced along the defect toward the outer AF and degenerates. By 18 months, the disk height decreases moderately and osteophytes form on both vertebrae adjacent to the lesion. A study has been completed of the chemical changes in this model during the disk degeneration.[79] The NP PG (hexuronate) content decreases after 6 months, falling to values close to those in the AF by 18 months. The NP collagen (hydroxyproline) and water contents also fall by 12 months and remain low. The proportion of NP PG present as aggregates decreases by 4 months but returns to normal by 6 months; the similar changes in the AF are delayed. At 4 months, the hydrodynamic volume of the nonaggregating PG decreases but reverts to normal by 8 to 12 months. The extractable protein increases between 8 and 12 months, then falls again by 18 months. Similar but less striking changes are seen in the disks adjacent to the operated disks, although no morphologically or histologically detectable anomalies are seen in the former.

Chondrodystrophoid dogs, notably the beagle, undergo a series of morphologic changes in their intervertebral disks during their first year that resemble closely those seen during aging of the human intervertebral disk.[80] During 40 months, chemical and histochemical analyses showed that the hexuronate of the NP decreased with age, whereas the KS/ChS ratio increased[81]; in a nonchondrodystrophoid breed, the greyhound, the hexuronate content was higher, and similar changes did not occur until much later in life.[82] The beagle NP contains five times more collagen than that of the greyhound.[83] The hexuronate content of AF and NP in the young beagle was higher than that in the old; the proportion of aggregate was higher in both tissues, and the glcN/galN in the nonaggregating fraction was higher.[84] Thus, the changes in the beagle disk with aging correspond closely to those reported for human disks.

The disks of the aging sand rat (*Psammomys obesus*) develop morphologic changes similar to those seen in human disks: radiologically detectable narrowing of the disk space and subchondral endplate sclerosis (more prevalent in the lower lumbar spine), formation of bony/cartilaginous spurs anteriorly, loss of physaliform cells and chondrocyte replication in the NP, and ligamentous calcification. In some respects, the findings also resemble diffuse idiopathic skeletal hyperostosis. The skeletal findings did not correlate with the diabetes mellitus that develops frequently in these animals.[85] With aging, the water content, fixed charge density (a measure of PG), and ability to resist changes in colloid osmotic pressure decreased.[86]

Regeneration can be induced in the canine intervertebral disk after the injection of chymopapain in vivo.[87] Two weeks after injection, the disk space is narrow, and PG is absent from NP, AF, and CEP as judged by safranin-O staining. Restoration is partial by 3 months and nearly complete by 6 months. The properties of the PG were similar to controls 3 and 6 months after the injection, but showed extensive degradation at 2 weeks. However, at the latter time, sulfate and glcN were actively incorporated into the ChS-rich PG. Studies of disk tissues exposed to chymopapain in vitro showed that some cells survive a 24-hour exposure to therapeutic levels of the enzyme.[88]

Mechanisms of Aging and Degeneration

Although many aspects of the effects of aging on disk morphology and biochemistry remain unclear, others have been well documented in studies using human and animal models. The conversion of the NP from a translucent gel to fibrocartilage is associated with loss of PG, a decreased proportion of PG aggregate, the degradation of the link protein, the accumulation of PG fragments in the tissue, and the loss of PG-associated water. Proteases capable of producing these chemical changes are present in the tissue, mainly as inactive precursors, and are held inactive by the presence of inhibitors. Certain as yet ill-defined agents are capable of activating the degradative process; one candidate is interleukin-1. The AF is less sensitive to breakdown but responds to mechanical stress by the local synthesis of type I collagen. The response to chymopapain in dogs suggests that, at least in the early stages of degeneration, the disk is capable of regeneration both morphologically and chemically. Thus, two major unresolved questions are (1) the mechanisms by which the degradative process is initiated and (2) the exact roles of degradative enzymes and their inhibitors in the modulation of this process.

The PG of the newborn disk differ in both GAG composition and in the proportion of aggregate from disks at 6 months of age and strikingly from mature adult disks. These changes could be effected either by changes in the biosynthetic processes, perhaps associated with the replacement of notochordal cells by chondrocytes, or by accumulation of degradation products in the tissue, or both.

Three hypotheses concerning the initiation of disk degeneration are supported by a large body of data and form the basis of much current research. First, both human postmortem specimens and induced anular lesions in sheep have demonstrated that the morphologic and chemical features of 'degeneration' can be initiated by a rim lesion in the AF. Second, severely degenerate disks usually are found only in spines in which all disks have low PG content. Third, inadequate nutrition may initiate the degeneration process (see the chapter by Urban on the effect of physical factors on disk cell metabolism). The relative importance of these mechanisms and of others as yet unrecognized remains to be demonstrated.

Acknowledgments

This research has been supported by Operating Grants from the Arthritis Society. The preparation of this review was aided by the provision of manuscripts prior to publication by Prof. Barrie

Vernon-Roberts of the University of Adelaide and Dr. Peter Ghosh of the University of Sydney.

References

1. Eyre D, Benya P, Buckwalter J, et al: The intervertebral disk. Part B. Basic science perspectives, in Frymoyer JW, Gordon SL (eds): *New Perspectives on Low Back Pain*. Park Ridge, IL, American Academy of Orthopaedic Surgeons, 1989, chap 5, pp 147-207.
2. Kohn RR: *Aging*. Kalamazoo, MI, Upjohn Co, 1973.
3. Vernon-Roberts B: Disk pathology and disease states, in Ghosh P (ed): *The Biology of the Intervertebral Disk*. Boca Raton, FL, CRC Press, 1988, vol II, chap 11, pp 73-119.
4. Schmorl G, Junghanns H, (translated by Besemann EF): *The Human Spine in Health and Disease*, ed 2. New York, NY, Grune & Stratton, 1971.
5. Vernon-Roberts B: The normal aging of the spine: Degeneration and arthritis, in Andersson GBJ, McNeill TW (eds): *Lumbar Spinal Stenosis*. St. Louis, MO, Mosby Year Book, 1992, chap 6, pp 57-75.
6. Vernon-Roberts B: Age-related and degenerative pathology of intervertebral disks and apophyseal joints, in Jayson MIV (ed): *The Lumbar Spine and Back Pain*, ed 4. Edinburgh, Churchill Livingstone, 1992, chap 2, pp 17-41.
7. Miller JA, Schmatz C, Schultz AB: Lumbar disk degeneration: Correlation with age, sex, and spine level in 600 autopsy specimens. *Spine* 1988;13:173-178.
8. Roberts S, Menage J, Urban JP: Biochemical and structural properties of the cartilage end-plate and its relation to the intervertebral disk. *Spine* 1989;14:166-174.
9. Hormel SE, Eyre DR: Collagen in the aging human intervertebral disk: An increase in covalently bound fluorophores and chromophores. *Biochim Biophys Acta* 1991;1078:243-250.
10. Osti OL, Vernon-Roberts B, Moore R, et al: Annular tears and disc degeneration in the human lumbar spine: A post-mortem study of 135 discs. *J Bone Joint Surg* 1992;74B:678-682.
11. Pritzker KP: Aging and degeneration in the lumbar intervertebral disc. *Orthop Clin North Am* 1977;8:65-77.
12. Bernick S, Cailliet R: Vertebral end-plate changes with aging of human vertebrae. *Spine* 1982;7:97-102.
13. Nachemson A: Lumbar intradiscal pressure: Experimental studies on post-mortem material. *Acta Orthop Scand Suppl* 1960;43:1-104.
14. Thompson JP, Pearce RH, Schechter MT, et al: Preliminary evaluation of a scheme for grading the gross morphology of the human intervertebral disk. *Spine* 1990;15:411-415.
15. Buckwalter JA: The fine structure of human intervertebral disk, in White AA, Gordon SL (eds): *American Academy of Orthopaedic Surgeons Symposium on Idiopathic Low Back Pain*. St. Louis, MO, CV Mosby, 1982, chap 9, pp 108-143.
16. Trout JJ, Buckwalter JA, Moore KC, et al: Ultrastructure of the human intervertebral disk: I. Changes in notochordal cells with age. *Tissue Cell* 1982;14:359-369.
17. Trout JJ, Buckwalter JA, Moore KC: Ultrastructure of the human intervertebral disk: II. Cells of the nucleus pulposus. *Anat Rec* 1982;204:307-314.
18. Buckwalter JA, Maynard JA, Cooper RR: Banded structures in human nucleus pulposus. *Clin Orthop* 1979;139:259-266.

19. Cornah MS, Meachim G, Parry EW: Banded structures in the matrix of human and rabbit nucleus pulposus. *J Anat* 1970;107:351-362.

20. Wu JJ, Eyre DR, Slayter HS: Type VI collagen of the intervertebral disk: Biochemical and electron-microscopic characterization of the native protein. *Biochem J* 1987;248:373-381.

21. Buckwalter JA, Cooper RR, Maynard JA: Elastic fibers in human intervertebral disks. *J Bone Joint Surg* 1976;58A:73-76.

22. Eyre DR: Biochemistry of the intervertebral disk, in Hall DA, Jackson DS (eds): *International Review of Connective Tissue Research.* New York, NY, Academic Press; 1979, vol 8, pp 227-291.

23. Ayad S, Weiss JB: Biochemistry of the intervertebral disk, in Jayson MIV (ed): *The Lumbar Spine and Back Pain*, ed 3. Edinburgh, Churchill Livingstone, 1987, chap 5, pp 100-137.

24. Püschel J: Der Wassergehalt normaler und degenerierter Zwischenwirbelscheiben. *Beitr Path Anat* 1930;84:123-130.

25. Gower WE, Pedrini V: Age-related variations in proteinpolysaccharides from human nucleus pulposus, annulus fibrosus, and costal cartilage. *J Bone Joint Surg* 1969;51A:1154-1162.

26. Urban JPG, Maroudas A: The measurement of fixed charge density in the intervertebral disc. *Biochim Biophys Acta* 1979;586:166-178.

27. Pearce RH, Grimmer BJ, Adams ME: Degeneration and the chemical composition of the human lumbar intervertebral disc. *J Orthop Res* 1987;5:198-205.

28. Urban JP, McMullin JF: Swelling pressure of the lumbar intervertebral disks: Influence of age, spinal level, composition, and degeneration. *Spine* 1988;13:179-187.

29. Johnstone B, Urban JP, Roberts S, et al: The fluid content of the human intervertebral disc: Comparisons between fluid content and swelling pressure profiles of discs removed at surgery and those taken postmortem. *Spine* 1992;17:412-416.

30. Eyre DR: Collagens of the disc, in Ghosh P (ed): *The Biology of the Intervertebral Disk.* Boca Raton, FL, CRC Press, 1988, vol I, chap 7, pp 171-188.

31. Adams P, Eyre DR, Muir H: Biochemical aspects of development and aging of human lumbar intervertebral discs. *Rheumatol Rehab* 1977;16:22-29.

32. Wu JJ, Niyibizi C, Eyre DR: A novel collagenous protein in annulus fibrosus of the disc. *Trans Orthop Res Soc* 1991;16:349.

33. Eyre DR, Muir H: Types I and II collagens in intervertebral disc: Interchanging radial distributions in annulus fibrosus. *Biochem J* 1976;157:267-270.

34. Eyre DR, Muir H: Quantitative analysis of types I and II collagens in human intervertebral discs at various ages. *Biochim Biophys Acta* 1977;492:29-42.

35. Brickley-Parsons D, Glimcher MJ: 1983 Volvo award in basic science: Is the chemistry of collagen in intervertebral discs an expression of Wolff's Law?: A study of the human lumbar spine. *Spine* 1984;9: 148-163.

36. Roberts S, Menage J, Duance V, et al: 1991 Volvo award in basic science: Collagen types around the cells of the intervertebral disc and cartilage end plate: An immunolocalization study. *Spine* 1991;16: 1030-1038.

37. Roberts S, Ayad S, Menage PJ: Immunolocalisation of type VI collagen in the intervertebral disc. *Ann Rheum Dis* 1991;50:787-791.

38. McDevitt CA: Proteoglycans of the intervertebral disc, in Ghosh P (ed): *The Biology of the Intervertebral Disk*. Boca Raton, FL, CRC Press, 1988, vol I, chap 6, pp 151-170.
39. Hallén A: Hexosamine and ester sulphate content of the human nucleus pulposus at different ages. *Acta Chem Scand* 1958;12:1869-1872.
40. Lyons G, Eisenstein SM, Sweet MB: Biochemical changes in intervertebral disc degeneration. *Biochim Biophys Acta* 1981;673:443-453.
41. Adams P, Muir H: Qualitative changes with age of proteoglycans of human lumbar discs. *Ann Rheum Dis* 1976;35:289-296.
42. Stevens RL, Ewins RJ, Revell PA et al: Proteoglycans of the intervertebral disk: Homology of structure with laryngeal proteoglycans. *Biochem J* 1979;179:561-572.
43. Pedrini-Mille A, Pedrini V, O'Connor R, et al: Age related changes in the proteoglycans of human intervertebral disc. *Orthop Trans* 1980;4:221.
44. Tengblad A, Pearce RH, Grimmer BJ: Demonstration of link protein in proteoglycan aggregates from human intervertebral disc. *Biochem J* 1984;222:85-92.
45. Pedrini-Mille A, Pedrini V, Tudisko C, et al: Proteoglycans of human scoliotic intervertebral disc. *J Bone Joint Surg* 1983;65A:815-823.
46. Buckwalter JA, Pedrini-Mille A, Pedrini V, et al: Proteoglycans of human infant intervertebral disc: Electron microscopic and biochemical studies. *J Bone Joint Surg* 1985;67A:284-294.
47. Donohue PJ, Jahnke MR, Blaha JD, et al: Characterization of link protein(s) from human intervertebral-disc tissues. *Biochem J* 1988;251:739-747.
48. Hallén A: The collagen and ground substance of human intervertebral disc at different ages. *Acta Chem Scand* 1962;16:705-710.
49. Stevens RL, Ryvar R, Robertson WR, et al: Biological changes in the annulus fibrosus in patients with low back pain. *Spine* 1982;7:223-233.
50. Hardingham TE, Adams P: A method for the determination of hyaluronate in the presence of other glycosaminoglycans and its application to human intervertebral disc. *Biochem J* 1976;159:143-147.
51. DiFabio JL, Pearce RH, Caterson B, et al: The heterogeneity of the non-aggregating proteoglycans of the human intervertebral disc. *Biochem J* 1987;244:27-33.
52. Ghosh P, Bushell GR, Taylor TK, et al: Distribution of glycosaminoglycans across the normal and the scoliotic disc. *Spine* 1980;5:310-317.
53. Bebault GM, Pearce RH: Assay of keratan sulfate as anion-exchanger bound hexose. *Connect Tissue Res* 1991;25:281-293.
54. Oegema TR Jr, Bradford DS, Cooper KM: Aggregated proteoglycan synthesis in organ cultures of human nucleus pulposus. *J Biol Chem* 1979;254:10579-10581.
55. McDevitt CA, Billingham MEJ, Muir H: In-vivo metabolism of proteoglycans in experimental osteoarthritic and normal canine articular cartilage and the intervertebral disc. *Semin Arthritis Rheum* 1981;11(Suppl 1):17-18.
56. Bayliss MT, Urban JP, Johnstone B, et al: In vitro method for measuring synthesis rates in the intervertebral disc. *J Orthop Res* 1986;4:10-17.
57. Bayliss MT, Johnstone B, O'Brien JP: 1988 Volvo award in basic science: Proteoglycan synthesis in the human intervertebral disc: Variation with age, region and pathology. *Spine* 1988;13:972-981.

58. Jahnke MR, McDevitt CA: Proteoglycans of the human intervertebral disc: Electrophoretic heterogeneity of the aggregating proteoglycans of the nucleus pulposus. *Biochem J* 1988;251:347-356.
59. Melrose J, Gurr KR, Cole T-C, et al: The influence of scoliosis and aging on proteoglycan heterogeneity in the human intervertebral disc. *J Orthop Res* 1991;9:68-77.
60. Pearce RH, Mathieson JM, Mort JS, et al: Effect of age on the abundance and fragmentation of link protein of the human intervertebral disc. *J Orthop Res* 1989;7:861-867.
61. Melrose J, Ghosh P: The non-collagenous proteins of the intervertebral disc, in Ghosh P (ed): *The Biology of the Intervertebral Disc*. Boca Raton, FL, CRC Press, 1988, vol 1, chap 8, pp 189-237.
62. Sedowofia KA, Tomlinson IW, Weiss JB, et al: Collagenolytic enzyme systems in human intervertebral disc: Their control, mechanism, and their possible role in the initiation of biomechanical failure. *Spine* 1982;7:213-222.
63. Melrose J, Ghosh P, Taylor TK: Neutral proteinases of the human intervertebral disc. *Biochim Biophys Acta* 1987;923:483-495.
64. Liu J, Roughley PJ, Mort JS: Identification of human intervertebral disk stromelysin and its involvement in matrix degradation. *J Orthop Res* 1991;9:568-575.
65. Knight JA, Stephens RW, Bushell GR, et al: Neutral protease inhibitors from human intervertebral disc and femoral head articular cartilage. *Biochim Biophys Acta* 1979;584:304-310.
66. Andrews JL, Melrose J, Ghosh P: A comparative study of the low-molecular mass serine proteinase inhibitors of human connective tissues. *Biol Chem Hoppe Seyler* 1992;373:111-118.
67. Melrose J, Ghosh P, Taylor TKF, et al: The serine proteinase inhibitory proteins of the human intervertebral disc: Their isolation, characterisation and variation with aging and degeneration. *Matrix* 1992;12:56-75.
68. Dickson IR, Happey F, Pearson CH, et al: Variations in the protein components of human intervertebral disc with age. *Nature* 1967; 215:52-53.
69. Sorce DJ, McDevitt CA, Greenwald RA, et al: Protein and lysozyme content of adult human nucleus pulposus. *Experientia* 1986;42: 1157-1158.
70. Melrose J, Ghosh P, Taylor TK: Lysozyme, a major low-molecular-weight cationic protein of the intervertebral disc, which increases with aging and degeneration. *Gerontology* 1989;35:173-180.
71. Miller RR, McDevitt CA: Thrombospondin in ligament, meniscus and intervertebral disc. *Biochim Biophys Acta* 1991;1115:85-88.
72. Fricke R: Serum proteins in connective tissues, in Peeters H (ed): *Protides of the Biological Fluids. Proceedings of Ninth Colloquium, Bruges*. New York, Pergamon Press, 1961, pp 249-252.
73. Bert JL, Pearce RH: Transport of fluids and solutes in the interstitium, in Renkin EM, Michel CC (eds): *Handbook of Physiology. Section 2. The Cardiovascular System. Volume IV. Microcirculation. Part 1*. Bethesda, MD, American Physiological Society, 1984, pp 521-547.
74. Smith JW, Walmsley R: Experimental incision of the intervertebral disc. *J Bone Joint Surg* 1951;33B:612-625.
75. Lipson SJ, Muir H: Experimental intervertebral disc degeneration: Morphologic and proteoglycan changes over time. *Arthritis Rheum* 1981;24:12-21.
76. Lipson SJ, Muir H: 1980 Volvo award in basic science: Proteoglycans in experimental intervertebral disc degeneration. *Spine* 1981;6:194-210.

77. Lipson SJ, Muir H: Vertebral osteophyte formation in experimental disc degeneration: Morphologic and proteoglycan changes over time. *Arthritis Rheum* 1980;23:319-324.
78. Osti OL, Vernon-Roberts B, Fraser RD: 1990 Volvo award in experimental studies: Anulus tears and intervertebral disc degeneration: An experimental study using an animal model. *Spine* 1990;15:762-767.
79. Melrose J, Ghosh P, Taylor TK, et al: A longitudinal study of the matrix changes induced in the intervertebral disc by surgical damage to the annulus fibrosus. *J Orthop Res* 1992;10:665-676.
80. Hansen HJ: A pathologic-anatomical study on disc degeneration in dog: With special reference to the so-called enchondrosis intervertebralis. *Acta Orthop Scand* 1952;11(Suppl):1-117.
81. Ghosh P, Taylor TK, Braund KG: The variation of the glycosaminoglycans of the canine intervertebral disc with aging: I. Chondrodystrophoid breed. *Gerontology* 1977;23:87-98.
82. Ghosh P, Taylor TK, Braund KG: Variation of the glycosaminoglycans of the intervertebral disc with aging: II. Non-chondrodystrophoid breed. *Gerontology* 1977; 23:99-109.
83. Ghosh P, Taylor TK, Braund KG, et al: The collagenous and non-collagenous protein of the canine intervertebral disc and their variation with age, spinal level and breed. *Gerontology* 1976;22:124-134.
84. Cole TC, Ghosh P, Taylor TK: Variations of the proteoglycans of the canine intervertebral disc with aging. *Biochim Biophys Acta* 1986;880:209-219.
85. Moskowitz RW, Ziv I, Denko CW, et al: Spondylosis in sand rats: A model of intervertebral disc degeneration and hyperostosis. *J Orthop Res* 1990;8:401-411.
86. Ziv I, Moskowitz RW, Kraise I, et al: Physicochemical properties of the aging and diabetic sand rat intervertebral disc. *J Orthop Res* 1992;10:205-210.
87. Bradford DS, Cooper KM, Oegema TR Jr: Chymopapain, chemonucleolysis, and nucleus pulposus regeneration. *J Bone Joint Surg* 1983;65A:1220-1231.
88. Suguro T, Oegema TR Jr, Bradford DS: Ultrastructural study of the short-term effects of chymopapain on the intervertebral disc. *J Orthop Res* 1986;4:281-287.

Chapter 28

Intervertebral Disk: Mechanical Changes That Occur With Age

Malcolm H. Pope, DMSc, PhD

The Intervertebral Disk as an Osmotic System

The disk "system," the disk and its adjacent structures, can be regarded as an osmotic system for the exchange of ions and fluids. The outer layers are semipermeable membranes and show size and charge selectivity. For example, Holm and associates[1,2] and Roberts and associates[3] have shown that negatively charged sulfate ions diffuse into the nucleus pulposus largely through the annulus, whereas neutral glucose diffuses mainly through the end plate. The hydroscopic capacity of the macromolecules in the interior of the disk (chondroitin sulfate and keratan sulfate) is so high that they can take up fluid even when under pressure. Osmotic fluid flow takes place against any pressure resulting from external loading until the osmotic and loading pressures are in equilibrium.

Biochemical Changes With Age and Their Mechanical Consequence

Through use of magnetic resonance imaging (MRI), Sether and associates[4] found a correlation between the decrease in signal intensity and age, although signal intensity decreased less than 6% in 80 years. The decrease in signal intensity is concomitant with decreases in disk water and chondroitin sulfate content and increases in disk collagen.[5] Chiang[6] reported changes in chondroitin sulfate chain length with aging, which could lead to decreased viscosity of both the annulus and nucleus in aged disks.[7,8]

Because the disk is in osmotic balance, biochemical changes are crucial to its mechanical behavior. The water content decreases in a nonlinear fashion with increasing age. This decrease is pronounced within the second to fourth decades and is minimal thereafter.[9] Urban and McMullin[10] examined how the water content of the nucleus pulposus varies with applied pressure for disks of various ages. Hydration decreased as pressure increased, but the equilibrium level of hydration depended on the relative amounts of collagen and proteoglycan in the tissue. Aged disks often have a low proteoglycan to collagen ratio, their equilibrium hydration tends to be low, and a far larger proportion of the total water is associated with the collagen than in the younger disks. Some insight into these relationships comes from the work of Melrose and associates,[11] who found that the normal pattern of age changes in the proteoglycans from scoliotic disks appears to be disturbed, probably because of the abnormal biomechanical forces intrinsic to the scoliosis. Thus, altered mechanical forces may have a similar effect in the normal disk.

Gross Structural Changes With Age

Both Koeller and associates[9] and DeCandido and associates[12] found that degenerative disk changes increased with age until the fifth decade of life, after which the disk remained unchanged. There was a positive relationship between disk degeneration and prolapse or herniation. Degeneration was more prevalent at the lower lumbar levels.[12,13] Butler and associates[13] found that both disk degeneration and facet osteoarthritis increase with increasing age. They concluded that disk degeneration occurs before facet joint osteoarthritis, which may be secondary to mechanical changes in the loading of the facet joints. Thus, the change of loading in one part of the spine will affect another part.

The cross-sectional area of lumbar and thoracic intervertebral disks increases with age; the lumbar disk increases are larger than the thoracic in all cases. The increase after the sixth decade is mainly a result of spondylitic changes. Changes in the heights and anteroposterior diameters of human lumbar intervertebral disks have also been studied by Amonoo-Kuofi.[14] There is an overall increase in the various dimensions of the disk with age. Growth of the disks apparently does not follow a linear pattern; there are alternating periods of overgrowth and thinning. After the fifth decade of life, there is an appreciable decline of disk height. Twomey and Taylor[15] report that the actual average disk height increases with age as the disks "sink" into the vertebrae. The loss of transverse trabeculae of lumbar vertebrae is primarily responsible for the change in shape of both vertebrae and disks in the elderly.

Histologic examination of aging lumbar spines showed that the cartilaginous end plates degenerated and were replaced by subchondral bone proliferation (endochondral bone formation).[16-18] The end-plate change correlated positively with disk-space narrowing and degeneration of the nucleus pulposus. Harris and MacNab[19] reported that the nucleus becomes fibrotic and shows chondroid changes and diffuse calcification. Such changes can be expected to have major effects on the mechanical behavior of the disk and, subsequently, of the motion segment.

Microstructural Changes

Marchand and Ahmed[20] investigated the structure of the lumbar disk annulus fibrosus using a layer-by-layer peeling technique and microscopic examination of various cut surfaces. They found that the annulus, excluding the transition zone, consists of 15 to 25 distinct layers. Nearly half of the layers terminate or originate in any 20° circumferential sector, thereby causing local irregularities. Increasing age was found to significantly reduce the number of distinct layers, to increase the thickness of individual layers, and to increase the interbundle spacing within an individual layer. The reduction in the number of distinct layers in older disks was said to be related to both the thickening of the transition zone and the gradual loss of organized fiber structures of the inner layers. The net width of the annulus is not reduced because of a concomitant increase in the thickness of the remaining layers. Increasing age also causes more irregular distribution of bundles within a layer. This may affect the strength of the annulus, particularly under complex loads.

Bernick and associates[21] also studied age changes in the annulus fibrosus. The annular laminas from individuals younger than 40 years of age consisted of obliquely oriented collagen fibers exhibiting a pennate arrangement. Beginning during middle age and continuing into the eighth decade, there was a progressive degeneration of the laminae characterized by fraying, splitting, and loss of collagen fibers. The continual deposition of chondroid substance in the annulus of the aging disks was not seen in the young disks. The collagen may become stiffer and the annulus and nucleus less viscous in the aging disks. These age-related changes may lead to a loss of local structural strength, which may be a factor in disk pathology.

Johnson and associates[22] examined lumbar intervertebral disks of various ages. Approximately 10% of the matrix of the annulus fibrosus consisted of elastic fibers, and the number of fibers tended to decrease slightly with increased age. Such changes may affect the mechanical behavior (stiffness) of the aging disk.

Yasuma and associates[23] found that, in the sixth decade of life or later, the orientation of the inner fiber bundles of the annulus fibrosus was reversed, so that they bulged inward. The reversal appeared to be the result of degeneration of the middle fibers of the annulus fibrosus (where the stresses are highest), atrophy of the nucleus, and narrowing of the disk space. This reversal may be the reason for the predominance of protrusions of the nucleus pulposus in patients who are younger than 60 years and of herniation in patients who are older than 60 years. The aging disk appears to behave like a thick-walled cylinder rather than a pressurized vessel. Support for this view came from the work of Seroussi and associates,[24] who implanted markers in the disk in vitro. They found that denucleation, possibly an analog of degeneration, results in an inward bulge of the annulus under load. This behavior differed markedly from that of the intact disk, in which the annulus bulged outward under load. Failure of the aging disk appears to be a complex phenomenon that is undoubtedly influenced by asymmetric force distributors. Early in life, the nucleus pulposus and annulus fibrosus are strong, and protrusion or prolapse of the nucleus pulposus hardly ever occurs. In the young, a protrusion or prolapse occurs only when a large asymmetric force is applied to the disk.[25]

Harada and Nakahara[26] examined fragments from herniated lumbar disks of patients over 60 years of age. In 70% of disks from patients between the ages of 60 and 69 and in 80% of disks from patients over 70, the fragments were composed of the annulus fibrosus and the cartilaginous end plate. The authors concluded that the cartilaginous end plate had avulsed from the vertebral body and herniated with the annulus fibrosus in these cases. This type of herniation in elderly patients is obviously a result of advanced disk degeneration. However, more work is needed to understand the precise mechanics of disk herniation in the degenerated disk.

Diurnal Height Changes

Diurnal changes in disk height are an indirect manifestation of osmotic flow. DePuky[27] was the first to make such measurements. Eklund and Corlett[28] optimized the measurement by means of an apparatus that carefully controls posture. Hindle and Murray-Leslie[29] found that the diurnal stature loss was markedly reduced in patients with ankylosing spondylitis. This reduction was assumed to result from the osseous changes in the disk and the ligaments. Krämer[30] reports that, in a well-controlled series of experiments, the overall loss of height during a day was nearly 18 mm for a group under the age of 30, but in a 50-year-old group, this figure averaged only 13 mm. This reduction in diurnal height loss may be caused by a decrease in disk hydration with age, which results in the disk not being able to recover fully from daily compression. Diurnal changes lead to different load distributions on the other elements of the motion segment. Therefore, these findings may have implications in the injury risk to aging manual workers, but more work is needed to understand this phenomenon.

Mechanical Changes With Age

Plaue and associates,[31] testing disks under axial compression, found no variation of compressive strength and modulus of elasticity with age. Berkson[32] and Nachemson and associates[33] also found no significant influence of age on the deformation behavior of human disks subjected to flexion, extension, or lateral bending. However, Farfan and associates[34] found that the average failure torque was 25% higher for undegenerated versus degenerated disks. They proposed torsion as a mode of disk failure. Horst and Brinckmann[35] measured the distribution of normal stresses at the interface between the vertebral body and the disk. They found an asymmetric distribution when degenerated or denucleated disks were subjected to eccentric axial compression and a symmetric distribution when young disks were tested.

Virgin[36] reported that hysteresis is highest in young disks, less in old (degenerated) disks, and least in disks of the middle aged. Hysteresis is related to energy absorption and, thus, may have implications in injury under impact loadings. Twomey and Taylor[37] report a marked decrease in initial flexion deformation and a slight increase in flexion creep deformation of elderly specimens. Kazarian,[38] however, found that in degenerated disks, creep occurs more rapidly than in nondegenerated young disks under compressive or cyclic compressive load. Findings were similar in specimens treated with chymopapain.[39,40] Degenerated disks have less ability to attenuate shock and vibration.

Koeller and associates[9] found that, from the first decade to the middle of the third decade, axial deformability decreases within the thoracic region and remains almost constant within the lumbar spine. Afterwards, axial deformability remains unchanged with age. The ratio of axial deformation to mean disk height is independent of age; values for the thoracic disks are generally higher than those for the lumbar disks. Up to the third decade, creep decreases in both regions as age increases. From the middle of the third decade to the beginning of the sixth, creep behavior remains fairly constant within the lumbar spine. After the sixth decade, increased disk bulge is reported.[41-43]

The clinical significance of these mechanical changes is not clear in times of changed kinematics. Soini and associates[44] studied the association between disk degeneration observed in plain radiographs and diskograms, and instability expressed as abnormal angular movement of lumbar vertebrae. They found that disk degeneration seldom results in abnormal angular movement and instability of the lumbar spine. Therefore, flexion-extension radiography may have only limited diagnostic value.[44] However, Benini[45] suggests that abnormal movements occur with disk narrowing and can lead to instability. Knutsson[46] and van Akkerveeken and associates[47] also report increased mobility in degenerated disks. More recently, Kaigle and associates[48] have developed a means of accurately making in vivo measurements that may clarify these contradictory findings.

Disk herniation cannot be created in vitro by a single compressive loading. Findings were similar under dynamic conditions. When lumbar motion segments were tested under a cyclic axial compressive load, the fractures observed were all in the end-plate region; no lesions were observed in the outer fibers of the annulus fibrosus.[49,50] Farfan and associates[34] showed that disk injury can be produced by slowly applied rotation in amounts within the range of normal lumbar movement. They suggest that impairment of the function of the posterior elements may result in a higher risk of disk degeneration. However, Adams and Hutton[51] showed that torsion of the lumbar spine is resisted by the joint that is in compression; the compression facet is the first structure to yield at the limit of torsion. These authors concluded that torsion seems unimportant in the etiology of disk degeneration and prolapse. Liu and associates[52] investigated the effect of cyclic torsional loads. Failure occurred in the end plates, facets, laminae, and capsular ligaments. However, disk herniations have been caused in combined flexion, compression, and lateral bending.[53] These complex loadings appear to be the key to the production of a herniation. Combined lateral bend, flexion, and axial rotation vibration loading was able to cause tracking tears proceeding from the nucleus through the posterolateral region of the annulus (in the aged disk, the degenerated nucleus gradually tracked through the annulus).[54]

Cyclic loading appears to have other damaging effects. It suggests that mechanical changes leading to instability of the motion segment is a mechanism for disk herniation. Studies conducted using a porcine model have encompassed disk nutritional changes under vibration[2] and creep and vibration.[55] Holm and Nachemson[2] demonstrated a loss of nutrition to the intervertebral disk at the first natural frequency and suggested disks could degenerate with prolonged exposure. High pressures were found in the disk nucleus at the first natural frequency. When vibration was applied directly to the porcine motion segment in vivo, accentuated creep was noted.[55]

Mathematical Models

Lin and associates[56] used a three-dimensional finite element model of the lumbar intervertebral joint to simulate axial loading tests on cadaveric material. They computed Young's moduli and shear moduli of the annulus fibrosus and found a linear decrease with increasing degree of degeneration. Shirazi-Adl and associates[57] demonstrated that in the degenerated disk the end plates are subjected to less pressure in the center and loads are distributed around the periphery (thick-walled cylinder). The axial stress is compressive, and the circumferential stress is close to zero.

Monroe and associates[58] suggested that the lamina separation seen in degenerated disks may result from an increase in the interlaminar shear stresses predicted in an injured disk, compared to an intact disk. Goel and associates[59] have shown that, in a normal motion segment, both the disk and facets transmit loads in the presence of active muscles. As muscle function deteriorates with age, the disk begins to play a bigger role. This supports the concept that disk degeneration occurs before facet joint osteoarthritis.

Discussions

More work is needed to quantify the mechanical changes in the aging versus the degenerated disk and to establish the consequence of these changes on the whole functional spinal unit. What is the temporal nature of such changes? What are the kinematic consequences, and can they be better quantified in-vivo?

The relationship between degenerative changes, hydration, and disk morphology is unknown at the level of the functional spinal unit. These changes have profound effects on the diurnal changes in vivo, and the implications remain to be explored.

More work is necessary to fully understand the implications of the aging and degenerative changes on disk protrusion or prolapse. The mechanics of failure under these circumstances and the consequent prevention in the workplace remains elusive.

References

1. Holm S, Maroudas A, Urban JP, et al: Nutrition of the intervertebral disc: Solute transport and metabolism. *Connect Tissue Res* 1981; 8:101-119.
2. Holm S, Nachemson A: Nutrition of the intervertebral disc: Effects induced by vibrations. *Orthop Trans* 1985;9:525.
3. Roberts S, Menage J, Urban JP: Biomechanical and structural properties of the cartilage end-plate and its relation to the intervertebral disc. *Spine* 1989;14:166-169.
4. Sether LA, Yu S, Haughton VM, et al: Intervertebral disc: Normal age-related changes in MR signal intensity. *Radiology* 1990;177:385-388.
5. Cole TC, Ghosh P, Taylor TK: Variations of the proteoglycans of the canine intervertebral disc with ageing. *Biochim Biophys Acta* 1986; 880:209-219.
6. Chiang YL: A study on topographical change of proteoglycans in human lumbar disk. *Nippon Seikeigeka Gakkai Zasshi* 1983;57:539-551.
7. Nachemson A, Lewin T, Maroudas A, et al: In vitro diffusion of dye through the end-plates and the annulus fibrosus of human lumbar intervertebral discs. *Acta Orthop Scand* 1970;41:589-607.
8. Peereboom JWC: Some biochemical and histochemical properties of the age pigment in the human intervertebral disc. *Histochemistry* 1973;37:119-130.
9. Koeller W, Muehlhaus S, Meier W, et al: Biomechanical properties of human intervertebral discs subjected to axial dynamic compression: Influence of age and degeneration. *J Biomech* 1986;19:807-816.
10. Urban JP, McMullin JF: Swelling pressure of the intervertebral disc: Influence of proteoglycan and collagen contents. *Biorheology* 1985;22:145-157.
11. Melrose J, Gurr KR, Cole T-C, et al: The influence of scoliosis and ageing on proteoglycan heterogeneity in the human intervertebral disc. *J Orthop Res* 1991;9:68-77.
12. DeCandido P, Reinig JW, Dwyer AJ, et al: Magnetic resonance assessment of the distribution of lumbar spine disc degenerative changes. *J Spinal Disord* 1988;1:9-15.
13. Butler D, Trafimow JH, Andersson GB, et al: Discs degenerate before facets. *Spine* 1990;15:111-113.
14. Amonoo-Kuofi HS: Morphometric changes in the heights and anteroposterior diameters of the lumbar intervertebral discs with age. *J Anat* 1991;175:159-168.
15. Twomey L, Taylor J: Age changes in lumbar intervertebral discs. *Acta Orthop Scand* 1985;56:496-499.

16. Aoki J, Yamamoto I, Kitamura N, et al: End plate of the discovertebral joint: Degenerative change in the elderly adult. *Radiology* 1987; 164:411-414.
17. Peereboom JW: Age-dependent changes in the human intervertebral disc: Fluorescent substances and amino acids in the anulus fibrosus. *Gerontologia* 1970;16:352-367.
18. van den Hooff A: Histological age changes in the anulus fibrosus of the human intervertebral disk with a discussion of the problem of disk herniation. *Gerontologia* 1964;9:136-149.
19. Harris RI, MacNab I: Structural changes in the lumbar intervertebral discs: Their relationship to low back pain and sciatica. *J Bone Joint Surg* 1954;36B:304-322.
20. Marchand F, Ahmed AM: Investigation of the laminate structure of lumbar disc anulus fibrosus. *Spine* 1990;15:402-410.
21. Bernick S, Walker JM, Paule WJ: Age changes to the anulus fibrosus in human intervertebral discs. *Spine* 1991;16:520-524.
22. Johnson EF, Berryman H, Mitchell R, et al: Elastic fibres in the anulus fibrosus of the adult human lumbar intervertebral disc: A preliminary report. *J Anat* 1985;143:57-63.
23. Yasuma T, Koh S, Okamura T, et al: Histological changes in aging lumbar intervertebral discs: Their role in protrusions and prolapses. *J Bone Joint Surg* 1990;72A:220-229.
24. Seroussi RE, Krag MH, Muller DL, et al: Internal deformations of intact and denucleated human lumbar discs subjected to compression, flexion, and extension loads. *J Orthop Res* 1989;7:122-131.
25. Yasuma T, Makino E, Saito S, et al: Clinico-pathological study on lumbar intervertebral disc herniation: 1. Changes in the intervertebral disc relation with age including Schmorl's node. *Sikei Saigaigeka* 1986;29:1565-1578.
26. Harada Y, Nakahara S: A pathologic study of lumbar disc herniation in the elderly. *Spine* 1989;14:1020-1024.
27. DePuky P: The physiological oscillation of the length of the body. *Acta Orthop Scand* 1935;6:338-348.
28. Eklund JA, Corlett EN: Shrinkage as a measure of the effect of load on the spine. *Spine* 1984;9:189-194.
29. Hindle RJ, Murray-Leslie C: Diurnal stature variation in ankylosing spondylitis. *Clin Biomech* 1987;2:152-157.
30. Krämer J: *Intervertebral Disc Diseases: Causes, Diagnosis, Treatment and Prophylaxis*, ed 2. New York, NY, Thieme Medical Publishers, 1990.
31. Plaue R, Gerner HJ, Salditt R: Das elastomechanische Verhalten menschlicher Bandscheiben unter statischem Druck. *Arch Orthop Unfallchir* 1974;79:139-148.
32. Berkson MH: Mechanical properties of the human lumbar spine: Flexibilities, intradiscal pressures, posterior element influences. *Proc Inst Med Chicago* 1977;31:138-143.
33. Nachemson AL, Schultz AB, Berkson MH: Mechanical properties of human lumbar spine motion segments: Influence of age, sex, disc level, and degeneration. *Spine* 1979;4:1-8.
34. Farfan HF, Cossette JW, Robertson GH, et al: The effects of torsion on the lumbar intervertebral joints: The role of torsion in the production of disc degeneration. *J Bone Joint Surg* 1970;52A:468-497.
35. Horst M, Brinckmann P: Measurement of the distribution of axial stress on the end-plate of the vertebral body. *Spine* 1981;6:217-232.
36. Virgin WJ: Experimental investigations into the physical properties of the intervertebral disc. *J Bone Joint Surg* 1951;33B:607-611.

37. Twomey L, Taylor J: Flexion creep deformation and hysteresis in the lumbar vertebral column. *Spine* 1982;7:116-122.
38. Kazarian LE: Creep characteristics of the human spinal column. *Orthop Clin North Am* 1975;6:3-18.
39. Bradford DS, Oegema TR Jr, Cooper KM, et al: Chymopapain, chemonucleolysis and nucleus pulposus regeneration: A biochemical and biomechanical study. *Spine* 1984;9:135-147.
40. Wakano K, Kasman R, Chao EY et al: Biomechanical analysis of canine intervertebral discs after chymopapain injection: A preliminary report. *Spine* 1983;8:59-68.
41. Hirsch C, Nachemson A: New observations on the mechanical behavior of lumbar discs. *Acta Orthop Scand* 1954;23:254-283.
42. Lin HS, Liu YK, Adams KH: Mechanical response of the lumbar intervertebral joint under physiological (complex) loading. *J Bone Joint Surg* 1978;60A:41-55.
43. Reuber M, Schultz A, Denis F, et al: Bulging of lumbar intervertebral discs. *J Biomech Eng* 1982;104:187-192.
44. Soini J, Antti-Poika I, Tallroth K, et al: Disc degeneration and angular movement of the lumbar spine: Comparative study using plain and flexion-extension radiography and discography. *J Spinal Disord* 1991;4:183-187.
45. Benini A: *Ischias ohne Bandscheibenvorfall.* Bern, Huber, 1976.
46. Knutsson F: The instability associated with disk degeneration in the lumbar spine. *Acta Radiol* 1944;25:593-609.
47. van Akkerveeken PF, O'Brien JP, Park WM: Experimentally induced hypermobility in the lumbar spine: A pathologic and radiologic study of the posterior ligament and annulus fibrosus. *Spine* 1979;4:236-241.
48. Kaigle AM, Pope MH, Fleming BC, et al: A method for the intravital measurement of interspinous kinematics. *J Biomech* 1992;25:451-456.
49. Hansson TH, Keller TS, Spengler DM: Mechanical behavior of the human lumbar spine: II. Fatigue strength during dynamic compressive loading. *J Orthop Res* 1987;5:479-487.
50. Brinckmann P, Johannleweling N, Hilweg D, et al: Fatigue fracture of human lumbar vertebrae. *Clin Biomech* 1987;12:94-96.
51. Adams MA, Hutton WC: The relevance of torsion to the mechanical derangement of the lumbar spine. *Spine* 1981;6:241-248.
52. Liu YK, Goel VK, Dejong A, et al: Torsional fatigue of the lumbar intervertebral joints. *Spine* 1985;10:894-900.
53. Adams MA, Hutton WC: Gradual disc prolapse. *Spine* 1985;10:524-531.
54. Wilder DG, Pope MH, Frymoyer JW: The biomechanics of lumbar disc herniation and the effect of overload and instability. *J Spinal Disord* 1988;1:16-32.
55. Keller TS, Hansson TH, Holm SH, et al: In vivo creep behavior of the normal and degenerated porcine intervertebral disk: A preliminary report. *J Spinal Disord* 1988;1:267-278.
56. Lin HS, Liu YK, Ray G, et al: Systems identification for material properties of the intervertebral joint. *J Biomech* 1978;11:1-14.
57. Shirazi-Adl SA, Shrivastava SC, Ahmed AM: Stress analysis of the lumbar disc-body unit in compression: A three-dimensional nonlinear finite element study. *Spine* 1984;9:120-134.
58. Monroe B, Goel VK, Gilbertson L, et al: Role of disc injury in producing disc degeneration: A finite element composite approach. Proceedings of WAM-ASME. Anaheim, CA, Nov. 8-11, 1992.

59. Goel VK, Konj WZ, Han JS, et al: Role of muscles in lumbar spine mechanics: A finite element investigation. Proceedings of the Annual Meeting of the Orthopaedic Research Society, San Francisco, CA, Feb. 18-22, 1993.

Chapter 29

The Effect of Physical Factors on Disk Cell Metabolism

Jill P. G. Urban, PhD

Introduction

The composition and organization of the disk matrix, and, therefore, its mechanical behavior, ultimately depend on the balance between the rate at which cells lay down their macromolecular components, such as collagen and proteoglycans, and the rate at which these components of the matrix are broken down and lost from the tissue. As long as the rates of matrix production and breakdown are in balance, the composition of the disk will remain constant. However, if, for example, the rate for proteoglycan synthesis decreases or its rate of breakdown increases, the inevitable result will be a decrease in the proteoglycan content of the matrix. Proteoglycan loss, found in disk degeneration,[1-4] must result from some such change in metabolic activity.

There have been relatively few studies on cell metabolism in the disk. However, in vivo radio-tracer studies in guinea pigs,[5] rabbits,[6,7] mice,[8,9] and dogs[10,11] have shown that proteoglycans are synthesized well into adult life. Turnover appeared to be around 30 days in young animals,[5,7,9] but it was estimated at around 18 to 30 months in adult dogs and rabbits.[7,10,12] In vivo labeling studies showed that a large aggregating proteoglycan is synthesized, but after synthesis it becomes progressively smaller with time, and it loses its ability to bind hyaluronic acid, possibly because of proteolytic removal of the hyaluronic acid-binding region.[13] Little is known about metabolism of other matrix constituents such as collagen. There is indirect evidence, however, that the annulus may remodel; changes in crosslink pattern indicating new collagen synthesis have been seen in disk herniations[14] and in disks adjacent to degenerated disks.[15] In addition, the ratio of collagen I/II in the annulus also appears to alter with age.[16] Very few in vitro studies of disk metabolism have been reported, probably because of experimental difficulties.[17] However, such studies have shown that the disk cells synthesize not only matrix components,[18] but also proteinases active in the degradation of matrix components.[19] New methods for disk explant culture[20,21] and for culture of disk cells[22,23] have been developed recently, and these allow the study of matrix synthesis in human disks taken during surgical procedures. Initial studies show that in human adult disks, turnover of proteoglycans is slow, with a turnover time of two to four years.[24]

The factors that regulate disk cell metabolism are not well understood, but, like those of other cartilages, disk cells appear to respond to growth factors and cytokines.[25-27] The physical environment of the disk cells may also influence metabolism, and may have a detrimental effect under some conditions, ultimately leading to disk degeneration. In particular, failure of nutrition may cause cell necrosis and decay of the matrix.[28-30] In addition, the disk is also constantly under external loads resulting from body weight and muscle activity; the effect of load on the matrix and on the blood supply may also be important in governing cellular activity.[31,32] This chapter will discuss the effect of these two physical factors on cellular activity and the possible changes with age that may lead to the degenerative changes seen in the disk.

Mechanical Loads on the Disk

It is now known that mechanical loads are potent modifiers of matrix composition in cartilaginous tissues. In articular cartilage in particular, many studies have found that cartilage thickens and proteoglycan concentration increases in areas of cartilage that are habitually subjected to high loads in vivo.[33,34] Conversely, if loads are removed from a joint, cartilage thins and proteoglycans are lost.[35,36] The few studies performed on intervertebral disks have shown similar responses to changes in mechanical loads, with increased long-term exercise leading to increases in disk metabolic rate, overloading, as in bipedal mice, causing structural alterations, and fusion leading to changes in proteoglycan structure and, in some instances, leading to degeneration.[11,37-41] Mechanical factors may influence the path to degeneration; injury coupled to normal loading appears to lead to matrix degeneration in an animal model.[42] These mechanically induced modifications to cartilage or disk composition could result from alterations to systemic factors, such as lactate, potassium, or hormone levels as the result of exercise; from changes in blood flow to the joint, which affect cell nutrition; or from cellular responses to load in the disk itself. Thus, from such in vivo studies it is difficult to determine the mechanism responsible. In vitro work, however, has shown that cartilaginous cells respond directly to mechanical stress and that load can either stimulate or depress synthesis, depending on the nature of the load and of the cartilage.[31] These studies have also shown that the cells respond, not to load itself, but to the changes in their physical environment induced by applied mechanical loads.

The Physical Environment of the Chondrocyte

The cells of the adult disk are isolated from one another and are embedded in a matrix made up of a high concentration of polyanionic proteoglycans (Fig. 1). The fixed negative charges on these proteoglycans dictate the extracellular ionic composition of the disk. In accordance with the Gibbs Donnan equilibrium conditions, the tissue thus contains a high concentration of free cations and a low concentration of anions,[7,43] the local concentration of which varies with proteoglycan concentration. Table 1 summarizes the range of extracellular ionic concentrations found in the disk. These concentrations of inorganic ions impart a high osmotic pressure to the tissue, considerably higher than that of the surrounding plasma. As a result of this osmotic pressure difference, fluid tends to be imbibed by the disk matrix while swelling is resisted by the network of collagen fibers and the load on the disk itself.

In addition to controlling the concentration of major ions around the cells and extracellular osmolality, an increase in fixed charge density leads to an increase in $H+$ concentrations and thus a drop in pH of about 0.5 pH units in the adult nucleus.[7] However pH is far more influenced by the lactic acid produced by the cells; the disk is avascular, oxygen concentrations are consequently low and the disk's metabolism is mainly anaerobic.[45]

The cells of the disk are constantly exposed to varying mechanical loads, which result from changes in posture and from muscle activity. Loading deforms the disk, causing hydrostatic pressures to rise.[46] This increase in hydrostatic pressure disturbs osmotic equilibrium. If the load is maintained, fluid is expressed, which increases proteoglycan concentration and fixed charge density and, consequently, increases extracellular cation concentrations and osmolality (Fig. 2).[47] The disk cells thus routinely experience changes to their physical environment when loads are applied, and it is apparent that some of these changes can affect cellular metabolism. While there are indications that both fluid movement itself, possibly through the generation of streaming potentials,[31] and also cell deformation,[48] can affect metabolism, these signals have not been studied systematically. However, changes in both fluid content and hydrostatic pressure have been shown to alter cell behavior.

The Effect of Fluid Loss on Matrix Metabolism When a static load is applied to the disk or to cartilage, fluid is expressed, and the amount of fluid loss is directly related to the magnitude of the load and to the composition of the tissue (Fig. 2).[49] The invariable finding in vitro is that fluid expression decreases matrix synthesis rates proportionally.[50] Cartilage swells little when excised and suspended in incubation medium,[7] and thus little is known about the response of cartilage chondrocytes to swelling. The disk, however, swells extensively in vitro because the osmotic equilibrium is disturbed by removal of the load normally on the disk; swelling can be prevented

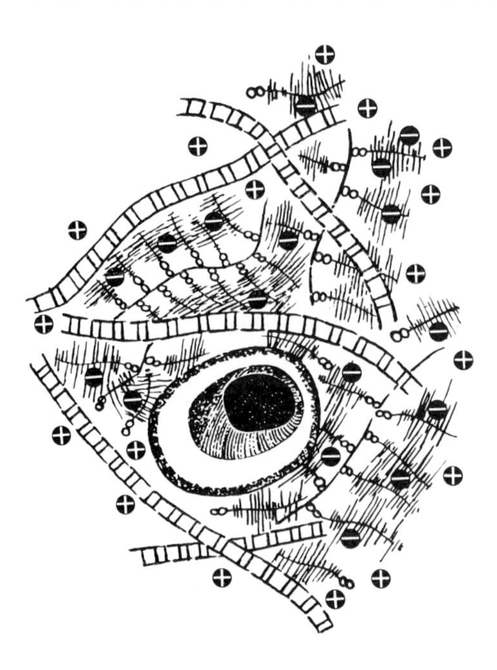

Fig. 1 *The environment of disk cells.*

Table 1 Extracellular ionic composition of the disk

Region	Na+ mM	Cl− mM	K+ mM	Ca++ mM
Nucleus	330-400	30-70	11-14	13-20
Outer annulus	220-280	50-110	6-9	4-10
Endplate	230-270	70-100	6-9	4-8
Serum	120-140	140-150	5	1.5-2.5

(Adapted with permission from Urban JP, Maroudas A: The measurement of fixed charge density in the intervertebral disc. *Biochim Biophys Acta* 1979;586:166-178.)

by applying either an osmotic or mechanical load.[51] The effect of hydration on disk metabolism has thus been studied over a large range; maximum synthesis rates appear to be at the hydration closest to that found in vivo, and synthesis rates fall quickly if the tissue swells or loses fluid (Fig. 3).[20,51]

Fluid Loss Changes Extracellular Osmolality When fluid is lost from the disk, the proteoglycan concentration increases, leading to an increase in extracellular cation concentrations and osmolality. The effect of osmolality on cell metabolism can be studied independently of changes in hydration by altering the ionic strength of the medium in which disk slices are suspended.[47] The fall in synthesis rates induced by increasing extracellular osmolality in this way is very similar to that induced by compressing the cartilage, thus caus-

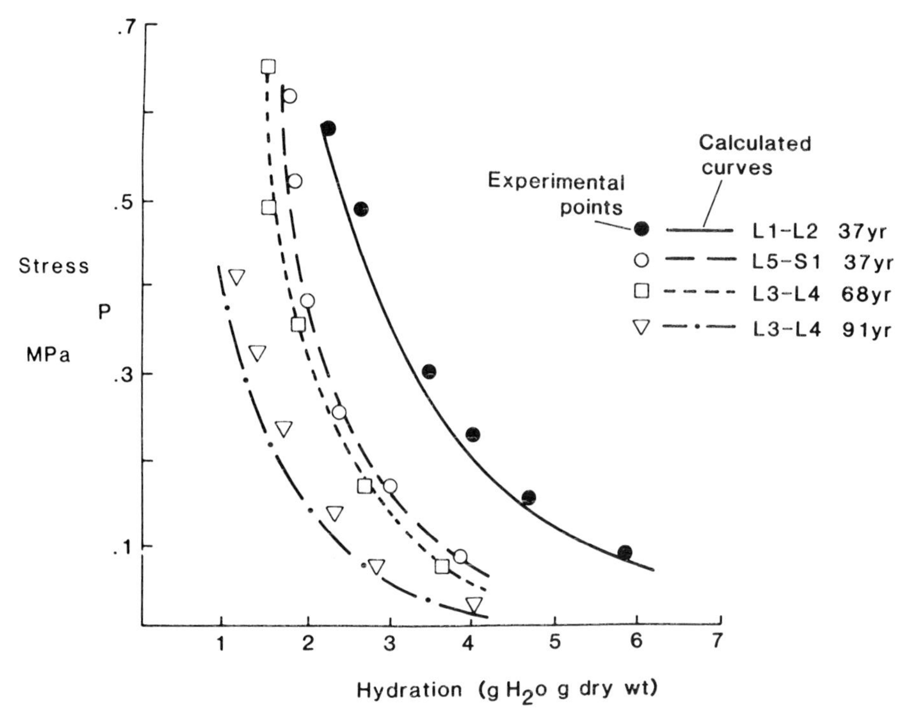

Fig. 2 *The effect of applied stress on the equilibrium hydration of nucleus slices taken from disks of various ages. (Reproduced with permission from Urban JP, McMullin JF: Swelling pressure of the lumbar intervertebral discs: Influence of age, spinal level, composition, and degeneration.* Spine *1988;13:179-187.)*

ing fluid loss and also increasing osmolality (Fig. 2).[50,52,53] It appears that the cells respond to changes in extracellular ion concentrations and osmolality rather than to changes in hydration as such. This result has been used recently to maintain disk metabolism at in vivo rates. If the disk is bathed in extracellular medium the tissue swells and synthesis rates fall rapidly. If, however, sucrose or sodium are added to the medium to return extracellular osmolality to that found in vivo, even though the tissue still swells the cells are retained at their in vivo volumes. Synthesis rates then appear to be very similar to those seen when the tissue itself is prevented from swelling by osmotic or mechanical loads.[54]

Extracellular osmolality could affect synthesis through its effect on the intracellular composition. Water crosses the plasma membrane of most cells readily; any change in external osmolality leads to a rapid flux of water across the cell membrane and thus a change in cell volume.[55] Such a change in cell volume will of course lead to alterations in the composition of ions and other components of the cytoplasm and could affect synthesis rates, by altering enzyme activities,[56] and, also, gene expression.[57]

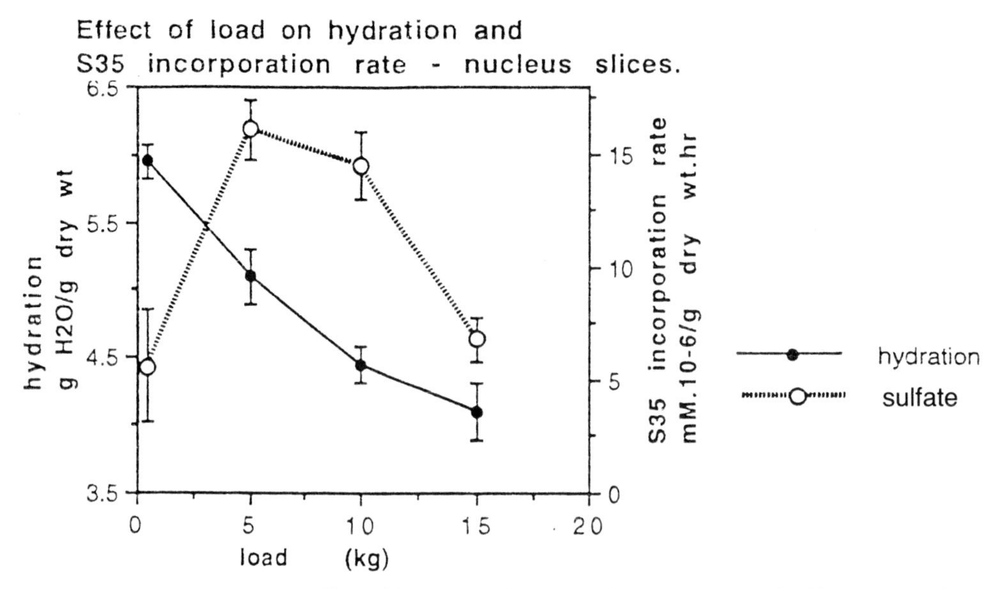

Fig. 3 *The effect of load on hydration and incorporation of sulfate into nucleus slices (bovine). (Adapted with permission from Ohshima H, Urban JP, Bergel DH: Measurement of metabolic profiles in the intervertebral disc in vitro by a new perfusion technique.* Trans Orthop Res Soc *1991;16:98.)*

Hydrostatic Pressure Also Influences Disk Metabolism The effects of hydrostatic pressure on cartilage metabolism can be examined independently of other changes to the tissue. Hydrostatic pressure can be applied uniformly, does not cause fluid flow, and does not deform the tissue macroscopically because water, the major component of cartilage and disk, is virtually incompressible over the physiologic ranges of pressure.

Hydrostatic pressures of the orders seen in cartilage and disk can affect many cellular processes.[58] It is, thus, not surprising that the changes in hydrostatic pressure seen on loading also affect metabolism in cartilaginous tissues.[59] In the disk, 20-second applications of hydrostatic pressures (10 to 25 atm) cause an increase in the rate of sulfate and amino acid incorporation in the nucleus and inner annulus over the following two hours (Fig. 4); a short mechanical stimulus that leads to a prolonged cellular response has also been reported in bone,[60] in the compression-resistant zone of the flexor tendon,[61] in osteoblasts[62] and in lung cells.[63] A longer application of pressure appears to decrease the stimulatory effect, and may even depress synthesis. These results suggest that hydrostatic pressure has at least two different effects on the synthetic pathway, one stimulatory and the other inhibitory; the net effect depends on the relative effect of each component.

It is of interest to note the different effect of the same pressure applied to different regions of the disk. The inner annulus is stimu-

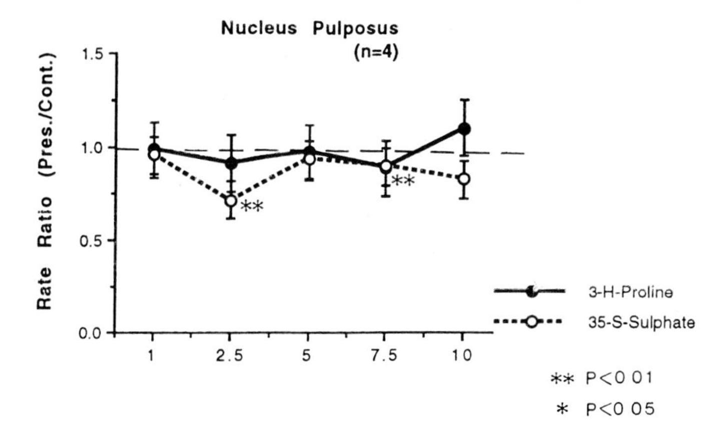

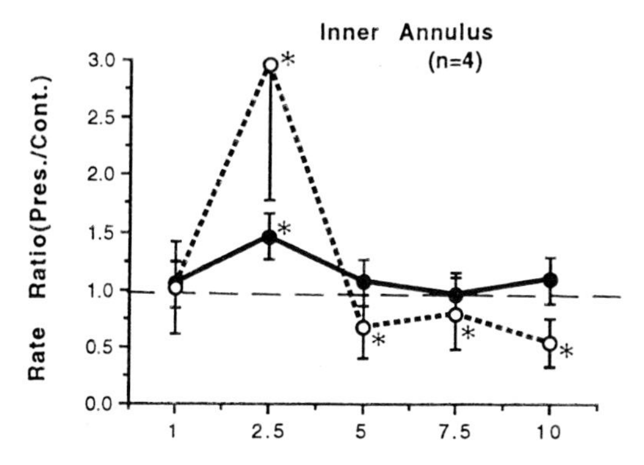

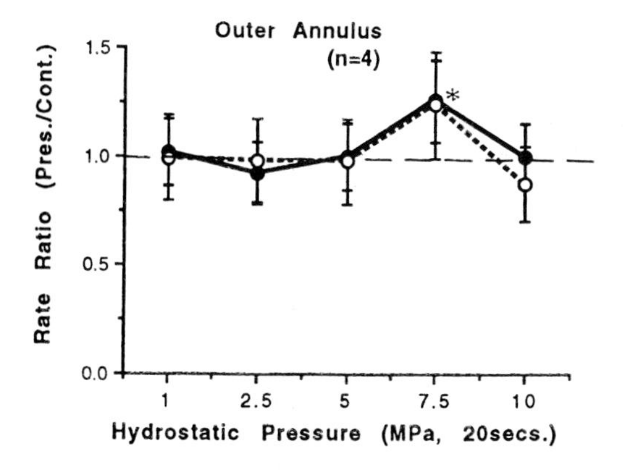

Fig. 4 *The effect of hydrostatic pressure on synthesis rates (S^{35} (o), proline (•)) in the nucleus, inner annulus and outer annulus of bovine disks.*

lated far more than the nucleus by application of physiologic pressures, while the outer annulus cells, which may be more fibroblastic than chondrocytic, do not appear to respond to hydrostatic pressure at all.

Effect of Aging on the Response to Mechanical Loads

The composition of the disk changes significantly with age. Hydration falls,[64,65] proteoglycan/collagen ratios decrease, and the glycosaminoglycan composition changes, with an increase in the proportion of keratan sulfate relative to chondroitin sulfate, particularly in the nucleus.[1-3] The collagen network of the annulus and nucleus becomes more disorganized,[2,3,30] and nonenzymic glycation[66] increases. All these changes may affect the tissue's response to mechanical load. The fall in fixed charge density, for example, will alter the extracellular ionic composition and osmotic pressure, and thus affect tissue turgor. Disks of low proteoglycan content have a higher hydraulic permeability and lower swelling pressure, and thus an increased rate and extent of fluid loss under applied loads.[44] The extent of the extracellular changes in osmolality on loading are thus related to tissue properties. Pressure rises on loading may alter as the result of changes in the organization of the annulus; changes in pressure profiles have been observed with degeneration.[67] It is not known how the cell's metabolism is affected by such changes, but, as discussed above, because matrix synthesis appears to vary in proportion to changes in pressure and osmolality, changes seen in the disk matrix with age may further alter cellular metabolism.

Disk Nutrition

The disk is the largest avascular tissue in the body. Cells in the center of human lumbar disks may lie 7 to 8 mm away from the nearest blood supply.[68] Nutrients must therefore move from the blood vessels at the margins of the disk, through the endplate and through the matrix to these cells (Fig. 5), and, thus, the actual pericellular concentration of any nutrient or metabolic by-product is governed by factors that influence transport.

Transport Into the Disk From the Surrounding Blood Vessels

The disk appears to be fed by two distinct routes—from the blood vessels at the disk endplate interface and from the rich capillary bed embedded in the soft tissue at the periphery of the disk.[29,69] This second route is not thought to be at risk; however, it has been shown in several studies that the route through the endplate is affected by degeneration or injury and diminishes with aging.[28,70-74]

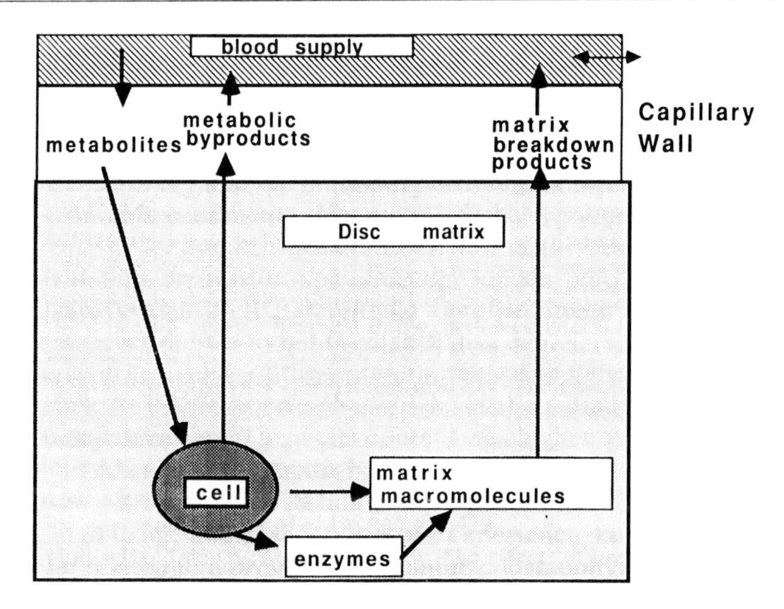

Fig. 5 *Transport of solutes from circulation to the disk cells. (Adapted with permission from Urban JP: Solute transport between tissue and environment, in Maroudas A, Kuettner K (eds):* Methods in Cartilage Resources. *London, Academic Press, 1990, sect 10, pp 241-273.)*

The Blood Supply to the Disk From the Vertebral Bodies The adult disk is avascular and exchanges nutrients and wastes via vascular buds, or marrow spaces which penetrate the bony endplate and are in direct contact with the cartilaginous endplate, and via small capillaries, which terminate at the cartilage endplate.[27,69] The exchange area formed from these contacts occupies about 10% to 20% of the endplate in human adults.[75,76] Factors such as smoking[77] and vibration,[32] which may reduce blood flow to these capillaries, have been shown to lead to a fall in transport of low molecular weight solutes into the disk, and are correlated with an increase in disk degeneration and low back pain.[78,79]

The Cartilaginous Endplate Once nutrients leave the blood vessels, they diffuse through the cartilaginous endplate into the disk. The endplate consists of a thin (0.1 to 1.6 mm) layer of hyaline cartilage separated from the bone at the outer annulus by a tidemark and a layer of calcified cartilage.[72,75] Because the composition and thus transport properties of the endplate cartilage are similar to that of other hyaline cartilages, the cartilaginous endplate should not ordinarily provide a barrier to the transport of substances into or out of the disk. However, any increase in calcification of this cartilage could severely limit nutrient transport as has been shown in degenerate disks.[26]

Transport Through the Matrix

Nutrients reach the surface of the disk from the blood supply, equilibrate with the disk matrix, and then diffuse several millimeters through the dense proteoglycan-rich matrix to reach the cells in the center of the nucleus. The factors that influence partition and diffusion of solutes through the matrix have been extensively reviewed.[7,80,81] In general, small neutral solutes such as oxygen and amino acids (around 100 MW), can enter the matrix freely. However, even slightly larger solutes, such as glucose (180 MW), are sterically excluded to some extent, and the degree of exclusion increases steeply with increase in solute size and increase in proteoglycan concentration.[7] The concentration of a growth factor, such as IGF-1, in cartilage, for example, could be only 4% to 10% of that present in the serum.[82]

Small nutrients move through the disk matrix to the cells mainly by diffusion under gradients set up by consumption or production of the solute by the cells. For small solutes, such as amino acids or glucose, the diffusivity in the matrix is 30% to 50% of that in solution, and varies with tissue hydration.[7,81] Diffusivity is thus affected by the changes in fluid content that occur under load.[83]

Effect of Mechanical Loads on Transport of Nutrients Through the Matrix It has often been suggested that the flow of fluid in and out of the disk, which results from pressure gradients during loading, is essential for the supply of nutrients. However, this suggestion is not supported either experimentally or theoretically. In dogs in vivo the movement of radioactive tracers into the disk was similar to that expected from diffusion whether net fluid movement was into or out of the disk.[84] In vitro studies on disk[85] and on cartilage[86] also showed that diffusion was the major mechanism for transport of nutrients, such as oxygen, glucose, and amino acids. This result is not surprising, because diffusivity for small solutes is much greater than the hydraulic permeability (which governs the rate of fluid flow through the tissue).[81] Also, in humans the overall direction of fluid flow is out of the disk for about 16 hours of 24,[44,87] and it is thus difficult to envisage how the cells could survive during such a prolonged period of fluid loss if they relied on fluid transport to supply basic nutrients, such as oxygen and glucose, or to remove lactate. Fluid movement, however, may be very important for the movement of larger solutes, such as growth factors, cytokines, and enzymes, because convection can significantly enhance the rate of transport of such molecules through the matrix of cartilaginous tissues.[86]

While fluid flow may have no effect on the movement of small nutrients through the matrix itself, loading may affect nutrition because of other factors. Loading causes shape changes which affect diffusion distances and thus influence the overall rate of diffusion into a disk.[88] A considerable degree of fluid loss occurs during the day's activities, and the consequent change in proteoglycan concentration could affect transport through the matrix and thus alter pericellular concentration of nutrients.[83] Loading may also affect the blood supply to the disk; experimental work suggests that vibration, for instance, diminishes the blood supply to the disk and thus affects nutrition in this manner.[32,89,90]

Cellular Activity

Because transport of small solutes takes place mainly by diffusion under concentration gradients set up by cellular activities, the rates of cell metabolism have a strong influence on nutrient transport and the development of nutrient concentration gradients throughout the disk.

Cellular Metabolic Activity Several studies have shown that the cells of cartilage and the disk use both glycolysis and oxidative phosphorylation to obtain energy.[45,91-94] However, a recent study indicated that virtually all glucose was consumed by glycolysis.[95] These cells do, however, consume oxygen at a rate independent of oxygen tension until relatively low oxygen concentrations are reached.[45] Below about 5% oxygen, the rate of consumption of oxygen falls with fall in tension, and lactic acid production rises steeply; the disk thus appears to exhibit a strong Pasteur effect (Fig. 6).[45,54] The glycolysis rates and oxygen consumption rates per cell have been shown to be relatively similar for all cartilaginous tissues, although there is some variation with age and source of cartilage.[45] However, because the cell density in the disk varies with disk thickness, as shown in Figure 7, the actual consumption or production of metabolites per volume of tissue can vary dramatically. For example, although the oxygen consumption rate per cell in young bovine cartilage is only 15% greater than that of an adult bovine cartilage, higher cell densities cause the oxygen consumption rate per gram of tissue to be nearly four times as high.

Concentration Profiles of Oxygen, Lactate, and pH The gradients in oxygen and lactate concentration that occur in the disk are set up by the metabolic activity of the cells; in a dead disk there are no oxygen concentration gradients. The concentration profile thus depends on the balance between the rate of transport through the matrix to the cells and the rate of consumption by the cells. Because oxygen consumption is high relative to its transport rate, steep concentration gradients develop and measured oxygen concentrations in the center of the disk may be only 2% to 4% of those at the periphery.[45,96] However, because sulfate consumption is low, the sulfate gradient is relatively flat, even though sulfate diffusivity and partition coefficients are lower than those for oxygen.[10] Because lactate is produced by the cells and the rate of production is greatest at low oxygen concentrations, lactate concentrations are highest in

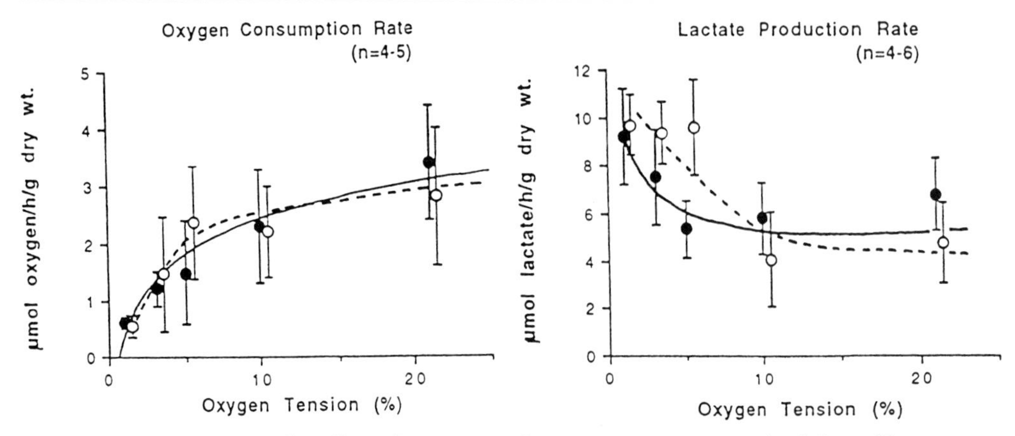

Fig. 6 *The effect of oxygen tension on oxygen consumption* **left**, *and lactate production* **right**, *rates in the nucleus (•) and outer annulus (o) of bovine disks. (Adapted with permission from Ishihara H, Urban JP: The effect of low oxygen concentrations and metabolic inhibitors on protein and proteoglycan synthesis rates in the intervertebral disc.* Trans Orthop Res Soc *1992.)*

the center of the nucleus, where they can rise to eight to ten times those at the periphery.[45] As a result, pH is lowest in this region (Fig. 8).

Does Supply of Nutrients Govern Cell Density? Stockwell[97] investigated the relationship between cell density and cartilage thickness over a large range of species and joint sizes. He found that the number of cells varied inversely in relation to cartilage thickness and was nearly constant per unit surface area. This relationship also exists in the disk, but, for equivalent thicknesses, the cell density of cartilage is always greater than that of the disk (Fig. 7). This difference could arise because the exchange area is always lower in the disk than in articular cartilage, which receives nutrients through its entire surface from the synovial fluid.[7,75] Because exchange area is an important factor in determining flux of nutrients,[98] articular cartilage is better supplied with nutrients than is the disk. The relationship between cell density, cartilage or disk thickness, and exchange area suggests that cell number could depend on nutrient supply.[96] Although this mechanism is speculative in the disk, nutritional factors have been shown to control cell density in tumor spheroids.[99]

Effect of Oxygen and Lactate Concentrations on Matrix Production in the Disk As can be seen from Figure 8 there are steep concentration gradients in both oxygen and pH throughout the disk. Both oxygen concentrations and extracellular pH have been found to affect matrix metabolism in other cartilaginous tissues[100] and in the disk. The maximum synthesis rate occurs not when the lactic acid concentration is zero but when it is 6 to 8 mM, which is approxi-

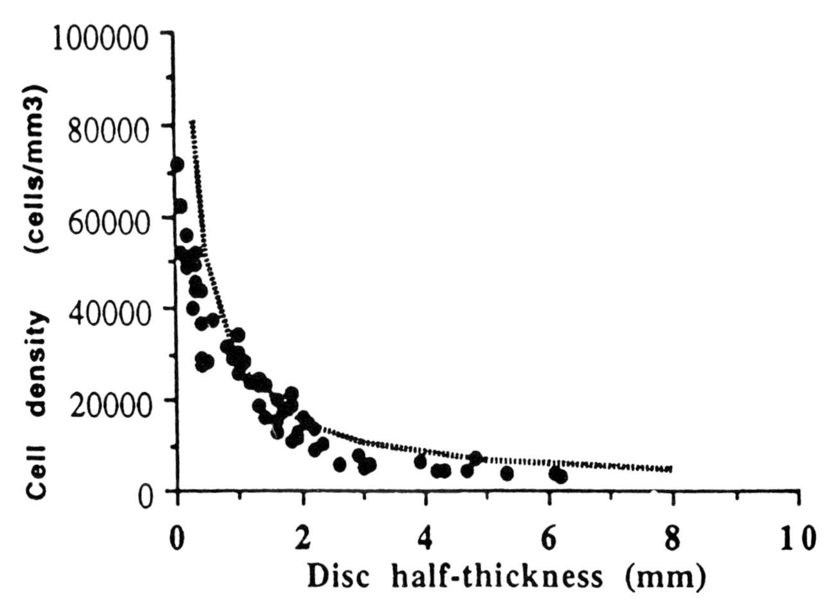

Fig. 7 *The variation of cell density with disk thickness. Samples were taken from mouse, rat, rabbit, pig, dog, and human disks. (Adapted with permission from Stairmand JW, Holm S, Urban JP: Factors influencing oxygen concentration gradients in the intervertebral disc: A theoretical analysis.* Spine *1991;16:444-449.)*

mately the concentration found in the center of the nucleus. However, once the lactic acid concentration rises beyond 10 mM, and pH falls below pH6.8, matrix synthesis rates fall steeply.[101] Low values of pH (below pH6.5) have been seen in some degenerate disks at surgery.[102-104] Such low values of pH lead to a significant reduction in matrix synthesis rates (Fig. 9) and could be a cause of the observed degeneration.

Oxygen concentrations also influence matrix synthesis rates, particularly in the nucleus;[54] below 5% oxygen, sulfate incorporation rates (a measure of proteoglycan synthesis), fall steeply in the nucleus but are less affected in the annulus. It has been suggested that the increase in the ratio of keratan sulfate to chondroitin sulfate, seen in the nucleus with maturation and age, results from alterations in glycosaminoglycan metabolism that arise as the result of a fall in oxygen tension.[105] There is no direct evidence to support this assumption. Present evidence, from tracer labeling studies performed in vivo and in vitro, suggests that the aggrecan molecule synthesized in the disk is similar to that found in other cartilages, but that it breaks down to smaller monomers within 30 to 60 days.[13] A decrease in oxygen concentrations below some critical level (around 5% in the bovine disk), thus not only affects energy metabolism in the disk by increasing lactate concentrations and lowering pH, but

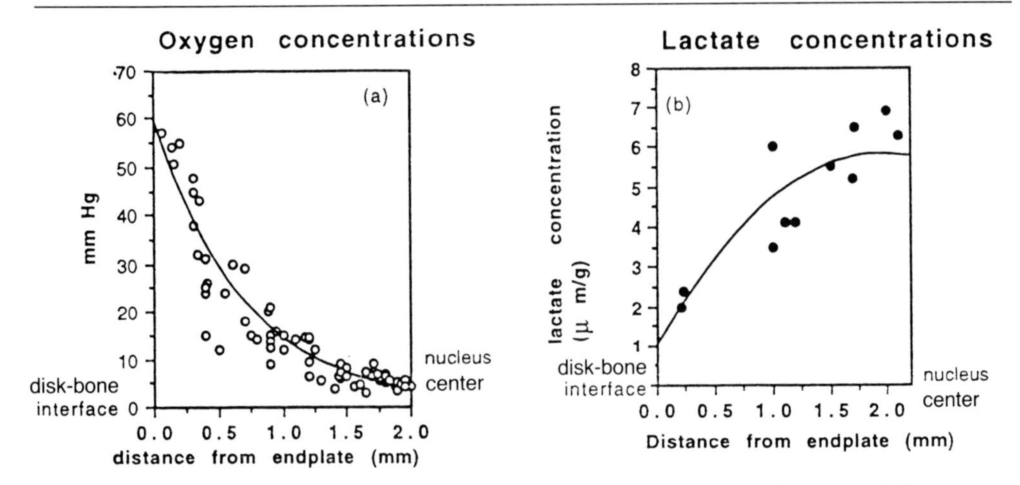

Fig. 8 *Profiles of O_2 and lactate concentrations across the canine disk. (Adapted with permission from Holm S, Maroudas A, Urban JP: Nutrition of the intervertebral disc: Solute transport and metabolism.* Connect Tissue Res *1981;8:101-119.)*

also leads directly to a fall in matrix production and may alter the type of matrix produced.

The results discussed here therefore suggest that a fall in the supply of nutrients, leading to a fall in oxygen levels and, consequently, a rise in lactate concentrations and a fall in pH, would eventually lead to a loss of proteoglycans and, possibly, to degeneration.

Effects of Aging on Disk Nutrition

Many of the pathways shown in Figure 5 change with age. In middle age, as the result of degenerative arterial disease, there is a rapid decline in the number of arteries that feed the disk, particularly around the fifth lumbar vertebral body.[106,107] This decline in functioning arteries could affect the blood supply to the disk. The permeability of the endplate has been found to decrease with age in several species,[28,51,72,74,108] both because the marrow spaces through the endplate become partly occluded, limiting the exchange area between disk and blood supply, and because the endplate calcifies, restricting diffusion of nutrients. As the disk loses hydration with aging, solute diffusivity decreases and thus transport slows. All these changes adversely affect the transport of nutrients to the cells of the disk. It is not known how cellular metabolism is affected, but, as discussed above, matrix synthesis rates fall if oxygen concentration or pH falls below a critical level. Presumably, the loss of other nutrients, such as glucose, also has deleterious effects on the matrix composition. If, as suggested, cell density is controlled by levels of nutrient supply, the concentration of viable cells would decrease as nutrient supplies fell.

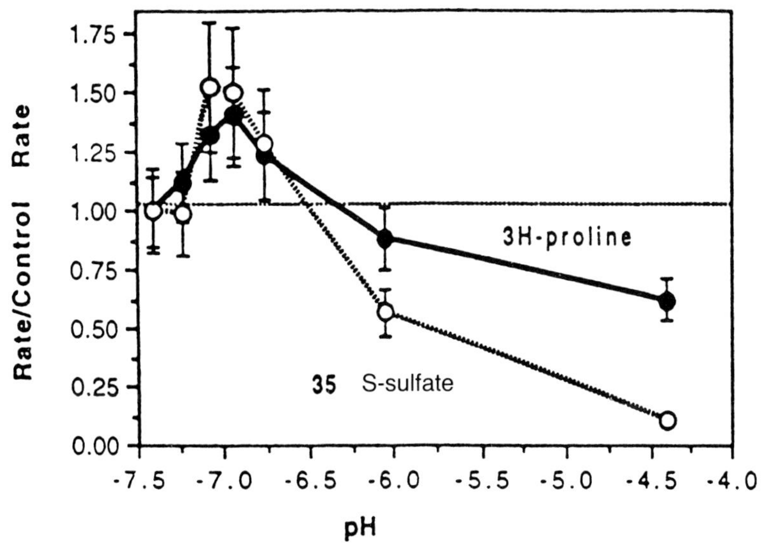

Fig. 9 *The effect of pH on incorporation of sulfate and proline into the nucleus of bovine disks. (Reproduced with permission from Ohshima H, Urban JP: The effect of lactate and pH on proteoglycan and protein synthesis rates in the intervertebral disc.* Spine *1992;17:1079-1082.)*

Conclusions

Disk cells are now known to make matrix components, as well as factors responsible for matrix breakdown.[18,19,24] A change in either of these activities may lead to alterations in matrix composition and properties. Little is known about how the balance between these factors is regulated. It is clear, however, that any change to the cell's physical microenvironment, such as an increase in extracellular osmolality or hydrostatic pressure, or a fall in oxygen levels or in pH, can lead to a rapid alteration in matrix synthesis rates,[51,53,54,109] and, in articular cartilage at least, can affect the pattern of proteinase-inhibitor release.[110] Indeed, recent work on isolated chondrocytes has suggested that changes in osmolality may uncouple the balance between matrix synthesis and breakdown.[111]

The composition of the disk matrix changes considerably with age, leading to profound changes in the microenvironment of the disk cells. From results obtained to date, these changes would significantly alter cellular metabolism; because of cellular heterogeneity,[23] some regions of the disk would appear particularly sensitive to these changes. However, present information has only shown the immediate effects of environmental change on cellular behavior. Initial studies have indicated that chondrocytes can adapt, at least partially, to a wide range of extracellular conditions. Indeed, chondrocytes in culture are in a very abnormal physical environment and yet some are able to synthesize a normal-appearing matrix around themselves.[112] In the disk, too, the cells of the nucleus appear able to rebuild a matrix even after the proteoglycans have been almost completely removed by chymopapain.[113-115] It should be noted, however, that these latter experiments were performed in previously healthy animal disks, containing viable cells. In the human, many cells (up to 50% in the nucleus)[116] appear necrotic in middle age and beyond, and the changes with age seen in the disk may be the inevitable result if too few cells remain to maintain the matrix.

Acknowledgements

I am grateful to the Arthritis and Rheumatism Council for financial support.

References

1. Pearce RH, Grimmer BJ, Adams ME: Degeneration and the chemical composition of the human lumbar intervertebral disc. *J Orthop Res* 1987;5:198-205.
2. Eyre DR: Biochemistry of the intervertebral disc. *Int Rev Connect Tissue Res* 1979;8:227-281.
3. Pritzker KP: Aging and degeneration in the lumbar intervertebral disc. *Orthop Clin North Am* 1977;8:66-77.
4. Lipson SJ, Muir H: 1980 Volvo award in basic science. Proteoglycans in experimental intervertebral disc degeneration. *Spine* 1981;6:194-210.
5. Lohmander S, Antonopoulos CA, Friberg U: Chemical and metabolic heterogeneity of chondroitin sulfate and keratin sulfate in guinea pig cartilage and nucleus pulposus. *Biochim Biophys Acta* 1973;304: 430-448.
6. Souter WA, Taylor TK: Sulphate acid mucopolysaccharide metabolism in the rabbit intervertebral disc. *J Bone Joint Surg* 1970;52B:371-384.
7. Maroudas A: Physical chemistry of articular cartilage and the intervertebral disc, in Sokoloff L (ed): *The Joints and Synovial Fluid.* New York, NY, Academic Press, 1980, pp 240-293.
8. Venn G, Mason RM: Biosynthesis and metabolism in vivo of intervertebral-disc proteoglycans in the mouse. *Biochem J* 1983;215:217-225.
9. Venn G, Mason RM: Changes in mouse intervertebral-disc proteoglycan synthesis with age: Hereditary kyphoscoliosis is associated with elevated synthesis. *Biochem J* 1986;234:475-479.
10. Urban JP, Holm S, Maroudas A: Diffusion of small solutes into the intervertebral disc: An in vivo study. *Biorheology* 1978;15:203-221.
11. Cole TC, Ghosh P, Hannan NJ, et al: The response of the canine intervertebral disc to immobilization produced by spinal arthrodesis is dependent on constitutional factors. *J Orthop Res* 1987;5:337-347.
12. Katsura N, Davidson EA: Metabolism of connective tissue polysaccharides in vivo: IV. The sulphate group. *Biochim Biophys Acta* 1966;121:134-142.
13. McDevitt CA: Proteoglycans of the intervertebral disc, in Ghosh P (ed): *The Biology of the Intervertebral Disc.* Boca Raton, FL, CRC Press, 1988, pp 151-170.

14. Lipson SJ: Metaplastic proliferative fibrocartilage as an alternative concept to herniated intervertebral disc. *Spine* 1988;13:1055-1060.
15. Herbert CM, Lindberg KA, Jayson MIV, et al: Changes in the collagen of human intervertebral discs during ageing and degenerative disc disease. *J Mol Med* 1975;1:79-91.
16. Brickley-Parsons D, Glimcher MJ: Is the chemistry of collagen in intervertebral discs an expression of Wolff's law?: A study of the human lumbar spine. *Spine* 1984;9:148-163.
17. Urban JP, Maroudas A: Swelling of the intervertebral disc in vitro. *Connect Tissue Res* 1981;9:1-10.
18. Oegema TR Jr, Bradford DS, Cooper KM: Aggregated proteoglycan synthesis in organ cultures of human nucleus pulposus. *J Biol Chem* 1979;254:10579-10581.
19. Liu J, Roughley PJ, Mort JS: Identification of human intervertebral disc stromelysin and its involvement in matrix degradation. *J Orthop Res* 1991;9:568-575.
20. Bayliss MT, Urban JP, Johnstone B, et al: In vitro method for measuring synthesis rates in the intervertebral disc. *J Orthop Res* 1986;4:10-17.
21. Johnstone B: Explant culture of the intervertebral disc, in Maroudas A, Kuettner K (eds): *Methods in Cartilage Research*. London, Academic Press, 1990, chap 32, pp 123-129.
22. Guo JF, Jourdian GW, MacCallum DK: Culture and growth characteristics of chondrocytes encapsulated in alginate beads. *Connect Tissue Res* 1989;19:277-297.
23. Banks GM, Chelberg MK, Oegema TR Jr: Identification of heterogeneous cell populations in normal intervertebral disc. *Trans Orthop Res Soc* 1992;17:681.
24. Bayliss MT, Johnstone B, O'Brien JP: 1988 Volvo award in basic science. Proteoglycan synthesis in the human intervertebral disc: Variation with age, region and pathology. *Spine* 1988;13:972-981.
25. Thompson JP, Oegema TR Jr, Bradford DS: Stimulation of mature canine intervertebral disc by growth factors. *Spine* 1991;16:253-260.
26. Shinmei M, Kikuchi T, Yamagishi M, et al: The role of interleukin-1 on proteoglycan metabolism of rabbit annulus fibrosus cells cultured in vitro. *Spine* 1988;13:1284-1290.
27. Davidson EA, Small W: Metabolism in vivo of connective-tissue mucopolysaccharides: I. Chondroitin sulfate and keratosulfate of nucleus pulposus. *Biochim Biophys Acta* 1963;69:445-452.
28. Nachemson A, Lewin T, Maroudas A, et al: In vitro diffusion of dye through the end-plates and the annulus fibrosus of human lumbar intervertebral discs. *Acta Orthop Scand* 1970;41:589-607.
29. Brodin H: Paths of nutrition in articular cartilage and intervertebral discs. *Acta Orthop Scand* 1955;24:177-183.
30. Eyre DR, Mooney V, Caterson B, et al: The intervertebral disc, in Frymoyer JW, Gordon SL (eds): *New Perspectives on Low Back Pain*. Park Ridge, IL, American Academy of Orthopaedic Surgeons, 1989, chap 5, pp 131-214.
31. Sah RL, Grodzinsky AJ, Plawas AHK, et al: Effects of static and dynamic compression on matrix metabolism in cartilage explants, in Kuettner KE, Schleyerbach R, Peyron JG, et al (eds): *Articular Cartilage and Osteoarthritis*. New York, NY, Raven Press, 1992, chap 26, pp 373-392.
32. Hirano N, Tsuji H, Ohshima H, et al: Analysis of rabbit intervertebral disc physiology based on water metabolism: II. Changes in normal intervertebral discs under axial vibratory load. *Spine* 1988;13: 1297-1302.

33. Oláh EH, Kostenszky KS: Effect of loading and prednisolone treatment on the glycosaminoglycan content of articular cartilage in dogs. *Scand J Rheumatol* 1976;5:49-52.
34. Saamanen A-M, Tammi M, Kiviranta I, et al: Maturation of proteoglycan matrix in articular cartilage under increased and decreased joint loading: A study in young rabbits. *Connect Tissue Res* 1987;16:163-175.
35. Palmoski M, Perricone E, Brandt KD: Development and reversal of a proteoglycan aggregation defect in normal canine knee cartilage after immobilization. *Arthritis Rheum* 1979;22:508-517.
36. Caterson B, Lowther DA: Changes in the metabolism of the proteoglycans from sheep articular cartilage in response to mechanical stress. *Biochim Biophys Acta* 1978;540:412-422.
37. Holm S, Nachemson A: Variations in the nutrition of the canine intervertebral disc induced by motion. *Spine* 1983;8:866-874.
38. Holm S, Nachemson A: Nutritional changes in the canine intervertebral disc after spinal fusion. *Clin Orthop* 1982;169:243-258.
39. Higuchi M, Abe K, Kaneda K: Changes in the nucleus pulposus of the intervertebral disc in bipedal mice: A light and electron microscopic study. *Clin Orthop* 1983;175:251-257.
40. Neufeld JH: Induced narrowing and back adaptation of lumbar intervertebral discs in biomechanically stressed rats. *Spine* 1992;17: 811-816.
41. Cole TC, Burkhardt D, Ghosh P, et al: Effects of spinal fusion on the proteoglycans of the canine intervertebral disc. *J Orthop Res* 1985;3:277-291.
42. Osti OL, Vernon-Roberts B, Fraser RD: 1990 Volvo Award in experimental studies. Anulus tears and intervertebral disc degeneration: An experimental study using an animal model. *Spine* 1990;15:762-767.
43. Urban JP, Maroudas A: The measurement of fixed charge density in the intervertebral disc. *Biochim Biophys Acta* 1979;586:166-178.
44. Urban JP: Factors influencing the fluid content of intervertebral discs. in Staub N (ed): *Advances in Microcirculation 13*, New York, New York, Springer-Verlag, pp 160-170.
45. Holm S, Maroudas A, Urban JP, et al: Nutrition of the intervertebral disc: Solute transport and metabolism. *Connect Tissue Res* 1981;8: 101-119.
46. Nachemson A, Elfström G: Intravital dynamic pressure measurements in lumbar discs: A study of common movements, maneuvers and exercises. *Scand J Rehabil Med* 1970;(Suppl):1-40.
47. Urban JP, Hall AS: Changes in cartilage osmotic pressure in response to loads and their effects on chondrocyte metabolism, in Karalis TK (ed): *Swelling Mechanics: From Clays to Living Cells and Tissues.* Berlin, Springer-Verlag, 1992; pp 513-526.
48. Ingber D: Integrins as mechanochemical transducers. *Curr Opin Cell Biol* 1991;3:841-848.
49. Urban JP, McMullin JF: Swelling pressure of the lumbar intervertebral discs: Influence of age, spinal level, composition, and degeneration. *Spine* 1988;13:179-187.
50. Gray ML, Pizzanelli AM, Lee RC, et al: Kinetics of the chondrocyte biosynthetic response to compressive load and release. *Biochim Biophys Acta* 1989;991:415-425.
51. Ohshima H, Urban JP, Bergel DH: Measurement of metabolic profiles in the intervertebral disc in vitro by a new perfusion technique. *Trans Orthop Res Soc* 1991;16:98.

52. Urban JP, Bayliss MT: Regulation of proteoglycan synthesis rate in cartilage in vitro: Influence of extracellular ionic composition. *Biochim Biophys Acta* 1989;992:59-65.
53. Urban JP, Hall A: Physical modifiers of cartilage metabolism, in Kuettner KE, Schleyerbach R, Peyron JG, et al (eds): *Articular Cartilage and Osteoarthritis*. New York, NY, Raven Press, 1992, chap 27, pp 393-406.
54. Ishihara H, Urban JP: The effect of low oxygen concentrations and metabolic inhibitors on protein and proteoglycan synthesis rates in the intervertebral disc. *Trans Orthop Res Soc* 1992;18:208.
55. Hoffmann EK, Saimonsen LO: Membrane mechanisms in volume and pH regulation in vertebrate cells. *Physiol Rev* 1989;69:315-382.
56. Horowitz SB, Lau Y-T: A function that relates protein synthetic rates to potassium activity in vivo. *J Cell Physiol* 1988;135:425-434.
57. Burns CP, Rozengurt E: Extracellular Na+ and initiation of DNA synthesis: Role of intracellular pH and K+. *J Cell Biol* 1984;98: 1082-1089.
58. Jannasch HW, Marquis RE, Zimmerman AM (eds): *Current Perspectives in High Pressure Biology*. London, Academic Press, 1987.
59. Hall AC, Urban JP, Gehl KA: The effects of hydrostatic pressure on matrix synthesis in articular cartilage. *J Orthop Res* 1991;9:1-10.
60. Rubin CT, Lanyon LE: Kappa Delta Award paper. Osteoregulatory nature of mechanical stimuli: Function as a determinant for adaptive remodeling in bone. *J Orthop Res* 1987;5:300-310.
61. Koob TJ, Vogel KG, Thurmond FA: Compression loading in vitro regulates proteoglycan synthesis by fibrocartilage in tendon. *Trans Orthop Res Soc* 1991;16:49.
62. Jones DB, Nolte H, Scholubbers JG, et al: Biochemical signal transduction of mechanical strain in osteoblast-like cells. *Biomaterials* 1991;12:101-110.
63. Wirtz HR, Dobbs LG: Calcium mobilization and exocytosis after one mechanical stretch of lung epithelial cells. *Science* 1990;250:1266-1269.
64. Sah RL, Kim YJ, Doong J-Y, et al: Biosynthetic response of cartilage explants to dynamic compression. *J Orthop Res* 1989;7:619-636.
65. De Puky P: The physiological oscillation of the length of the body. *Acta Orthop Scand* 1935;6:338-347.
66. Hormel SE, Eyre DR: Collagen in the ageing human intervertebral disc: An increase in covalently bound fluorophores and chromophores. *Biochim Biophys Acta* 1991;1078:243-250.
67. McNally DS, Adams MA: Internal intervertebral disc mechanics as revealed by stress profilometry. *Spine* 1992;17:66-73.
68. Maroudas A, Stockwell RA, Nachemason A, et al: Factors involved in the nutrition of the human lumbar intervertebral disc: Cellularity and diffusion of glucose in vitro. *J Anat* 1975;120:113-130.
69. Crock HV, Goldwasser M, Yoshizawa H: Vascular anatomy related to the intervertebral disc, in Ghosh P (ed): *Biology of the Intervertebral Disc*. Baton Rouge, LA, CRC Press, 1988, pp 109-133.
70. McFadden KD, Taylor JR: End-plate lesions of the lumbar spine. *Spine* 1989;14:867-869.
71. Edelson JG, Nathan H: Stages in the natural history of the vertebral end-plates. *Spine* 1988;13:21-26.
72. Bernick S, Cailliet R: Vertebral end-plate changes with aging of human vertebrae. *Spine* 1982;7:97-102.
73. Torner M, Holm S: Studies of the lumbar vertebra end-plate region in the pig. *Ups J Med Sci* 1985;90:243-258.

74. Brown MD, Tsaltas TT: Studies on the permeability of the intervertebral disc during skeletal maturation. *Spine* 1976;1:240-244.
75. Roberts S, Menage J, Urban JP: Biochemical and structural properties of the cartilage end-plate and its relation to the intervertebral disc. *Spine* 1989;14:166-174.
76. Pope M, Wilder D, Booth J: The biomechanics of low back pain, in White AA III, Gordon SL (eds): American Academy of Orthopaedic Surgeons *Symposium on Idiopathic Low Back Pain*. St. Louis, MO, CV Mosby, 1982, chap 14, pp 252-295.
77. Holm S, Nachemson A: Nutrition of the intervertebral disc: Acute effects of cigarette smoking: An experimental animal study. *Ups J Med Sci* 1988;93:91-99.
78. Deyo RA, Bass JE: Lifestyle and low-back pain: The influence of smoking and obesity. *Spine* 1989;14:501-506.
79. Heliovaara M, Makela M, Knekt P, et al: Determinants of sciatica and low-back pain. *Spine* 1991;16:608-614.
80. Urban JP: Solute transport between tissue and environment, in Maroudas A, Keuttner K (eds): *Methods in Cartilage Research*. London, Academic Press, 1990, sect 10, pp 241-273.
81. Urban JP: Solute transport in articular cartilage and the intervertebral disc, in Hukins DWL (ed): *Connective Tissue Matrix, 2*. Boca Raton, FL, CRC Press, 1990, pp 44-65.
82. Maroudas A, Popper O, Grushko G: Partition coefficients of IGF-1 between cartilage and external medium in the presence and absence of FCS. *Trans Orthop Res Soc* 1991;16:398.
83. Koh K, Kusaka Y, Mifune T, et al: Self diffusion coefficient of water and its anisotropic property in bovine intervertebral discs analyzed by pulsed gradient nmr method. *Trans Orthop Res Soc* 1991;16:355.
84. Urban JP, Holm S, Maroudas A, et al: Nutrition of the intervertebral disc: Effect of fluid flow on solute transport. *Clin Orthop* 1982;170: 296-302.
85. Katz MM, Hargens AR, Garfin SR: Intervertebral disc nutrition: Diffusion verus convection. *Clin Orthop* 1986;210:243-245.
86. O'Hara BP, Urban JP, Maroudas A: Influence of cyclic loading on the nutrition of articular cartilage. *Ann Rheum Dis* 1990;49:536-539.
87. Adams MA, Dolan P, Hutton WC, et al: Diurnal changes in spinal mechanics and their clinical significance. *J Bone Joint Surg* 1990;72B:266-270.
88. Adams MA, Hutton WC: The effect of posture on diffusion into lumbar intervertebral discs. *J Anat* 1986;147:121-134.
89. Hirano N, Tsuji H, Ohshima H, et al: Analysis of rabbit intervertebral disc physiology based on water metabolism: 1. Factors influencing metabolism of the normal intervertebral discs. *Spine* 1988;13:1291-1296.
90. Ishihara H, Tsuji H, Hirano N, et al: Effects of continuous quantitative vibration on rheologic and biological behaviors of the intervertebral disc. *Spine* 1992;17(3 Suppl):S7-S12.
91. Marcus RE: The effect of low oxygen concentraiton on growth, glycolysis, and sulfate incorporation by articular chondrocytes in monolayer culture. *Arthritis Rheum* 1973;16:646-656.
92. Marcus RE, Srivastava, VM: Effect of low oxygen tensions on glucose-metabolizing enzymes in cultured articular chondrocytes. *Proc Soc Exp Biol Med* 1973;143:488-491.
93. Bywaters EGL: Metabolism of joint tissues. *J Pathol Bacteriol* 1937;44:247-268.
94. Rosenthal O, Bowie MA, Wagoner G: Studies on the metabolism of articular cartilage. *J Cell Physiol* 1917;221:233.

95. Henderson G, Mason R: Effect of oxygen tensions on synthesis of glycosaminoglycans in cartilage explant cultures. *Biochem Trans* 1991;19:364.
96. Kofoed H, Levander B: Respiratory gas pressures in the spine: Measurements in goats. *Acta Orthop Scand* 1987;58:415-418.
97. Stockwell RA: The interrelationship of cell density and cartilage thickness in mammalian articular cartilage. *J Anat* 1971;109:411-421.
98. Stairmand JW, Holm S, Urban JP: Factors influencing oxygen concentration gradients in the intervertebral disc: A theoretical analysis. *Spine* 1991;16:444-449.
99. Freyer JP, Sutherland RM: Regulation of growth saturation and development of necrosis in EMT6/Ro multicellular spheroids by the glucose and oxygen supply. *Cancer Res* 1986;46:3504-3512.
100. Gray ML, Pizzanelli AM, Grodzinsky AJ, et al: Mechanical and physiochemical determinants of the chondrocyte biosynthetic response. *J Orthop Res* 1988;6:777-792.
101. Ohshima H, Urban JP: The effect of lactate and pH on proteoglycan and protein synthesis rates in the intervertebral disc. *Spine* 1992;17:1079-1082.
102. Nachemson A: Intradiscal measurements of pH in patients with lumbar rhizopathies. *Acta Orthop Scand* 1969;40:23-42.
103. Diamant B, Karlsson J, Machemson A: Correlation between lactate levels and pH in discs of patients with lumbar rhizopathies. *Experientia* 1968;24:1195-1196.
104. Mooney V: A perspective on the future of low back research. *Spine* 1989;3:173-183.
105. Bosman T, Scott JE: Changes with age of KS and CS in the human intervertebral disc. Presented at the Federation of European Connective Tissue Societies Davos meeting 1992.
106. Ratcliffe JF: The anatomy of the fourth and fifth lumbar arteries in humans: An arteriographic study on one hundred live subjects. *J Anat* 1982;135:753-761.
107. Ratcliffe JF: The arterial anatomy of the adult human lumbar vertebral body: A microarteriographic study. *J Anat* 1980;131:57-79.
108. Hassler O: The human intervertebral disc: A micro-angiographical study on its vascular supply at various ages. *Acta Orthop Scad* 1969;40: 765-772.
109. Ishihara H, Urban JP, Hall AS: The effect of physiological hydrostatic pressures on synthesis in different regions of the intervertebral disc. *Orthop Trans* 1992.
110. Maciewicz RA, Hall AC, Urban JP: Regulation of the degradative potential of chondrocytes by load: The effects of hydrostatic pressure and osmotic pressure on the secretion of cysteine proteinases and their inhibitors from cartilage. *Biochem Soc Trans* 1992;19:396.
111. Aydelotte MB, Mok SS, Michal L, et al: Influence of changes in environmental osmotic pressure on the synthesis and accumulation of proteoglycans in matrix assembled by cultured articular chondrocytes. *Trans Orthop Res Soc* 1992;18:14.
112. Aydelotte MB, Schumacher BL, Kuettner KE: Heterogeneity of articular chondrocytes, in Kuettner KE, Schleyerbach R, Peyron JG, et al (ed): *Articular Cartilage and Osteoarthritis*. New York, NY, Raven Press, 1992, chap 16, pp 237-250.
113. Kitano S, Tsuji H, Hirano N, et al: Water, fixed charge density, protein contents, and lysine incorporation into protein in chymopapain-digested intervertebral disc of rabbit. *Spine* 1989;14:1226-1233.

114. Garvin PJ, Jennings RB: Long-term effects of chymopapain on intervertebral disks of dogs. *Clin Orthop* 1973;92:281-295.
115. Bradford DS, Oegema TR Jr, Cooper KM, et al: Chymopapain, chemonucleolysis and nucleus pulposus regeneration: A biochemical and biomechanical study. *Spine* 1984;9:135-147.
116. Trout JJ, Buckwalter JA, Moore KC: Ultrastructure of the human intervertebral disc: II. Cells of the nucleus pulposus. *Anat Record* 1982;204:307-314.

Future Directions

Determine the relationship between individuals' activity-related and environmental factors and disk degeneration.

Epidemiologic studies have provided some evidence about these relationships, but the main effects of different factors, including the aging processes, remain elusive. Improved knowledge is important to design appropriate preventive interventions. The use of such imaging techniques as magnetic resonance imaging (MRI) permits limited determinations of disk degeneration by noninvasive techniques, and the degree of degeneration may be possible to quantify. Cross-sectional studies of carefully selected populations could provide indications of specific relationships, which can then be explored experimentally. Longitudinal studies of populations with varied exposures to possible effectors would be particularly helpful. Intervention studies may also be possible when relationships become better clarified. Studies of twins have provided interesting data and should be pursued.

Determine the impact of the degree of disk degeneration on the development of disk herniation.

Clinical information indicates that there is an age interval during which the risk of herniation is particularly high. This age interval appears to vary from one disk level to another, and some disks herniate with regularity, whereas others rarely herniate. Other information suggests that in younger individuals, the nucleus pulposus material herniates, whereas with aging, the herniated tissue consists mostly of annular and endplate material.

Ideally, longitudinal studies would be used to address this topic, but information can also be obtained by specifically studying populations with disk herniations.

Determine the relationship between disk degeneration and osteoarthritis (OA) of extremity joints.

There is widespread belief that OA can exist as a generalized disease affecting multiple joints. The relationship of this condition to disk degeneration, including spondylosis, is not well established; there is a need to determine if there is a common underlying process that causes both conditions.

Epidemiologic techniques would be used to address this question, followed by specific biologic and biomechanical studies.

Determine at which levels disks degenerate as a function of aging.

Clinical information indicates that disks degenerate earlier at some disk levels than others, and that degeneration is rare at some levels. Careful documentation of this process using MRI combined with biomechanical analysis of stresses and strains would better define the role of mechanical factors in this process.

Epidemiologic techniques can be used to determine which levels degenerate and when. Biomechanical techniques can then be used to interpret the epidemiologic findings.

Develop clinical markers of disk degeneration (DD).

MRI, the only currently available technique to study DD in vivo, is expensive and at present is mainly qualitative. New imaging modalities are needed that are sensitive to matrix and cellular changes. Biochemical markers in body fluids could provide less expensive, quantitative measures of the presence of DD as well as the process of leading to the development of DD.

Define the age-related changes in disk gross morphology, including changes in shape, volume, and height, and the age-related changes in disk microstructure.

Although human disk volume and shape are known to change with age, no complete quantitative description of the age-related changes in disk morphology exists. A quantitative morphologic study should include the cartilage endplates, the annulus fibrosus, the nucleus pulposus, and the anterior and posterior spinal ligaments for each disk level. This information is essential for detailed studies of the changes in spinal mechanics with age and would provide a better understanding of the relationships between changes in disk morphology and changes in spine motion and loading. Ideally, this work should initially be based on human autopsy specimens and later should be correlated with high resolution three-dimensional imaging such as MRI or CT scans. It might then be possible to perform large scale studies using imaging alone.

The age-related changes in disk microstructure, including cell types, cell density, matrix compartments, and matrix macromolecules, have not been well defined. Modern quantitative studies should be undertaken on human tissue for each spinal region and for each decade and should include the annulus fibrosus, nucleus pulposus, cartilage endplate, subchondral bone, and adjacent tissues, including the spinal ligaments. Cell phenotypes should be identified by histochemistry and in situ hybridization.

Investigate the age-related changes in disk repair.

The effects of age on the ability of the disk to respond to tissue damage are currently unknown. The ability of disks to restore matrix and cells following either specific surgical disruptions of the annulus fibrosus or enzymatic digestion of the nucleus pulposus should be assessed for all separate disk tissues taken from donors representing various age groups. Assessment of repair should include morphology, biochemistry, and biomechanics.

Develop experimental models of the pathogenetic mechanisms of disk degeneration.

Data are now available concerning the relationship between annular rim lesions and degeneration of disks in sheep. Similar information is needed concerning two other possible pathogenetic mechanisms of disk degeneration: proteoglycan loss and impaired disk nutrition. The former might be achieved by the introduction of chronic enzymic depletion of nucleus proteoglycan; the latter, by interference with transport between the circulation and the nucleus pulposus or by metabolic inhibitors.

Determine if denucleated disk is an appropriate biomechanical or biologic model for a degenerated or aging disk.

The denucleated disk has been used in biomechanical studies as an analog for the degenerated disk, although the biomechanical and biochemical differences between the two states have not been quantified. It is unclear how differences in disk height, pressure, endplate changes, and others affect interpretations of the behavior of the degenerated disk. The relative merit of an immediate or time-dependent animal model should be assessed. Models of fibrotic and fragmented nucleus should also be created.

Determine the relative role of mechanical forces and force repetitions versus time in disk degeneration.

Animal models may show how interventions can change the force distribution and, thus, degeneration. The diurnal variation of such forces must be established, and the effect of these forces on the disk and on other tissues should be evaluated as a function of the age of the animal. The ability of mathematical models to simulate temporal changes is presently limited. Material properties for such models should be determined. The effect of disk degeneration on other structures must be established.

Determine the mechanism of failure of the intervertebral disk at different ages.

The intervertebral disk prolapses or cracks at different forces at different ages. The relationship between failure and changes in the structural or material properties of the disk needs further study. Does part of the endplate avulse; if so, how? What combination of forces and material properties is likely to cause a prolapse or crack propagation in different age groups? The experiments and models should include the variety of cracks noted in pathologic specimens, and concepts used in the study of fractures of other connective tissues should be examined for their relevance to disk tissues.

Develop methods for in vivo detection of the onset of disk degeneration and the consequences thereof.

Most current methods are limited in their ability to assess the effect of disk degeneration on kinematics and/or load distribution in vivo. New methods that could detect kinematic and coupling changes in six degrees of freedom would make it possible to create a model able to predict the stress distribution changes that accompany disk degeneration.

Develop and evaluate animal models of disk degeneration in terms of mechanical environment, biochemistry, metabolism, and morphology, and develop mathematical models of the disk as part of the functional spinal unit, which includes muscles.

Animal models are not developed in a systematic manner and tend to be predicated on the desired outcome rather than on the mechanisms. The sophistication of mathematical models will enable the investigators to model the effects of disk degeneration on adjacent structures and on surgical and other interventions as a function of aging. Such models may make it possible to include the effect of muscles on stabilization and to include temporal effects such as remodeling.

Determine the physical properties of disk tissue in different regions, at macroscopic and microscopic structural levels.

The quality, structure, and function of disk matrix changes with age. The resulting changes in the mechanical, electromechanical, and physicochemical properties may be important in the understanding of disk degeneration and the effect of the local matrix environment on cellular metabolism. Therefore, these properties should be evaluated using material taken from different regions of the disk at different ages. In addition, to better examine microstructure-function relationships as a function of age, methods are needed to measure material properties at smaller structural levels that exhibit local anisotropies and heterogeneities. Ultimately, properties at the cellular level may be of interest.

Elucidate mechanisms for the degradation of macromolecules in the disk and determine why the disk does not maintain homeostasis matrix macromolecules at different sites.

A large proportion of the molecules found in the disk at any given time appears to represent fragments of matrix constituents, such as aggrecan fragments. A study of these fragments should permit identification of the particular proteolytic activities (eg, enzymes, free radicals, glycosylation) responsible for this degradation. Because many of the fragments may be retained for a longer time in the disk, it represents a particularly suitable model for studies of such mechanisms. Determination of age-dependent alterations in these processes is of particular interest.

Altered matrix composition in disk degeneration results from fragmentation and retention of select molecules/fragments. Understanding of factors governing this retention and the capacity of the cells to replace lost molecules in the correct location and orientation in the matrix are of key importance in understanding disk degeneration. For example, the method of amino acid racemization can be explored to compare residence times of various macromolecules/fragments in the disk.

Characterize alterations in matrix macromolecules and identify age-dependent alterations in posttranslational modifications.

Examples are alterations in glycosylation, phosphorylation, and sulfation as well as altered cross-link formation. Typical age changes in glycosaminoglycan chain structure have been recorded in other tissues. Such changes may profoundly influence the functions of these macromolecules and alter their susceptibility to degradation.

Identify tissue macromolecules unique to or selectively enriched in the intervertebral disk structures.

Such molecules can be used to identify processes occurring in aging whether they are released from the tissue or retained as fragments. Released fragments can be used as systemic markers for disk processes and may permit the early identification of degenerative processes and assist in monitoring therapy. Studies of such fragments may also aid in understanding the mechanisms for tissue retention of fragments by distinguishing fragments that interact with other matrix components from those that are physically retained in the tissue.

Elucidate the role of notochordal cells in establishing and maintaining the gelatinous nature of the nucleus pulposus.

The disappearance of notochordal cells from the human nucleus pulposus always occurs before fibrotic changes are observed; therefore, important information can be obtained by identifying factors that influence metabolism of neighboring cells or matrix assembly. Chondroma cells in culture may represent an alternative way of studying the cell population.

Use quantitative or semiquantitative methods at the light microscopic and ultrastructural level to determine the distribution of matrix proteins, including proteoglycans and collagens, in annulus, nucleus, endplate, and perispinal structures.

Knowledge of the distribution of such proteins will help in understanding the organization and assembly of the matrix and will aid in the identification and characterization of tissue compartments. Identification of variations related to different spinal levels and to variations in disk shape will help in understanding why they occur with different pathologies and will be important in understanding mechanisms for tissue failure.

Characterize the cells in disk and surrounding structures in relation to site and age by histologic methods designed to assess variations in cell structure, cell-matrix structure, and metabolic status of cells.

Little is known about the nature of cells in the disk. There are obvious differences in the appearance and in the environment of cells from different localities. Histologic methods allow internal comparison of the properties of different populations of cells within the same disk and surrounding tissues without removing cells from their environment.

Characterize metabolism of cells from disks of different ages in vitro in relationship to their physical and nutritional environment and site of origin.

The cells of the disk exist in an environment in which the supply of nutrients and extracellular pH vary with distance from the blood supply and in which the physical environment alters cyclically in response to diurnal loading patterns. The effect of these changes on matrix production and energy metabolism can be studied using newly developed in vitro culture techniques.

Examine the degree of cell necrosis in relation to disk aging and degeneration.

Some reports suggest that a significant proportion of cells in the adult human disk are necrotic. This work needs to be followed up in more detail, with the proportion of necrotic cells in different regions of the disk related to the condition of the matrix. The ultimate aim of this work would be to determine what proportion of cells need to survive to maintain matrix integrity.

Index

Page numbers in italics refer to figures or figure legends.

N

O

P

R

S